Understanding

BASIC STATISTICS

Understanding
BASIC STATISTICS

Harvey W. Kushner

C. W. Post College / Long Island University

Gerald De Maio

Baruch College / City University of New York

Holden-Day, Inc.

San Francisco · London · Dusseldorf · Singapore
Sydney · Tokyo · New Delhi · Mexico City

To Meredith Hope

Preface

These days, most programs in the social and behavioral sciences, as well as education, offer an introductory course in statistics as part of the curriculum. A student in any of these fields needs to be literate in statistics—that is, to understand it and to speak its unique language.

In the course of teaching statistics to social science students, we have observed that basically two general types of statistics texts are available. The first type is the basic elementary statistics textbook. This does help the student to master some basic skills, but in illustrating statistical techniques it tends to use some rather unrealistic examples—such as horse races, card games, and whether undergraduate males prefer blonds, brunettes, or redheads. The second type of statistics text, usually categorized as "social statistics," has appeared within the last ten or fifteen years. In response to the social sciences' growing tendency to use quantitative techniques, these texts try to blend substantive social science concerns with statistics. But they don't quite fill the bill, either: often they use examples that may be difficult for the beginning student, or else they try to force relevancy into areas where it may be inappropriate. For example, some books deal with probabilistic considerations by referring to the number of Democrats or Republicans who walk through a door or are pulled out of a hat. We find this forced attempt at relevancy unnecessary. Relevancy must be kept in proper perspective.

So this text is meant to fill the gaps that other types of statistics texts leave open. Anyone using this book should be able to master basic statistical skills, and to develop an appreciation for how professional researchers in the student's substantive discipline use such skills. We believe that the best way to do this is to let the student work through hypothetical problems that simulate projects he or she may wish to carry out in his or her own course work, as well as to interpret research that's reported in professional journals. We also believe, however, that the examples need not come directly from the professional literature; such literature sometimes contains subtleties that hide the text's main purpose—namely, to master statistical techniques. However, since we want to teach statistics and its applicability to realistic research questions seriously rather than frivolously, we have presented some of the topics traditionally and others within the framework of a behavioral science application. Although beginning students often expect statistics to be dry and lifeless, actually it needn't be. Its intellectual history, for instance, is rather interesting; and we have included historical anecdotes wherever we could.

This text is meant to be used in a one-semester introductory statistics course, or a statistics-for-social-scientists course. We have designed it mainly for undergraduates. However, it can also be used profitably in a graduate first-semester quantitative techniques course. Recently, graduate programs in the applied social science fields of public administration, criminal justice, and urban studies have expanded. Many of these programs attract working professionals who have been away from college ten or more years, and who want to learn quantitative techniques but may doubt that they have enough mathematical training. Since this text is meant to be understood by people whose prior quantitative training is on the level of high school mathematics, re-entry graduate students should be able to use it to their advantage.

Like any other statistics text, this book treats all the standard topics that an introductory statistics course usually covers. But we also emphasize those techniques most likely to interest students in the social and behavioral sciences—namely, correlational analysis (including multivariate analysis) and nonparametrics. Most elementary texts do not detail these topics in such detail; but this one does—which gives the instructor some leeway. The instructor who wants to teach a conventional introductory statistics course can omit some of the topics covered in Chapters 15 and 16 without losing any continuity. It's our belief that any student who completes a course, using this text, will be able to understand professional social science literature with confidence, as well as to be competent in statistical analysis.

Understanding Basic Statistics is the creation of many people whose contributions we must acknowledge. At the beginning were our own teachers, S. Richard Maisel and Matthew P. Drennan, who taught us that statistics could be fun.

Both of us came to intellectual maturity during a period when quantitative applications in the social sciences were subjects of heated controversy. We are indebted to Steven J. Brams and Robert Burrowes for sensitizing us to these issues. The Inter-University Consortium for Political and Social Research (ICPSR) at the University of Michigan, through its institutes, further stimulated our awareness of the many uses of quantification in the social sciences.

This text was written largely because of the support and encouragement of Frederick H. Murphy. Also, it is a pleasure to acknowledge the assistance provided by Barbara Gordon and the rest of the staff at Holden-Day. Naomi Steinfeld, as well, provided invaluable editorial commentary and direction during the production of this text. And lastly, various colleagues and reviewers offered pertinent comments. The final product, of course, remains our own responsibility.

The senior author would like to thank his wife, Sara Kushner, who took time from her own responsibilities to listen, criticize, type, and support every moment of the writing of this text. She also made possible a congenial working atmosphere for both of us. And Meredith Hope Kushner, to whom this text is dedicated, not only provided many delightful moments but also deserves credit for being able to adapt to her father's unusual working hours.

We wish to thank our parents, Albert and Iryne M. Kushner and Salvatore and Antoinette C. De Maio, who taught us the value of love and knowledge. Mr. De Maio also assisted with the graphics.

Finally, we are grateful to the Literary Executor of the late Sir Ronald A. Fisher, F.R.S.; to Dr. Frank Yates, F.R.S.; and to Longman Group Ltd., London, for permission to reprint Tables III, IV, V and VI from their book, *Statistical Tables for Biological, Agricultural and Medical Research* (6th edition, 1974).

Harvey W. Kushner
Gerald De Maio

Contents

*The time may not be very remote when it will be
understood that for complete initiation as an efficient
citizen of one of the new great complex world wide
states that are now developing, it is as necessary to be
able to compute, to think in averages and maxima and
minima, as it is now to be able to read and write.*
H. G. Wells

Introduction to Basic Statistics

1.1 Introduction: Why Study Statistics?

We may or may not like it, but statistical findings have become basic to our thinking about social phenomena. Intuitively, we recognize that education and income are related, that different religious groupings show somewhat different political preferences, and that cigarettes are reputed to be a health hazard. But statistics are what we use to illustrate, prove, and justify our intuitive observations. Not only do we use statistical findings in our lives, but we share an intellectual vocabulary that contains elementary statistical concepts. Today's college student is familiar with the concepts of average and percentage, and most students don't need to consult a dictionary to define (even if not fully) such terms as *crime rate, percentile rank,* and perhaps even *statistical inference*.

But just being a citizen, these days, or a professional in our world means that we need more than a superficial knowledge of statistical concepts and findings. As our society produces more and more data, citizens need to understand statistical concepts more precisely, so that they can make intelligent decisions. Social scientists, as well as people in related professions, find that the literature relevant to their work is increasingly statistical in nature. Today's informed citizen must be able to think in even more advanced statistical terms than H. G. Wells, that nineteenth-century seer, prophesied. Therefore, if you're really interested in contemporary society you have to understand the concepts involved in the study of statistics. Then you can evaluate information accurately and make intelligent decisions.

1.2 What Is Statistics?

As early as the middle of the eighteenth century, many of the basics of modern statistics were already established—but, strangely, by people who wouldn't have labeled their work "statistical." A *statist,* in Elizabethan England, was a functionary of the state (a politician, of sorts) who collected and analyzed data (for instance, births and deaths) that the government required for effective planning, ruling, and tax collecting. Perhaps this explains why the term *statistician* is sometimes viewed with suspicion, especially by people who are afraid of government manipulation.

Today, however, the term *statistics* does not just refer to politics or state-craft (as its Latin root suggests). But knowing what something is *not* doesn't always help us know what it *is.* Statistics is still hard to define precisely, even in the most elegant and advanced treatises. It is generally recognized, however, that statistics has two distinct functions. One is called *descriptive statistics.* This is concerned with collecting and then manipulating data so that quantitative summarizing statements can be made—for example, "Of 100 television viewers polled, 40 watched the pilot film." The summarizing statement here is the percentage—40 percent watched the pilot instead of other programming.

But only rarely are we interested in the descriptive statistics on their own. The preferences per se of a hundred out of millions of viewers would hardly arouse any network executive's interest. However, if this sample could be considered to accurately represent the preferences of the entire viewing audience—that would be something else! So the second function of statistics concerns the process of making inferences from samples to populations in the face of uncertainty. Not surprisingly, this function is called *inferential statistics.*

This discussion has already included certain terms that might not be familiar to you; so before we go any further with the subject of statistics, let's take a look at some working definitions.

1.3 Some Basic Definitions

The following definitions should become part of your active, working vocabulary. Since you'll come across them often throughout the text as well as in the current literature of the social sciences, you might as well get comfortable with them now.

> *Variable: A characteristic that can take on a range of values.*

The height of males in the United States is a characteristic that takes on a range of values. Some males are short; others are average size; and still others are potential basketball stars. The age of females in the United States is another characteristic that is capable of taking on different values. Think of how many traits there are that you and your neighbors may differ about: age, income, political preference, and hair color, to name just a few. In contrast to a variable is a *constant,* whose value never changes (for example, $\pi \approx$ [nearly equal to] 3.1415926).

Data: The measurements collected as a result of observing a variable.

Data can take the form of *scores,* such as the intelligence quotients (IQs) of children; or of *ranks,* such as those given the finalists in a beauty pageant; or of *frequency counts,* such as the number of individuals voting for a winning candidate. Therefore, when we say that we have data on a particular characteristic, we mean that we have measured a variable.

Population: A complete set of things sharing some specified characteristic.

The number of blue-collar workers living in New York City can constitute a population. All the citizens of the United States are a population; so are all the residents of Sheboygan, Wisconsin. In short, a population is what you choose it to be for the purposes of your study.

Sample: A subset of a population.

A sample is any number of cases smaller than the population from which it was drawn. If we have a population of 100,000 voters, a sample might consist of only 10, 1,500, or 99,999 voters. Usually, however, samples consist of only a small or very small proportion of the population that is being investigated. Remember that at one time the ten voters may qualify as a sample, while at another they may constitute an entire population. The difference is whether your goal is to infer something about the larger group (as with a sample) or to describe only those you have selected (a population).

Parameter: A measurable characteristic of a population.

The proportion of registered Independents among Americans of voting age is a parameter. Other examples of parameters include the percentage of adults over eighteen who have completed high school in a given city, or the number of musicals currently playing on Broadway. As with a variable, a population has countless characteristics that are measurable. In this text, we follow the tradition of using Greek letters to represent population parameters. (For example, μ [small Greek mu] is used for the population mean or average.)

Statistic: A measurable characteristic of a sample.

The proportion of registered Independents for a sample of Americans of voting age can be a statistic, as can the percentage of Star Trek freaks in a sample of 1,000 science-fiction fans. Generally, we use a statistic to estimate the population parameter; for example, a sample of Independents can be used to estimate the proportion of Independents in the entire electorate. Here, too, we follow traditional practice and use Roman letters to represent sample statistics.* (For example, $\overline{X}$ [read X bar] is used for the sample mean or average.)

1.4 Caution: Two Abuses of Statistics

Both the camera and statistics are tools to serve humanity; both are used to capture an accurate picture of reality. Occasionally, the photographer may want to delete some portion of a photograph—after all, imagine seeing a mole on the Playmate of the Month! In statistics, however, such a deletion might well be a gross misrepresentation—in other words, abuse.

Take the TV commercial we have all seen hundreds of times. An actor is providing expert testimony on behalf of some product. At just the right moment, he or she points to a carefully sketched graph whose lines move in such a way that they appear to prove the validity of the testimony. Great care is taken never to label the axes. Now, this type of abuse may be infuriating—but because it is so obviously dishonest, we can laugh it off as being rather harmless. However, unintentional abuse—the abuse that occurs as a direct result of oversight (or ignorance)—is something else.

Unintentional abuse is unnecessary; it can be avoided through proper education in statistics. However, such education has not always been sought. The most glaring example of unintentional abuse was the 1936 *Literary Digest* poll that predicted that Alf Landon would trounce FDR. The outcome, of course, was that Roosevelt won all but two states, and many of them by a landslide. The *Digest* had made its error when it chose a sample from telephone directories and its own list of subscribers. But this sample was not random with respect to the population of voters; those voters who had no telephone and did not subscribe to the magazine had no chance of being included in the sample. Such people were more likely to be poor, which means that the poor were more or less systematically excluded from the poll. In this election, the poor overwhelmingly voted Democratic. Some years later, pollsters ob-

---

*The term *statistics* is often confusing, since there are two functions of statistics. In addition, inferential statistics makes a distinction between a statistic (characteristic of a sample) and a parameter (characteristic of a population). You can easily remember the difference between these two definitions if you ask yourself, "What is being described—sample or population?"

served the *Digest's* blunders and developed stratified sampling techniques to make sure that future samples would be more representative. (After that election, incidentally, the *Literary Digest* lost both its credibility and its readership, and ceased publication—probably because of its error.)

It's very likely that some individuals, caught up in the competitive world of private enterprise, do deliberately abuse statistics. Unfortunately, not all social scientists are innocent of deception, either. But since we believe that intentional abuses are not the common practice in scientific research, in this text we shall point out certain trouble areas which can plague even the most scrupulous researcher. We hope that these "Caution," or warning, sections will alert you to a possible unintentional abuse. These sections will help you avoid lying unintentionally; they will also make you aware of when researchers are "cooking" their results.

Are you interested in finding out more about the abuses (unintentional and otherwise) of statistics? If so, then race to your bookstore and invest in a copy of *How to Lie with Statistics,* by Huff (1954).* This little classic is reasonably priced, thoroughly enjoyable, and highly informative.

1.5 A Note to the Student

If you view statistics with anxiety and hostility, you are not alone. Despite the fact that we all have contact with various forms of statistics throughout our lives, most people feel intimidated and insecure about dealing with numbers in any form. *Understanding Basic Statistics* can help you to reduce your anxiety, reverse your hostility, and bolster your confidence in statistics.

In what we think is a friendly format, we shall introduce topics using nontechnical language wherever possible. We believe that you are more likely to master a statistical concept if it is removed from the world of mathematical symbols and instead is tied to your own experience. If you are terrified of mathematics, you can understand the term *mean* by reminiscing about an unforgettable ride on a seesaw, or the concept *probability* by thinking of games of chance, or the rationale for *statistical inference* by questioning the honesty of a coin.

However, you will need to be somewhat familiar with mathematics if you want a well-rounded understanding of statistics. Skimming through the text won't give you that degree of understanding—at times, you will have to read with pencil and paper in hand. Don't leave a topic until you have mastered it. Statistics is a cumulative course of study—try not to be the last in your class to

---

*In citing references throughout this text, ordinarily we shall give the author's name, the date of publication, and, when necessary, the page number. You will find complete publication facts in the "Bibliography" section at the end of the text.

find that out. Be the first to realize that the terms and symbols you fail to learn now will reappear later in the text in one form or another. Relax and even enjoy. Good luck.

Where We Stand

In this chapter we have distinguished between two functions of statistics. *Descriptive* statistics involves collecting and then manipulating data so that quantitative summary statements can be made. *Inferential* statistics concerns making inferences from samples to populations in the face of uncertainty.

The next six chapters are devoted solely to acquainting you with the use of statistics to analyze the data at hand—that is, with descriptive statistics. However, often data make up samples from a larger population; and we are interested in generalizing conclusions from samples to populations. In the later chapters of this text we will pay attention to those statistical techniques that help us make such inferences.

Terms and Symbols to Remember

A list of terms and symbols that have been introduced in a chapter follows each "Where We Stand" section. Since some of these terms and symbols are further defined in other chapters, they may appear more than once. You might find it helpful to explain, in your own words, what these terms and symbols mean. You also might benefit by explaining them to someone else (assuming, of course, that you can find someone who is willing to listen).

Statistics
Descriptive Statistics
Inferential Statistics
Variable
Constant
Data
Population
Sample
Parameter
Statistic

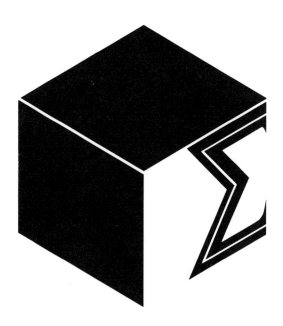

The most useful definition of measurement is the
assignment of numerals to things so as to represent facts
and conventions about them.
S. S. Stevens

Measurement

2.1 Introduction: What Is There to Know About Variables?

Statistics, as we said in Chapter 1, can be divided into two elements: description and inference. Descriptive statistics involves collecting and then manipulating data so that quantitative summarizing statements can be made; inferential statistics concerns making generalizations from data based on samples to more general populations. Both aspects of statistics implicitly assume that we can actually measure the variables of interest. Well, some we can—for instance, such variables as age, height, and weight. But it's harder to measure such widely used social science variables as power, social status, and prejudice. So before we discuss data description and statistical inference, let's consider how social science variables are measured.

In this chapter, we will treat the classification of variables according to levels of measurement, scale continuity, and their roles in research. In addition, we will discuss some basic statistical notation that is used to label and manipulate data. In short, we will present everything, or nearly everything, you ever wanted to know about variables (but never dared to ask).

2.2 The Concept of Measurement

In 1932, a select committee of the British Association for the Advancement of Science met to define the term *measurement*. After seven years of bitter debate, they failed to reach any kind of consensus. But their hard work was not in vain. In 1946, the renowned psychologist S. S. Stevens gave what is today the most widely accepted definition of measurement in the social sciences. Paraphrasing the final report of the British select committee, Stevens wrote that "measurement is the assignment of numerals to objects and events according to rules" (1946: 677). For our purposes, this statement expresses the basic nature of measurement.

Stevens' definition instructs us to take an object or event (or quality, or quantity, or whatever) and assign to that entity one or more symbols, such as 1, 2, 3, . . . , etc., which are called *numerals*.* According to this definition, we must assign numerals according to rules. In social science, we generally have three or four different sets of rules that we can use in our assignment of numerals. These different rules lead us to different *levels* of measurement that help classify variables and their data.

2.2.1 Nominal Level

The *nominal level of measurement* simply involves naming or labeling categories for a variable—that is, placing cases into categories and counting their frequency of occurrence. For example, the American Institute of Public Opinion (Gallup Poll) might use a nominal measure to indicate whether each respondent favors or opposes capital punishment. Specifically, interviewers might question a national sample of 1,500 adult respondents and determine that 800 favor capital punishment while 700 oppose it.

Other nominal level variables are sex (male or female), responses (yes or no), political party (Democrat, Republican, or Independent), and religion (Protestant, Catholic, or Jewish), to name just a few.

When dealing with nominal data, we must be able to place every case into a category. Suppose that in response to the Gallup Poll just mentioned, various respondents express "no opinion." In that case, we must expand the original category system to include a miscellaneous-type category in which unforeseen responses or exceptions can be placed. In short, the categories for a nominal variable must be *exhaustive* (there must be a category for *every* case) as well as *mutually exclusive* (no case must be in more than one category).

With nominal-level measurement, we never imply that one category is greater or better than another category. We limit our conclusions to such statements as, "There are more Democrats"—or more Republicans—or more of some category than another category. A nominal measure of sex does not, for example, signify whether males are "superior" or "inferior" to females.

---

*You can distinguish a numeral from a number by the fact that the numeral does not necessarily have any quantitative value attached to it.

We may use numbers arbitrarily as tags for different categories. We may assign "1" to Democrats, "2" to Republicans, and "3" to Independents; but this in no way justifies the use of the usual arithmetic operations (addition, subtraction, multiplication, and division) on these numbers. A "1" (Democrat) plus a "2" (Republican) does not equal a "3" (Independent). We label nominal data merely to group cases into separate categories so that we can indicate similarity or dissimilarity with respect to a given quality or characteristic.

The nominal level is really the crudest of all measurement systems. But this does not mean that it is not useful to us as social scientists. In many instances, nominal measures are the only appropriate measures to use. We shall see evidence of this in forthcoming chapters.

2.2.2 Ordinal Level

Sometimes we can order categories in terms of the degree to which they possess a certain quality, and still be unable to specify exactly how much of that quality they possess. It is even possible for us to rank data so precisely that none of them are located at the same point on a single continuum. But for the most part there will be many ties, and they will make it impossible for us to distinguish between certain data and will force us to lump these data together into a single category. We are able to say, however, that these data are *greater than* ($>$) or *less than* ($<$) other data. Thus, we may classify individuals according to socioeconomic status (such as upper class, middle class, and lower class), or political preference (such as strong Democrat and weak Democrat).

Since this type of measurement system, the *ordinal level of measurement,* allows us to separate categories and rank-order them as well, it is somewhat more sophisticated than the one used in obtaining a nominal scale. Examples of ordinal scales are social class, power, and the many attitudinal scales that range, for example, from "strongly agree" to "strongly disagree."

It is important to remember that ordinal-level measurement lets us state that one category is greater than or less than another, but it does not let us say anything about the magnitude of that difference. And, as with nominal measures, we cannot perform any of the familiar arithmetic operations that we usually attribute to numbers. Ordinal measurement does allow us a rank ordering that nominal-level data do not, but neither of these levels is quantitative in the usual sense.

2.2.3 Interval-Ratio Level

In the narrow sense of the word, measurement applies only to those instances when we can classify and rank-order data and determine the exact distance between data. When this is possible, we have attained the *interval level of measurement.* In order to have interval data, we need to obtain some sort of unit of measurement that will be agreed upon as standard and that is replicable—that is, one that can be used indefinitely and still produce the same results. This is what happens when we measure age in years, height in feet, or weight in pounds. Unfortunately, however, there are no comparable units for power, social status, and prejudice—none that can be agreed upon by all social scientists and can be assumed to be standard from one set of conditions to the next. Using a standard unit of measurement, we can say that the difference between, say, two scores is thirty units, or even that one score is three times as large as the other. This means that we can add or subtract scores in the same fashion that we can add or subtract distances or can subtract a foot from a board by sawing it into two. Similarly, we can add the incomes of husband and wife, but it would make no sense at all to add or subtract their ages.

To obtain the *ratio-level measurement,* we must be able to locate a true zero point on the scale. If we can do this, then we can compare scores by taking ratios; for example, twenty pounds is twice as heavy as ten pounds. But if there is no true zero point, such as in the case of IQ, taking ratios is meaningless. It would make no sense at all to state that the child with the high IQ is twice as intelligent as the brain-damaged child, even though we can state unequivocally that the unit differential between an IQ of 110 and an IQ of 115 is the same as the difference between an IQ of 125 and an IQ of 130.*

Actually, it is difficult to specify an interval scale that is not also ratio. Once we agree upon the size of the unit, we can conceive of zero units, regardless of whether we can achieve zero weight or attain zero velocity. Therefore, we refer only to interval scales, and we assume that in all instances where a standard unit of measurement is available, it's permissible for us to use all the familiar arithmetic operations that we assign to numbers.

2.2.4 Comparing the Levels of Measurement

As we have seen, there are several distinct levels of measurement. These various levels of measurement themselves form the cumulative scale shown in

---

*Some people still believe that the concept of temperature does not allow us to take ratios. But temperature is now measured on a ratio scale by the Kelvin thermometer. Not too long ago, it was measured on an interval scale by either the Centigrade or Fahrenheit thermometers. In these two, the zero point was selected arbitrarily, with reference to the freezing and boiling points of water. In contrast to these arbitrary zero points of the Centigrade and Fahrenheit thermometers, the Kelvin thermometer has a true zero point that represents the complete absence of heat. The development of the Kelvin scale demonstrates that the level of measurement that can be attained is not a function of the phenomenon itself, but rather is a function of our ability to develop instruments that are sensitive to how we conceptualize the phenomenon. Thus, the property we commonly refer to as "temperature" clearly illustrates that the level of measurement can be improved as the basic concept is redefined.

Table 2.1. Notice that the ordinal scale has all the characteristics of the nominal scale, plus ordinality. The interval scale has all the qualities of both nominal and ordinal scales, plus a standard unit of measurement. This means that we can always legitimately move down one or more levels of measurement in classifying data—but we lose information in doing so. For example, if we know that Bert has four years of college and Harry has two years of college, and if we make use only of the fact that Bert has *more* college, we in effect throw away the information telling us that the difference in years of college is two. Our point is that although we *can* reduce the level of measurement, it isn't *advisable*. If the data have been assembled in a way that involves interval measurement, then those data should be analyzed with the more powerful statistical techniques appropriate to the interval level. For example, under the rules given, you could treat a variable such as reaction time as a ratio variable; but in fact, reaction time might be treated more appropriately as an ordinal variable when used as a tool to measure some underlying process. Consider the case where a researcher is studying decision time and obtains a reaction time of four seconds under one set of conditions and two seconds under another. Is it justifiable to assume that it takes twice as long to decide under the first set? Probably not. Sometimes the voluntary loss of information is desirable; often it is necessary.

Table 2.1 Characteristics of the Different Levels of Measurement

Level	Mutually Exclusive Categories	Ordered Relationship Between Categories	Standard Unit of Measurement Between Categories
Interval-Ratio	Yes	Yes	Yes
Ordinal	Yes	Yes	No
Nominal	Yes	No	No

What can we say about the reverse process—that is, moving up the scale of measurement from, say, a nominal to an ordinal scale? For example, consider the "yes," "no," "absent," and "abstain" alternatives that an actor in a voting body may use. These voting options form clear-cut categories of a nominal variable. But can we assume that they are scaled for qualities such as "higher or lower," "more or less," or "superior or inferior"? If we do make this assumption, we should explain *why* we are assuming an ordered relationship, since it isn't obvious why any one option might be superior to another. We face a similar challenge when we try to rank-order the degree of severity of different crimes, or to cast social and political issues into ordinal or interval variables. Unless it is justified, *decreasing* the level of measurement destroys information. But *increasing* the level of measurement, without justification, adds noise.

2.3 Continuous and Discrete Measurement

Yet another distinction is made between variables: do they have a *continuous* or a *discrete (discontinuous)* scale of measurement? A continuous variable (see Figure 2.1) has an unlimited number of intermediate values. For example, suppose that a psychologist is recording the times it takes rats to run a maze. Obviously, the measurements that the psychologist records are only as accurate as the timing device that is used. Even if this device is accurate to the nearest second, it's always possible—or at least imaginable—that another timing device could be even more precise: say, to the nearest tenth or hundredth of a second. So we can think of *any* measure as only approximate, and as one that could be refined even more if a more sensitive measuring device were available. In other words, you can have an unlimited number of different values between any two values. This concept implies a *continuous* scale of possible measurement values. Age, intelligence, and certain measures of social class, such as income, are all examples of continuously measured variables.

A *discrete* variable, on the other hand, has only a limited number of possible values. In graphic terms (see Figure 2.1), it is a finite scale with a series of gaps between the data. Take the variable of "family size." When you realize that a family can have one, two, or more children, but not a *fraction* of a child, you also realize that "family size" is a discrete variable. When you have frequency-type data—for instance, the number of crimes committed in Detroit, or the number of delegates attending the Democratic National Convention—these data are discrete. Therefore, nominal variables are always discrete. However, "average family size," as reported by the Bureau of the Census, is not a discrete variable, although it might seem to be. Because it is based on data that are gathered from different households and are then averaged out, yielding a summary figure that can take on an unlimited number of values, "average family size" is a continuous variable.

2.4 Caution: The True Limits of Continuous Measurement

As we mentioned, continuously distributed variables can take on an unlimited number of intermediate values. That's why we said that numerical values of

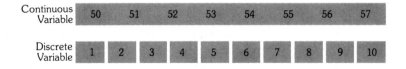

Figure 2.1 Classification of Variables by Scale Continuity

continuously distributed variables are always approximate. However, we can still specify the probable limits that the true value falls within. How? By assuming that the *true limits* of a value of a continuous variable are equal to that number plus or minus one-half of the unit measurement.

Suppose that you weigh yourself on your bathroom scale, which is calibrated in terms of pounds. Suppose further that the pointer appears to register your weight as 150 pounds. Thus, the *apparent value* is 150, but the *true value* of your weight falls somewhere between 149.5 and 150.5 pounds, assuming that your scale is accurate. If the same scale was calibrated in terms of tenths of pounds and the pointer registered 150 pounds, the true limits of your weight would be 149.95 and 150.05 pounds.

2.5 Independent Versus Dependent Variables

There is a third way in which variables are distinguished, and it is an important one—how they are used in research. For our purposes, we distinguish two basic types of variables: *independent* and *dependent*.

Why do we make this distinction? Social scientists are interested in the ways in which particular variables fluctuate from observation to observation in a population. For example, some individuals have higher incomes than others, different income groups show somewhat different political preferences, and certain crimes are committed more frequently than others. But why? To answer this question, the researcher looks for explanations and predictions of the variation in one variable, based on variations in another or several other variables. The similarity among variables generally becomes the substance of social science theory.

The variable that the researchers find the most interesting—the one whose variation is to be explained in the research—is called the *dependent variable*. The researcher sees the recorded differences between observations on this variable as being dependent on some other variable (or variables). For example, a student's grades in college may be dependent upon high school performance, hours spent studying, and basic intellectual capacity. Attaining a high social standing in the community may be dependent upon an individual's wealth, education, and religious persuasion. Voting for a Democrat may be dependent upon income bracket, education, and ethnic background.

Independent variables, on the other hand, are those that are thought to influence dependent variables. They serve an explanatory function; that is, they may help explain why a dependent variable varies as it does in some population. The independent variables associated with achieving high grades in college, in the above example, are the student's high school performance, the amount of time spent studying, and the basic intellectual capability. The independent variables for social status are wealth, education, and religious persuasion.

The variables you will use in your research may be dependent or independent. That's for *you* to decide, and your decision will be based on whether the variable *influences* another variable or whether it *is influenced* by another variable. Here, you must rely on your knowledge of how the two variables are related to each other. A variable can be dependent for some purposes and independent for others. For example, take religion as a variable. If you use it as an independent variable, it can help explain why people vote for particular parties. But if you want to explain the intensity of a person's religious devotion by a variety of factors such as type of education (parochial or public), ethnic background, and socioeconomic status, then religion becomes a dependent variable. Variables are not fixed as independent or dependent. How you use them has to do with the question you are trying to answer.

Later on in this book, we'll see that the role of independent and dependent variables is crucial. For now, it's important that you understand how they differ.

2.6 Statistical Notation

The field of statistics is full of symbols—not so much because they are necessary, but because they are useful. Once we really understand statistical notation, we can use it to express briefly something that otherwise would require a paragraph if it were written out in words. All the symbols and notational facts presented in this text* have more to do with the language of statistics than with the language of science. In order to understand statistics, we have to learn statistical notation; just as in order to understand English we have to learn the alphabet. If we master the notation used to label variables, data, and mathematical operations, it will be easier for us to understand the component parts of most statistical procedures.

2.6.1 The Subscript

Since the formulas that we will study in later chapters must be applied to different kinds of data, we must use some symbol such as X or Y to represent the data to which the formulas are to be applied. But we have to modify our symbolization, or else immediate confusion will result. For example, if we tried

---

*Different authors use different symbols—in other words, no standardized system of notation exists. However, there is some consensus; and, where possible, we try to use the statistical notation that is recognized most widely.

Proof:

$$\Sigma(X + k) = (X_1 + k) + (X_2 + k) + (X_3 + k) + \ldots + (X_N + k)$$
$$= (X_1 + X_2 + X_3 + \ldots + X_N) + (k + k + k + \ldots + k)$$
$$= \Sigma X + Nk$$

Rule II: The summation of the values of a variable minus a constant equals the summation of the values of the variable minus N times the constant.

Formally, the rule states:

$$\Sigma(X - k) = \Sigma X - Nk \quad \text{(see Rule I)}$$

Rule III: The summation of the sum of the values of two variables equals the sum of the individual summation.

Formally, the rule states:

$$\Sigma(X + Y) = \Sigma X + \Sigma Y$$

Proof:

$$\Sigma(X + Y) = (X_1 + Y_1) + (X_2 + Y_2) + (X_3 + Y_3) + \ldots + (X_N + Y_N)$$
$$= (X_1 + X_2 + X_3 + \ldots + X_N) + (Y_1 + Y_2 + Y_3 + \ldots + Y_N)$$
$$= \Sigma X + \Sigma Y$$

Rule IV: The summation of the differences between values of two variables equals the differences between the values of the individual summations.

Formally, the rule states:

$$\Sigma(X - Y) = \Sigma X - \Sigma Y \quad \text{(see Rule III)}$$

Rule V: The summation of a constant times a variable equals the constant times the summation of the values of the variable.

Formally, the rule states:

$$\Sigma kX = k\Sigma X$$

Proof:

$$\Sigma kX = kX_1 + kX_2 + kX_3 + \ldots + kX_N$$
$$= k(X_1 + X_2 + X_3 + \ldots + X_N)$$
$$= k\Sigma X$$

Rule VI: The summation of the values of a variable divided by a constant equals the reciprocal of the constant times the summation of the values of the variable.

Formally, the rule states:

$$\Sigma \frac{X}{k} = \frac{1}{k}\Sigma X$$

Proof:

$$\Sigma \frac{X}{k} = \frac{X_1}{k} + \frac{X_2}{k} + \frac{X_3}{k} + \ldots + \frac{X_N}{k}$$
$$= \frac{1}{k}(X_1 + X_2 + X_3 + \ldots + X_N)$$
$$= \frac{1}{k}\Sigma X$$

Rule VII: The summation of a binomial $(X - Y)$ raised to the second power equals the summation of the individual terms that result from expanding the binomial.

Formally, the rule states:

$$\Sigma(X - Y)^2 = \Sigma X^2 - \Sigma 2XY + \Sigma Y^2$$

Proof:

$$\Sigma(X - Y)^2 = \Sigma(X^2 - 2XY + Y^2)$$
$$= \Sigma X^2 - \Sigma 2XY + \Sigma Y^2 \quad \text{(see Rule III)}$$

2.6.4 Caution: The Use of Parentheses in Summation

Some differences may appear to be minor and therefore irrelevant in the expression of a mathematical operation, but they may make a great deal of difference in the end product. A case in point is the use of parentheses in the following operation:

$$\Sigma X^2 \text{ versus } (\Sigma X)^2$$

The operation ΣX^2 means that each value of X is squared (multiplied by itself) and all the results are added together:

$$\Sigma X^2 = X_1^2 + X_2^2 + X_3^2 + \ldots + X_N^2$$

In contrast, the operation $(\Sigma X)^2$ specifies that all the values of X are added, and then the sum is squared. That is:

$$(\Sigma X)^2 = (X_1 + X_2 + X_3 + \ldots + X_N)^2$$

When, for example, the operation ΣX^2 is applied to scores of 1, 2, and 3, the result is:

$$1^2 + 2^2 + 3^2 \text{ or } 14$$

Applying these same scores to the operation $(\Sigma X)^2$, we obtain:

$$(1 + 2 + 3)^2 \text{ or } 36$$

Notice that, by using the same values of X, the sum of all the squared values of X does not equal the squared sum of the values of X. In symbols:

$$\Sigma X^2 \neq (\Sigma X)^2$$

A similar source of confusion is the interpretation of ΣXY. The operation $\Sigma XY \neq (\Sigma X)(\Sigma Y)$ because ΣXY equals $X_1Y_1 + X_2Y_2 + X_3Y_3 + \ldots + X_NY_N$, and $(\Sigma X)(\Sigma Y)$ equals $(X_1 + X_2 + X_3 + \ldots X_N)(Y_1 + Y_2 + Y_3 + \ldots Y_N)$.

Where We Stand

One of the most important things that the student of social science statistics can learn is how to select the most appropriate statistical method from those that are available. You can always file away the computational procedures you learn and call on them when you need them; but what's really important is to develop the habit of asking (and answering) the question, "What statistical methods can I use, and what assumptions must I make in using each one?" A statistical method that is appropriate for one kind of variable may not be appropriate for another. That's why we distinguish variables by level of measurement, by scale continuity, and by their role in research. It's important that you understand each and every distinction. If you don't, you may find yourself using statistical techniques that are inappropriate for the analysis you wish to perform.

Terms and Symbols to Remember

Nominal Level
Ordinal Level
Interval Level
Ratio Level
Continuous Measurement
Discrete (Discontinuous) Measurement
True Limits
Apparent Value
Independent Variable
Dependent Variable
Subscript
Case
Summation Sign (Σ)

Exercises

1. At which level of measurement would you classify the following variables?
    (a) Education, measured in years
    (b) Education, measured as arts, business, social science, and science
    (c) Education, measured as elementary, high school, and college
    (d) Television viewing, measured in number of hours per month
    (e) Television viewing, measured in terms of program; for example, documentary, movie, news, situation comedy, and sports
    (f) Degree of agreement with the statement, "Social scientists should be more involved with public policy decisions": agree strongly, agree somewhat, disagree somewhat, and disagree strongly
    (g) Place of residence, measured in terms of size of community; for example, rural, suburban, urban, and inner city
    (h) Place of residence, measured as county of residence
    (i) Place of birth, measured as country
    (j) Personality, measured as likable and agreeable
2. In the following examples, determine the level of measurement and whether the variable is discrete or continuous.
    (a) The number of students majoring in sociology at Yale University
    (b) Zip codes

(c) The distance between New York and San Francisco

(d) The number of postal employees at varying ranks in the United States Postal Service

(e) Life expectancy for males in Norway

(f) The number of pages in this text

(g) The votes obtained by all the Democratic Party presidential candidates in the New Hampshire primary

(h) Yesterday's temperature

(i) The signers of the Declaration of Independence

(j) The average number of automobiles stolen every thirty minutes in the United States

(k) Tosses ot a coin

(l) The color of your eyes

(m) The number of automobiles sold during June in Maine

(n) Hat sizes

(o) The birth rate

(p) The number of unfinished automobiles left over at the end of a day on an assembly line in Detroit

(q) The top ten finalists in a pie-baking contest at a county fair

(r) Ten different telephone numbers

(s) A line drawn on a piece of graph paper

(t) Time, measured by an atomic clock

3. It is often argued that a variable such as income, if measured in dollars, is not really interval, since a difference of $500 will have different social meanings depending on whether the difference is between incomes of $10,000 and $10,500 or between incomes of $100,000 and $100,500. Do you agree? Explain your answer.

4. Is the severity of crime measurable? If yes, then how would you measure it?

5. What indicator would you suggest to select in constructing a measure for prestige? Specify the level of measurement for each indicator. Do the same for democracy, power, and prejudice.

6. List the problems involved in increasing the level of measurement. Why would it benefit the social scientist not to destroy information?

7. Reduce the following variables by at least one level, or by two, if possible. For example, take the variable, *education*. Measured in years, it's interval; measured as elementary school, high school, and college, it's ordinal; and measured as secular or religious, it's nominal.

(a) Income
(b) Age
(c) Crime
(d) Suicide
(e) Television consumption
(f) Political ideology
(g) Aggression
(h) Infant mortality
(i) Socioeconomic status
(j) Religion

8. Write the true limits for the following measurements:
   (a) 10.01
   (b) −10
   (c) 0
   (d) 1.6
   (e) 3.1

   (f) .5
   (g) 0.55
   (h) 0.5
   (i) 16
   (j) −35

9. Express each of the following sentences using statistical notation.
   (a) Sum all values of X from the first case to the last case and then divide by two.
   (b) Sum all values of X minus Y from the third case to the eighth case and then multiply by seven.
   (c) The sum of all values of X times a constant, k, equals k times the sum of all values of X.
   (d) Sum all values of X plus Y plus Z from the first case to the third case.
   (e) The mean ($\bar{X}$) of the X values is equal to the sum of the values divided by the number of values.

10. Write each of the following in summation notation.
    (a) $X_1 + X_2 + X_3 + X_4 + X_5 + X_6 + X_7$
    (b) $X_1 + X_2 + X_3 + \ldots + X_N$
    (c) $(X_1 + Y_1) + (X_2 + Y_2) + (X_3 + Y_3) + \ldots + (X_N + Y_N)$
    (d) $(X_2 + a) + (X_3 + a) + (X_4 + a) + (X_5 + a) + \ldots + (X_N + a)$
    (e) $(X_6 + Y_6 - k) + (X_7 + Y_7 - k) + (X_8 + Y_8 - k) + (X_9 + Y_9 - k)$
    (f) $X_1^2 Y_1^2 + X_2^2 Y_2^2 + X_3^2 Y_3^2 + X_4^2 Y_4^2 + X_5^2 Y_5^2 + X_6^2 Y_6^2$
    (g) $(X_1 - Y_1) + (X_2 - Y_2) + (X_3 - Y_3) + \ldots + (X_8 - Y_8)$
    (h) $(X_1 + Y_1) + (X_2 + Y_2) + \ldots + (X_N + Y_N)$
    (i) $(X_2 - Z_2) + (X_3 - Z_3) + (X_4 - Z_4)$
    (j) $[X_1 - (a + b)] + [X_2 - (a + b)] + \ldots + [X_N - (a + b)]$

11. Express each of the following without summation notation.

    (a) $\displaystyle\sum_{i=2}^{8} (X_i + Y_i)$

    (b) $\displaystyle\sum_{i=1}^{N} (X_i + k)$

    (c) $\displaystyle\sum_{j=2}^{4} (X_j - Y_j)$

    (f) $\displaystyle\sum_{i=1}^{3} (X_i + Y_i - k)$

    (g) $\displaystyle\sum_{i=5}^{9} X_i^2$

    (h) $\displaystyle\sum_{i=1}^{3} X_i^2 Y_i^2 Z_i^2$

(d) $\displaystyle\sum_{i=2}^{10} (X_i + 2)$ (i) $\displaystyle\sum_{i=2}^{3} (X_i + a^2)$

(e) $\displaystyle\sum_{i=4}^{7} (X_i + Y_i + k)$ (j) $\displaystyle\sum_{i=7}^{15} (X_i - Z_i + a)$

12. If $X_1 = 4$, $X_2 = 3$, $X_3 = 5$, $X_4 = 6$, $X_5 = 2$, $Y_1 = 2$, $Y_2 = 3$, $Y_3 = 1$, $Y_4 = 0.7$, $Y_5 = 6$, and $k = 5$, determine the value for the following.

(a) $\displaystyle\sum_{i=1}^{N} X_i$ (f) $\displaystyle\sum_{i=1}^{N} Y_i - Nk$

(b) $\displaystyle\sum_{1}^{3} Y_i$ (g) $\Sigma Y - \Sigma X - Nk$

(c) ΣXY (h) $\Sigma X + \Sigma Y$

(d) $\displaystyle\sum_{i=1}^{3} (X_iY_i + k)$ (i) $\displaystyle\sum_{i=2}^{4} X_i$

(e) $\Sigma(Y - k)$ (j) $\displaystyle\sum_{i=2}^{5} X_i$

13. Using the values for X in Exercise 12, demonstrate whether $\Sigma X^2 = (\Sigma X)^2$.

14. Using the values in Exercise 12, show whether $\Sigma XY = \Sigma X \Sigma Y$.

15. Prove that $\Sigma Xk = k\Sigma X$. Verify your proof by making up your own five values for X and a value for the constant k.

16. Prove that $\Sigma(X - Y + Z) = \Sigma X - \Sigma Y + \Sigma Z$. Verify your proof by making up six values for X, Y, and Z.

17. Demonstrate using summation Rule II that:
$\Sigma(X - \bar{X}) = \Sigma X - N\bar{X}$, where $\bar{X}$ (arithmetic mean) is a constant

18. If $X_1 = 1$, $X_2 = 2$, $X_3 = 4$, $X_5 = 5$, and $k = 10$, verify that:
$$\sum_{1}^{5} kX = k\sum_{1}^{5} X$$

19. Using the values for Exercise 18, verify that:
$$\sum_{1}^{5} \frac{X}{k} = \frac{1}{k}\Sigma X$$

20. Making up your own values for X and Y, convince yourself that:
$$\Sigma(X + Y)^2 = \Sigma X^2 + \Sigma 2XY + \Sigma Y^2$$

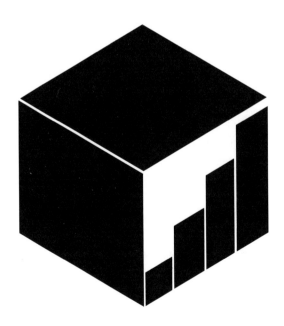

Order gave each thing view.
William Shakespeare

Organization of Data

3.1 Introduction: Why Organize Data?

When we collect information, the result is often unwieldy masses of data. If we are to understand them or present them effectively, we must summarize them in some manner. To create order out of this chaos, we can use either of two methods of presenting data. One method involves summarizing the numbers themselves and presenting them in some form, usually tabular; the other method involves presenting data in pictorial form—charts, graphs, or similar displays. Using either one of those methods to represent a set of data is part of the descriptive function; this should, in turn, help us understand the data set. Now let's see how tables and graphs can help us conceptualize whatever it is that we have found.

3.2 Grouping Data into Tabular Form

Suppose that all freshmen who enter Statistics State Teachers College immediately are required to take an intelligence test. So the freshman class of 150 students marches into Anxiety Hall to take the test. Now suppose that you, a work-study student assigned to the dean's office, are asked to transfer the individual scores to a master list. Further assume that you are instructed to organize these data in such a way that you can get an immediate feel for the distribution of IQ scores. Dutifully, you write the scores down on a piece of paper. The result is Table 3.1.

Table 3.1. IQ Scores of 150 College Freshmen

105	93	97	101	115	149	135	120	130	140
110	93	109	113	98	111	100	102	107	103
90	142	111	108	102	109	107	119	113	96
120	135	91	110	117	104	105	120	114	92
110	120	102	92	114	99	112	107	99	100
115	115	90	136	110	106	123	109	114	109
117	114	98	106	110	104	134	109	127	113
119	113	116	124	123	110	136	132	116	108
121	112	141	109	116	109	141	117	134	93
92	110	109	122	109	97	93	107	104	108
87	89	121	111	110	103	114	113	150	156
104	117	114	110	121	107	106	114	142	114
120	112	116	109	111	113	114	98	113	112
121	99	109	123	111	116	104	99	109	117
109	109	110	97	105	102	109	101	97	103

As you mull over these scores, you have to admit that your efforts up to this point are clearly insufficient. It is just about impossible to tell by eye the manner in which these 150 scores distribute themselves. You ponder what to do. Then, *eureka!* With pen in hand, you array all potential scores from lowest to highest, placing a tally mark alongside a score every time it occurs. This technique generates the frequency distribution presented in Table 3.2.

Table 3.2. Frequency Distribution of IQ Scores of 150 College Freshmen

Score	Tally	f	Score	Tally	f	Score	Tally	f
87	/	1	110	⧸⧸⧸⧸ ⧸⧸⧸⧸	10	133		0
88		0	111	⧸⧸⧸⧸	5	134	//	2
89	/	1	112	////	4	135	//	2
90	//	2	113	⧸⧸⧸⧸ //	7	136	//	2
91	/	1	114	⧸⧸⧸⧸ ////	9	137		0
92	///	3	115	///	3	138		0
93	////	4	116	⧸⧸⧸⧸	5	139		0
94		0	117	⧸⧸⧸⧸	5	140	/	1
95		0	118		0	141	//	2
96	/	1	119	//	2	142	//	2
97	////	4	120	⧸⧸⧸⧸	5	143		0
98	///	3	121	////	4	144		0
99	////	4	122	/	1	145		0
100	//	2	123	///	3	146		0
101	//	2	124	/	1	147		0
102	////	4	125		0	148		0
103	///	3	126		0	149	/	1
104	⧸⧸⧸⧸	5	127	/	1	150	/	1
105	///	3	128		0	151		0
106	///	3	129		0	152		0
107	⧸⧸⧸⧸	5	130	/	1	153		0
108	///	3	131		0	154		0
109	⧸⧸⧸⧸ ⧸⧸⧸⧸ ⧸⧸⧸⧸	15	132	/	1	155		0
						156	/	1

$$\Sigma f = N = 150$$

If we now inspect this table, we see that the scores are spread out widely, and that over half the scores have one or no tally mark. But even with this type of array, it is hard for any but the most experienced eye to see the shape of the distribution. Moreover, if we assume that many other arrays are even more complex than this one, it is obvious that something must be done in order to distinguish the various patterns that the data can take. In actual research, this would be accomplished by creating a *grouped frequency distribution* with appropriate class intervals to accommodate a number of different scores.

3.2.1 The Grouped Frequency Distribution

Anyone who has filled in a questionnaire, applied for credit, or read a newspaper has some familiarity with the grouped frequency distribution. What is a grouped frequency distribution? Take a peek at Table 3.3. Recognize it? Sure you do. Still, we'd lay odds that at some time you have wondered about its purpose and manner of construction. From the direction our present discussion is taking, you should be able to see the purpose of this distribution. Its manner of construction, on the other hand, will be explained shortly.

Before we show you how to construct the class intervals of the grouped frequency distribution, we must decide on the number of intervals to use. This decision is largely arbitrary; it depends on such factors as the range of the distribution and the data themselves. However, common sense tells us to select few enough intervals so that we can readily perceive the resulting distribution, but not so few that the group pattern becomes hidden and our ability to identify the individual scores becomes inhibited. Common sense also tells us to make each interval width, which is a function of the number of intervals, of sufficient size that it represents the data effectively. For example, if we were referring to the price of automobiles, intervals of 20 dollars would seem ridiculous and would produce too many intervals. If, on the other hand, we used the same interval width to summarize prices of socks, we would effectively destroy all the information, since virtually all prices would fall in the 0- to 20-dollar range. However, there is a solution to this thorny problem: we can restrict the number of class intervals to a minimum of *five* and a maximum of *fifteen*. This range has proven sufficiently accurate to accommodate most social science data.

If we keep this general rule-of-thumb in mind, the construction of the grouped frequency distribution becomes rather straightforward. In fact, all we need to remember are these three simple steps:

Step I: Find the difference between the highest and lowest score values contained in the original data. Add 1 to obtain the total number of potentially different score values.

Step II: Divide the above figure by the number of class intervals (you have decided on) to obtain the width (symbolized as i) for each interval.

Step III: Chart the class intervals by making the lowest score value the lower limit of the first class interval. Add to this $i - 1$ to obtain the score at the upper limit for the first class interval. Begin the next class interval at the number following the upper limit of the preceding interval. Follow these procedures for each successive interval until the scores are included in their appropriate class intervals.

Now we are ready to assign each score to the class interval to which it belongs. In making this assignment, it's a good idea to move systematically through the unordered data, making a tally mark next to the appropriate class interval whenever a score appears, rather than to jump around in an attempt to count all scores of a certain value. Then total the tally marks to obtain the class frequencies.

In the present example, our highest number is 156 and our lowest is 87. In accordance with the procedure outlined in Step I, we take the difference between the highest and lowest numbers and add 1, yielding 70. In this particular example, we have selected fourteen class intervals. The width for each interval is, therefore, 70/14 or 5 (Step II).* Since 87 is the lower limit for the first class interval, the upper limit is $87 + (5 - 1)$, or 91. The next interval would then start at 92 and end at 96; the next would start at 97 and end at 101; and so on for each interval (Step III). The grouped frequency distribution, excluding tally marks, for this example appears in Table 3.3.

From the grouped frequency distribution, we can obtain a better representation of the distribution of 150 IQ scores. Among other things, it shows that almost half the scores are located between 107–116, which lies approximately at the center of the distribution. As we can see, very few students scored at the extremes. Insights into the character of a distribution are frequently based on such simple findings.

Table 3.3. Grouped Frequency Distribution of IQ Scores of 150 College Freshmen

Class Interval	f	Class Interval	f
87–91	5	122–126	5
92–96	8	127–131	2
97–101	15	132–136	7
102–106	18	137–141	3
107–111	38	142–146	2
112–116	28	147–151	2
117–121	16	152–156	1
			$N = 150$

*Note that in computations involving classified distributions, the *midpoint* will be used to substitute for each score in the interval. For this reason, we recommend the choice of an odd number for i whenever possible. Nothing is sacred about this suggestion; it just makes the midpoint a whole number of units, thus simplifying computation.

3.2.2 The True Limits of a Class Interval

The grouped frequency distribution seems to create missing units between each class interval. For example, between the first interval of Table 3.3 and the next interval, there is a missing unit: the upper limit of the first interval is 91, while the lower limit of the second interval is 92. The unit between 91 and 92 appears to be missing and unaccounted for within either interval. Why did this happen? Because the value for the upper and lower limits for each interval are the *apparent limits,* not the *true limits.*

In Section 2.4, we said that the true limits in which a measurement falls are equal to the apparent numerical value plus and minus one-half of the unit of measurement. This is also true of these values even after we condense the data into class intervals. So even though we write the limits of the first interval as 87–91, the interval's *true* limits are 86.5–91.5. You can see that our missing unit is divided evenly between the classes: half the unit goes to the upper half of the first interval, and the other half of the unit goes to the lower half of the second interval. But remember—the true limits of the class interval are *not* the same as the true limits for a measurement or score.

3.2.3 The Midpoints of Class Intervals

We need to discuss one more feature of grouped frequency distributions. When we classify a set of data, we lose the ability to say exactly what the original numbers were. To illustrate, let's go on with our example. If a student in the group of 150 cases has an IQ of 107, he is simply counted as one of the 38 individuals who make up the frequency for the interval 107–111. We cannot tell from the grouped distribution exactly what the IQ of those 38 individuals are; we can only tell that they fall within the limits 107–111. What *are* the IQs of these individuals?

One solution is to say that they are all 109, the *midpoint* of the interval. The midpoint of any class is the number that falls exactly halfway among the numbers included in the interval. You can see that 109 falls halfway between the true limits of 106.5–111.5. Similarly, all five cases falling into the class interval 87–91 will be called by the midpoint 89; all eight cases falling into the class interval 92–96 will be called 94; and so on for each of the other class intervals.

The true limits of an interval can also be defined in terms of the midpoint. We can then write:

$$True\ Limits\ =\ Midpoint\ \pm\ \frac{1}{2}\,i$$

where i equals the interval width. Thus, if we were given the midpoints of a grouped frequency distribution, we should have no trouble finding the true limits.

3.3 The Cumulative Distribution

Arranging data into a *cumulative distribution* is really helpful. It allows us to obtain the number (or the proportion) of cases in a distribution below or above each class interval (or boundary), but not the cumulative frequencies (or percentages) below or above the countless intermediate values within each interval. In Chapter 4, we'll use the cumulative distribution to compute any position, according to the user's interests and needs.

Table 3.4 shows the process of constructing a cumulative tabulation. The cumulative frequency for a class interval stands for the frequency of cases below the upper true limit of the interval. For example, in the fourth interval from the top of the table, the entry 46 in the cumulative frequency column indicates that 46 freshmen scored lower than the upper true limit of that interval, which is 106.5. This cumulative frequency is obtained by combining successive class frequencies: that is, $5 + 8 + 15 + 18 = 46$. Note that the last entry in the cumulative frequency column must always equal N.

The cumulative percentage, also shown in the table, is computed by dividing the entry for the cumulative frequency column by the total number of cases (N) and multiplying by 100. Reading from the cumulative percentage column, we find that 3 percent of the cases are under 91.5; 9 percent of the cases are under 96.5; and so on for each of the class intervals. Note that the last entry must equal 100 percent, since all the scores must fall below the upper true limit of the last interval.

Table 3.4.   Cumulative Distribution of IQs of 150 College Freshmen

Class Interval	f	Cum f	Cum %
87–91	5	5	3
92–96	8	13	9
97–101	15	28	19
102–106	18	46	31
107–111	38	84	56
112–116	28	112	75
117–121	16	128	85
122–126	5	133	89
127–131	2	135	90
132–136	7	142	95
137–141	3	145	97
142–146	2	147	98
147–151	2	149	99
152–156	1	150	100
	$N = 150$		

3.4 Grouping Noninterval Data into Tabular Form

So far, we've only talked about intervally scaled data. But noninterval data are presented in tabular form, too. Table 3.5 shows this kind of *qualitative frequency distribution*. In this case, it contains the responses to a 1976 Gallup Poll that asked a national sample of 1,500 adults, "Would you favor or oppose requiring all young women to give one year of service to the nation, either in the military forces, or in nonmilitary work here or abroad, such as VISTA or the Peace Corps?"

The problem here does not concern the selection of the proper interval width for numerical distributions, but definition; that is, we must be exact about the criteria for each category. For example, do the categories in Table 3.5 give us sufficient information or not? Are there degrees of favorable responses to the question of national service for women? How are individuals who favor some forms of national service for women but who are vehemently opposed to military service classified? Would the "favor" category suffice? In short, are the categories too broad for an acceptable degree of precision?

Consider another example. Suppose that we were to classify respondents to a questionnaire on the basis of political ideology, and that we used the following categories: "conservative," "moderate conservative," "middle-of-the-road," "moderate liberal," and "liberal." In this case, how do we distinguish moderate conservatives and liberals from each other and from those calling themselves middle-of-the-road? Unless we define the categories explicitly, we'll have trouble classifying individuals on the basis of such qualitative criteria. Therefore, wherever possible, it's advisable to use standard criteria when creating qualitative classifications.

Table 3.5. National Service for Women

Categories	f
Oppose	765
Favor	600
No Opinion	135
	$N = 1,500$

Source: The American Institute of Public Opinion, *The Gallup Opinion Index* (February 1977): 20. Data by permission.

3.5 Graphing Techniques

There are always some people who would rather not read tables, who could understand the information better if it were presented in pictorial form. Our prehistoric ancestors undoubtedly knew this when they made the first cave drawings. Similarly, the Egyptians, Greeks, and Romans used drawings and sculpture to convey information about their respective societies. Thus, art was used to carry information throughout the ages. Art is also valuable to us in *describing* information.

Graphs, the pictorial forms that follow, are not meant to substitute for tabular construction. Rather they are meant as visual aids that help us to describe and think about the shape of the distribution. In fact, you cannot plan or construct a graph until you have prepared the corresponding table. The graphic forms shown here correspond to both qualitative and quantitative distributions. In the following chapter, we'll consider the graphic equivalent to the grouped cumulative distribution.

3.5.1 The Pie Chart

The *pie chart* is one of the more popular pictorial devices. It is used to convey an impression of the pattern of a qualitative distribution. The test of any graph is its ability to carry information clearly. So let's take a look at the pie chart, in view of this standard. Table 3.5 arranges our earlier Gallup Poll data. Although we can scan these data without a pie chart, it helps us to interpret the results of the survey more easily.

Figure 3.1 shows a pie chart. Using a protractor to lay off the angles, we make the pie chart by dividing the 360 degrees of a circle in proportion to the percentage of cases in each survey category. Thus, we assign 51 percent of 360 degrees, or 183.6 degrees, to the group that opposes national service for women. Similarly, we assign 144 degrees to the group that favors such service. We assign the remainder of the pie to those who hold no opinion.

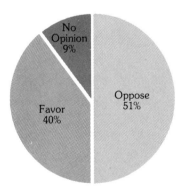

Figure 3.1. Pie Chart Illustrating a Survey on National Service for Women

Source: The American Institute of Public Opinion, *The Gallup Opinion Index,* (February 1977): 20. Data by permission.

We have placed the wedges next to one another, according to relative size. You don't necessarily need to follow this system, since *nominally* scaled variables provide no rank order for the categories. But when you are displaying *ordinally* scaled data, place the categories according to their ranks, regardless of the size of the wedge.

As Figure 3.1 illustrates, the pie chart shows qualitative data by means of a visual representation, which is often more helpful than tabular construction. The pie chart—and, for that matter, all graphs—can be made even more effective by carefully shading and coloring. The potential problem with pie charts is that they can't accommodate many categories easily. Generally speaking, if you have to divide the pie into more than seven pieces, the chart becomes rather difficult to interpret.

3.5.2 The Bar Graph

The *bar graph* (see Figure 3.2) is another pictorial device that can highlight the prominent features of a qualitative frequency distribution. We draw a vertical bar for each category; the height of the bar represents the number (or the percentage of cases) of observations for that category. The space between the bars emphasizes the discrete character of the data.

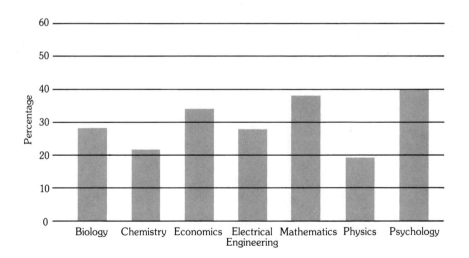

Figure 3.2. Young Doctorate Faculty* Investigators** as a Percentage of All Faculty Investigators (1974)

*Those who had held doctorates seven years or less at the time of each study.
**Spending 20 percent or more of their time in research.
Source: Adapted from National Science Foundation, *Science Indicators: 1974* (December 10, 1975): 129.

Since we are not assuming any order underlying the classes displayed in Figure 3.2, we can set them down along the horizontal axis in any order that seems plausible. In Figure 3.2, they are arranged alphabetically. Very often, the classes are positioned according to either their ascending or descending order of frequency.

Bar graphs can be displayed horizontally. In such a case, we place the frequencies or the proportions of cases along the horizontal axis, and place the categories themselves along the vertical axis. Figure 3.3, which shows college entrance rates determined by student ability and socioeconomic status of parents, illustrates this change with an ordinally scaled variable. Recall our preceding discussion on ordinal variables. Here, such variables are treated in the same way, except that the categories are placed in accordance with their rank order along the axis in question.

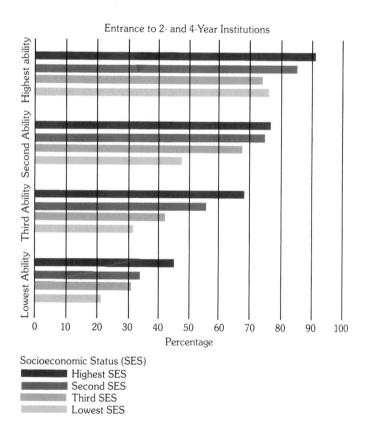

Figure 3.3.   College Entrance Rates, by Students' Ability and Socioeconomic Status of Parents (1967)

Source: Executive Office of the President: Office of Management and Budget, *Social Science Indicators: 1973* (1973): 88.

The fact that this graph depicts two variables illustrates the utility of pictorial devices in helping us to discern relationships among variables. As we can see from Figure 3.3, there appears to be an association between the two variables (student ability and socioeconomic status of parents) and the third variable (college entrance rates). Implicitly, entrance rates is the dependent variable and student ability and socioeconomic status of parents are the independent variables. Later on in this text more precise methods of determining the degree of association between variables will be discussed.

3.5.3 The Sliding-Bar Graph

When we present data, we are not limited to the format of the simple bar graph. In fact, some people who find bar graphs static, unimaginative, or boring substitute pictures of the things represented in the graph. The Bureau of the Census is especially fond of using human figures to represent populations. When we want to find the most effective display for our data, we might even design our own graph for our tabulations, if we don't think that any of the standard models is suitable. This flexibility is demonstrated by the *sliding-bar graph,* which is so named because the bars slide back and forth across a fixed origin.

Figure 3.4 shows how the Department of Justice graphically depicts the percentage of crimes against property that were cleared by arrest as compared

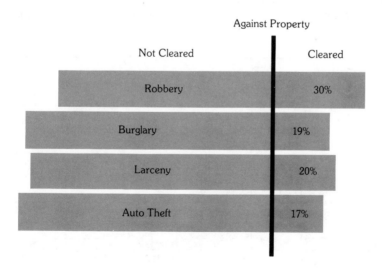

Figure 3.4. Crimes Cleared by Arrest for 1972

Source: Department of Justice, *Crimes in the United States: 1972* (August 8, 1973): 32.

to those that were not cleared. The graph, a succession of bars of equal length and width, is placed at right angles to the vertical axis, which represents the boundary between percentage cleared and percentage not cleared. There are as many bars as there are categories, and each bar has as many segments as there are category divisions; segments are proportional to division percentages. Each bar then represents 100 percent. The displacement of each bar relative to the vertical axis is proportional to category percentage, and the different displacements are supposed to emphasize the differences among the categories.

3.5.4 The Histogram

The *histogram* is the graphic equivalent of the grouped distribution for interval-level data. It consists of a set of adjacent bars whose heights are proportional to either the absolute frequencies or to the proportions of cases in each interval of the variable, as occurs in Figure 3.5.

The most noticeable feature of the histogram is its structural simplicity. Bars are understood more easily than numbers. The histogram shows the relative concentration of data in each interval as well as the shape of the distribution.

However, if you assume that the heights of bars are proportional to the frequencies (or the proportion of cases) in each class interval (which is often true), the histogram can be very misleading unless all the intervals are equal in width. If we combined the intervals 111.5–116.5 and 116.5–121.5, we would find a larger number of cases in the combined interval; therefore the histogram

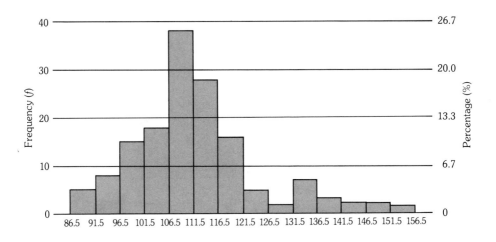

Figure 3.5. Histogram of IQ Scores of 150 College Freshmen

Source: Table 3.3

would not represent the data from Table 3.3 accurately. How can we compensate for this inconsistency in interval width? We can make the bar for the combined interval only half as high. Why? Because we have doubled the width and, on the average, have included twice as many cases in the combined interval as there would be in either of the two intervals of regular size. If we made this correction, we would get a histogram that was much more like the one we originally obtained (Figure 3.5). In general, then, we should always think in terms of areas rather than heights. We should keep the areas of bars proportional to the number of cases, as well. When all class intervals are of equal widths, the heights of bars will also be proportional to the frequencies (or percentages). If the width of each bar is then taken to be one unit, and if heights are represented as proportions, then the total area under the histogram will, of course, equal unity. Thus, for our present example:

$$1(5/150) + 1(8/150) + 1(15/150) + \ldots + 1(1/150) = 1.00$$

It will help you to think of the total area under the histogram as unity, since you'll run across this concept later in the text when we discuss the area under the normal curve.

3.5.5 The Polygon

It is easy to convert a histogram into the much-used *polygon*. All we need to do is connect the midpoints of the tops of the bars with straight lines (see Figure 3.6). Ordinarily, we would not use both types of figures in the same graph; however, by superimposing the polygon on the histogram, we see that the areas under both figures are identical. For every triangle that lies inside the polygon but outside the histogram, there is an identical triangle under the histogram but above the polygon (see inset in Figure 3.6). Thus, we can also take the area under the polygon to be unity.

When data are compressed into a relatively small number of broad intervals, the straight lines that are used to connect the midpoints of tops of bars form a jagged polygon. But we could reasonably suppose that this jagged appearance would become smoother if we increased the number of observations and used many more, rather small class-intervals. In the limit, the polygon that was drawn from the histogram would give us an approximation of the smooth curve that presumably would emerge if class intervals were made infinitesimally small and if the number of observations were unlimited. We use these curves throughout the text to show various hypothetical and theoretical probability distributions. But you should never use a smoothed polygon for a discrete variable, since it suggests a continuously scaled variable.

Polygons are particularly useful when we wish to present a comparison of two or more distributions on the same graph. They do not blur their respective

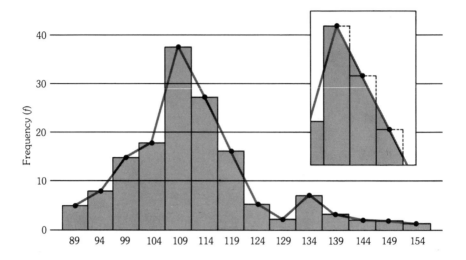

Figure 3.6. Histogram Transformed into Polygon

outlines, as histograms do. For an example, let's take the family incomes for blacks and whites in the United States in 1968. Suppose that we wanted to compare these two racial groups with respect to income. We would simply construct polygons for both distributions on the same graph (see Figure 3.7). The shaded section reveals the areas under the polygons that overlap and correspond to each other. Note the black-white and white-black overlap at the lower and upper income levels, respectively. The space depicted on the graph between the respective polygons gives us an intuitive feel for the data. In this way, we can use polygons as visual aids to help us think about the differences and similarities between distributions.

3.6 Caution: The Misuses of Graphing Techniques

A graph can be attractive, especially if it's drawn by an artist with a flair for the unusual. For example, take a look at Figure 3.8. Notice how the two graphs carry entirely different messages, even though both graphs are constructed from the same data. Notice, too, how they correspond along their horizontal axes but differ markedly with respect to the scale units on their vertical axes. Someone (perhaps a member of an irate citizens' group that was seeking more police protection) who wanted to show a dramatic increase in crime from June through September would naturally call upon the chart shown in Graph I (an example of what Huff calls the "Gee Whiz" chart). On the other hand, some-one (perhaps a member of the Chamber of Commerce, or the incumbent

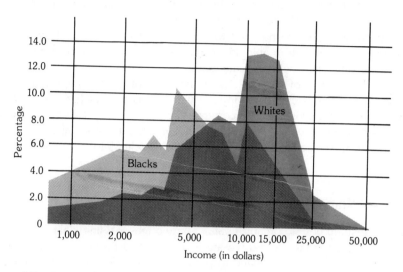

Figure 3.7. Income Distribution of White and Black Families in the United States (1968)

Source: Adapted from Bureau of the Census, *Measures of Overlap of Income Distributions of White and Negro Families in the United States* (1970): 9.

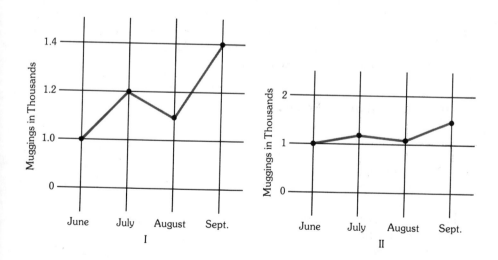

Figure 3.8. Varying the Vertical Units while Maintaining the Horizontal Units—An Illustration

mayor) who wanted to de-emphasize the increase would demonstrate that point with the "rubber band" effect produced by Graph II.

Bar graphs can also be used to accomplish similar types of distortions. For example, by cleverly manipulating the axes, it is possible to minimize or magnify the difference between categories.

Clearly, the use of such devices is not part of good "graphsmanship." However, it is equally clear that anyone who wants to stretch the scale units along the horizontal axis is just as correct as someone who wants to shrink the vertical axis, and vice versa. To avoid graphic anarchy, most social scientists lay out the vertical axis so that its height is approximately ⅔ to ¾ of the horizontal axis.

Where We Stand

Why do we study how to organize data? For two reasons: first, data often seem to be an unwieldy mass, and if we are to understand them we must organize them somehow. Second, we need to prepare the data for the inferential function of statistics. Since samples are sets of data, we need some method for describing samples so that we can use the information contained in them to make inferences about populations. We cannot make an inference about anything when we cannot describe the object of our interest.

To help us make such a description, we place data in either tabular or graphic form. Usually, we need to make tables so that we can create graphs from those tables; but both techniques are independent aids that help us to describe the object of our interest. For some purposes, tabular arrays and graphs are useful in themselves. However, they are also the step that comes before the development of numerical measures of data description. These measures allow us to make inferences from samples to populations. We will be taking a look at numerical measures—but first let's further investigate the cumulative distribution and its equivalent graphing technique.

Terms and Symbols to Remember

Frequency Distribution
Frequency (f)
Grouped Distribution
Class Interval
Midpoint
True Limits of the Class Interval
Cumulative Distribution
Qualitative Distribution
Pie Chart
Bar Graph

Sliding-Bar Graph
Histogram
Polygon

Exercises

1. The following are verbal Scholastic Aptitude Test scores of 200 high
    school seniors applying for admission to a college in New Jersey.

525	607	540	524	620	514	513	608	462	613
677	525	510	620	533	516	567	543	498	597
567	600	751	568	594	668	480	467	448	587
561	441	536	529	690	547	536	550	500	591
555	510	598	489	500	444	522	532	536	632
485	525	701	532	675	534	564	489	595	517
455	700	525	555	569	535	515	519	546	594
562	800	526	525	487	564	489	538	565	483
723	557	587	565	508	515	536	557	571	579
609	667	772	560	553	499	622	549	539	731
610	549	601	570	680	468	518	549	705	489
652	621	600	574	539	519	500	542	696	718
508	427	467	560	545	592	520	564	567	593
587	684	632	617	587	601	473	555	545	456
677	601	651	690	645	619	565	694	567	598
487	511	619	627	628	543	520	523	609	601
502	603	544	545	525	625	540	442	525	520
509	523	571	678	643	543	609	443	478	542
617	546	576	532	512	539	487	499	634	532
789	499	544	512	643	611	600	678	743	645

    (By just inspecting the data in their present form, can you really discern a
    pattern?)

    (a) Construct an ungrouped frequency distribution ($i = 1$) for the data.
    (b) Construct a grouped frequency distribution.
    (c) Justify why you selected the number of intervals you did.

2. Below are a political pressure group's cumulative percentage ratings for
    the 1st Session of the 92nd Congress. Each score represents a senator's
    percentage of agreement with the group's position for 200 roll-call votes.
    (a) Construct an ungrouped frequency distribution ($i = 1$) for the data.
    (b) Construct a grouped frequency distribution with five class intervals.
    (c) Construct a grouped frequency distribution with ten class intervals.
    (d) Construct a grouped frequency distribution with twenty class in-
         tervals.

(e) Discuss the advantages and the disadvantages of employing interval widths of $i = 20$, $i = 10$, and $i = 5$.

82	77	43	71	23	11	22	75	9	83
29	81	1	62	3	6	12	10	79	49
0	10	16	60	56	95	87	32	66	10
26	57	85	36	2	93	18	48	93	12
89	53	33	56	1	40	65	28	33	33
99	67	14	35	3	33	61	42	94	14
23	37	11	6	73	85	18	12	85	88
70	91	17	47	71	23	67	7	8	52
0	45	9	57	10	20	49	54	40	89
55	63	78	15	18	5	64	95	34	8

3. Given the low score and the high score, as well as the interval width, prepare the class intervals for the following distributions.
(a) 20, 91, $i = 9$ 　　　　　　　　　　(e) 1,201, 9,800, $i = 860$
(b) .05, .94, $i = .09$ 　　　　　　　　(f) 1, 100, $i = 5$
(c) 1.6, 11.5, $i = 2$ 　　　　　　　　(g) -101, $-1,000$, $i = 90$
(d) 21, 460, $i = 44$ 　　　　　　　　(h) .0021, .0080, $i = .006$

4. Specify the true limits for the first and the last class intervals for each of the distributions listed in Exercise 3.

5. Using the data in Exercise 2, use $i = 10$ for the class intervals.
(a) Prepare a cumulative frequency distribution.
(b) Prepare a cumulative percentage distribution.

6. The following are 100 reaction times (in seconds) for a psychological experiment.

44	71	68	66	74	66	71	67	41	66
66	63	68	72	72	71	67	58	66	62
67	60	74	65	70	58	74	59	68	67
70	68	62	68	72	50	76	59	69	62
63	62	65	69	62	70	40	62	74	66
55	60	65	63	57	66	68	67	64	60
71	67	62	55	60	72	73	44	61	67
63	62	61	71	60	68	67	67	59	69
64	71	61	66	67	67	74	76	63	52
76	63	71	76	55	81	55	68	51	45

(a) Group these data into a table that has the class intervals 40–44, 45–49, 50–54, etc.
(b) In order to show how the choice of different interval widths can alter the overall shape of a distribution (see also Exercise 19), regroup the data into tables that have:
(i) The true limits 29.5–40.5, 40.5–50.5, 50.5–60.5, etc.
(ii) The true limits 32.5–42.5, 42.5–52.5, 52.5–62.5, etc.

7. Take your university or college catalogue in which the teaching staff is listed, and construct a qualitative table showing how many teachers are professors, associate professors, assistant professors, and instructors. How else might you show how the teaching staff is distributed?

8. Using the data in the table, construct a pie chart for the adult population of the United States by religion.

Adult Population by Religion in the United States

Religion	Percent
Protestant	66
Roman Catholic	26
Jewish	3
Other religion	1
No religion	3
Religion not reported	1

9. Using the data in the table, construct a pie chart for the "rented" population by type of crime.

Household Crimes in Portland: Victimization Rates (per 100,000 households), by Type of Crime and Form of Tenure, 1971–1972

Type of Crime	Owned or Being Bought	Rented
Burglary	141.8	162.6
Household larceny	155.5	141.9
Motor-vehicle theft	30.0	38.8

10. Represent the tabled data below by a bar graph.
    (a) How should the bars be ordered? Explain.
    (b) Can the bars touch one another? Explain.

Distribution of Government Scientists

Agency	Percent
Commerce	7
DOD	45
HEW	4
Interior	8
NASA	7
USDA	15
Other	16

11. Construct a bar graph for the adult population of the United States by religion, using the data of Exercise 8.

12. Construct a bar graph for the "owned or being bought" population of Portland by type of crime, using the data of Exercise 9.

13. Construct percentage bar graphs in both vertical and horizontal positions from the data listed below.

American Class Structure	Percentage
Lower class	10
Working class	40
Lower middle class	40
Upper middle class	9
Upper class	1

14. Construct a bar graph showing percentage of armed robbery (any weapon) for each geographic region, and construct a sliding-bar graph for all geographic regions, locating the vertical axis between armed and strong-arm robbery.

	Northeastern States	North Central States	Southern States	Western States
Armed (any weapon)	70	64	67	61
Strong-arm (no weapon)	30	36	33	39

15. Draw a frequency histogram for the distribution obtained in Exercise 1.

16. Draw a percentage histogram for the distribution obtained in Exercise 1.

17. Draw both frequency and percentage histograms for the data in Exercise 6(a).

18. Draw a frequency polygon for the data in Exercise 6(a).

19. Draw frequency polygons for the distributions obtained in Exercise 6(b). Also construct percentage polygons for the same two distributions.

20. Does the histogram or the polygon adhere more closely to the grouped frequency table? Explain.

21. Suppose that you wished to compare two distributions by superimposing the graph of one on the graph of the other. Would you select the histogram or the polygon? Explain.

22. Using the data listed below, construct a graph:
    (a) Showing that the student grades are similar.
    (b) Showing that the student grades are dissimilar.
    (c) Showing the true relationship between the grades.

Student	Grade-Point Index	Student	Grade-Point Index
a	3.20	f	2.60
b	3.05	g	3.30
c	3.10	h	2.95
d	2.85	i	2.90
e	2.75	j	2.80

CHAPTER

4

Percentiles

4.1 Introduction: Why a Frame of Reference?

We've referred to scores and measurements in the last two chapters, and in the process we've taken for granted that a single score or measurement describes some characteristic of an individual (such as mathematical ability, IQ, or mechanical aptitude). But now it's time to emphasize that a score or measurement has meaning only in relation to the rest of the distribution of which it is part.

Suppose that we trapped a Loch Ness monster who measured fifty feet in length. Should we call our monster "big" or "small"? Clearly, we need more information before we can pass judgment on the creature's size. What kind of information? Well, for instance, just where does our Nessie stand in relation to the rest of its breed or to other Nessies of similar age and sex?

This is why standard scores are used. A standard score is a score that allows us to find out where it is located in a distribution. Referring back to our deep-sea monster, if we were to be told by zoologists the percentage of all Nessies who measure less than fifty feet, we would then have a frame of reference, a standard, for making comparisons among groups. In fact, what the zoologists cited would be the *percentile rank* of the Nessie's length. The percentile rank of a score, then, represents the percentage of cases in a distribution that achieved scores lower than the one cited. Therefore, if we say that a score has a percentile rank of, say, 95, we mean that 95 percent of the comparison group scored lower. But remember that a single score or measurement does not give us this information.

In this chapter, we will explore just one type of standard score: the percentile rank. We will discuss the other types later. Keep in mind that the use of a standard score such as a percentile rank is an attempt to make a single score or measurement more meaningful.

4.2 The Ogive

When a graph is used to present a cumulative percentage distribution, it is called an *ogive* (Figure 4.1). The term was coined by the English statistician and biologist, Sir Francis Galton (1822–1911), because of its resemblance to the curved rib of a Gothic vault, which the cumulative curve often resembles. The ogive is constructed on a pair of perpendicular axes, just like the polygon. The horizontal axis represents the values for the upper true limits of each class interval, and the vertical axis indicates the percentage of observations for each interval. A dot is then placed directly above the upper true limit of the class boundary, at whatever height is appropriate, to indicate the proportion of cases

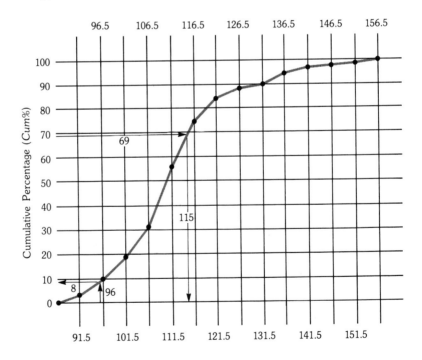

Figure 4.1. Ogive Showing IQ Scores of 150 College Freshmen

Source: Table 3.4

less than the upper true limit of the interval. After plotting all interval values with their corresponding percentages, the dots are joined by straight lines.

The ogive is used to obtain a reasonable approximation of the percentage of cases in a distribution below a certain value. To read a percentile rank—let's say a score of 96—from the ogive drawn in Figure 4.1, we first locate the value of interest, 96, along the horizontal axis. We then construct a line perpendicular to the horizontal axis at 96 so that it intercepts the ogive. Reading directly across on the scale to the left, we see that the percentile rank of 96 is approximately 8. Thus, only 8 percent of all students had IQ scores less than 96. If we reverse the procedure, we obtain the IQ score for any given *percentile*. For example, the IQ for a percentile of 69 is approximately 115.

It is helpful to know the form of the ogive, since it tells us whether the cumulative distribution has been cumulated up (as it has in this case) or down. To cumulate down, we simply determine the cumulative percentage for each interval—but in reverse order, moving from high to low value. The result is a diagonally shaped ogive, which has its highest point directly opposite the vertical axis, and which descends as we trace its path along the horizontal axis away from the origin. Thus, we have access to the cumulative values above the lower true limits of each class boundary. Then we report percentile ranks in terms of the percentage of the same comparison group above a particular value; for example, 92 percent of the students scored higher than 96 on the IQ test. Generally speaking, the "cumulating up" method is the more widely used of the two. This procedure is used to report standard examination scores such as those given for the National Merit tests. What high school junior or senior, about to embark for college, has not looked to see what percentage of other students nationwide scored lower than he/she did?

4.3 Computing Percentile Ranks

As we pointed out earlier, reading the percentile rank off an ogive is only an approximate procedure. A more accurate method of computing percentile ranks is by *linear interpolation* within the cumulative frequency column, to determine the cumulative frequency corresponding to a given score. Linear interpolation is an operation that's performed to find a value that is not known but that can be derived from those that are known.

Using the data in Table 4.1 (they originally appeared in Table 3.4), we once more find the percentile rank of an IQ of 96. We begin by noting that an IQ of 96 falls somewhere within the true limits of the interval 91.5–96.5. Since a percentile rank of a score is equal to the cumulative frequency of that score divided by the total number of cases in the distribution times 100,

$$PR = \frac{Cum\ f}{N}\ (100) \tag{4.1}$$

then we need to find the exact cumulative frequency corresponding to a score of 96. Inspecting Table 4.1, we can see that the cumulative frequency for a score of 96 is somewhere between the fifth and thirteenth case (that is, the cumulative frequencies at both extremes of the interval). Therefore, we are forced to interpolate within the interval 91.5–96.5 to find the exact cumulative frequency for an IQ of 96—or, more specifically, we have to determine the proportion of distance that we must move into the interval in order to locate the number of scores included up to a score of 96.

Table 4.1. Grouped Cumulative Frequency Distribution of IQs of 150 College Freshmen

Class Interval	f	Cum f
87–91	5	5
92–96	8	13
97–101	15	28
102–106	18	46
107–111	38	84
112–116	28	112
117–121	16	128
122–126	5	133
127–131	2	135
132–136	7	142
137–141	3	145
142–146	2	147
147–151	2	149
152–156	1	150
	N = 150	

To find the number of score units above the lower true limit of the interval, we subtract the score at the lower true limit 91.5 from our score of 96. We now know that a score of 96 is 4.5 score units above the lower true limit of the interval. Now, since there are five score units within the interval, a score of 96 is 4.5 fifths of the distance into the interval. So we must assume (lacking any other information) that the eight scores within the interval are distributed at equal distances throughout the interval. An IQ of 96 is, then, the 4.5/5th or the 7.2nd case (4.5/5 $[8]$ = 7.2) within the interval; or, in other words, the frequency 7.2 in the interval corresponds exactly to a score of 96. Adding this frequency to the number of cases below the lower limit of the interval brings us to a cumulative frequency of 12.2. You may appreciate this procedure more after you've examined Figure 4.2. In any case, we are ready to calculate the percentile rank for an IQ of 96, using Formula (4.1):

$$PR = \frac{12.2}{150} \ (100) = 8.13$$

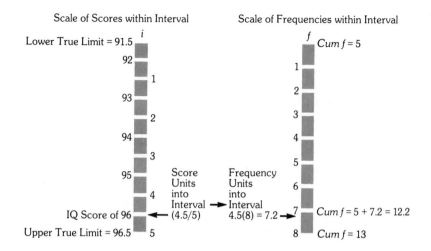

Figure 4.2. Linear Interpolation from Scale of Scores to Scale of Frequencies

Note that this answer, when rounded, agrees with the approximation obtained from the ogive.

This is the general formula for computing the percentile rank:

$$PR = \frac{Cum\, f_{\ell\ell} + \left(\dfrac{X - X_{\ell\ell}}{i}\right) f}{N}(100) \qquad (4.2)$$

where:

$Cum\, f_{\ell\ell}$ = the cumulative frequency at the lower true limit of the interval containing X,

X = the given score,

$X_{\ell\ell}$ = the score at the lower true limit of the interval containing X,

i = the width of the interval, and

f = the number of cases within the interval containing X.

To illustrate the application of this formula, we return to the values in Table 4.1 to compute the percentile rank for the IQ of 96:

$$PR = \frac{5 + \left(\dfrac{96 - 91.5}{5}\right) 8}{150}(100) = 8.13$$

8.66

But before we use this formula, let's see—so that we don't become too dependent on formulas—if we can understand the logic behind the linear interpolation from a scale of scores to a scale of frequencies.

4.4 Computing Percentiles

Sometimes we don't want to compute a percentile *rank* for a particular score, but to secure a score corresponding to a given *percentile.** This basically means that we have to reverse the procedure involved in the linear interpolation from the scale of scores to the scale of frequencies. To obtain the score, we simply interpolate from the cumulative frequency scale to the scale of scores. We begin the interpolation by first finding out the cumulative frequency corresponding to, for example, the 8.13th percentile. To do this we transpose Formula (4.1) into:

$$Cum \ f = \frac{(PR)(N)}{100} \tag{4.3}$$

Since we are interested in the score at the 8.13th percentile, the cumulative frequency of a score at that percentile is:

$$Cum \ f = \frac{(8.13)(150)}{100} = 12.2$$

If we look at Table 4.1, we can see that a cumulative frequency of 12.2 is found in the interval with the true limits of 91.5–96.5. The frequency is 7.2 frequencies into the interval, since the cumulative frequency at the lower true limit of the interval is 5, which is 7.2 less than 12.2. Since there are eight cases within the interval, the frequency 12.2 is 7.2 eighths of the way through an interval with a lower true limit of 91.5 and an upper true limit of 96.5. Actually, it is 7.2 eighths of the way into five score units. In terms of score units, then, it is (7.2/8)5 or 4.5 score units above the lower true limit of the interval. We then add together 4.5 and 91.5 to obtain the score at the 8.13th percentile, which is 96. Figure 4.3 depicts this linear interpolation graphically.

---

*If what you want to know is what score is the point below which a given percentage of the distribution falls, then what you are interested in is the *percentile*. If what you want to know is what percentage of the distribution falls below a given score, what you are interested in is the *percentile rank*.

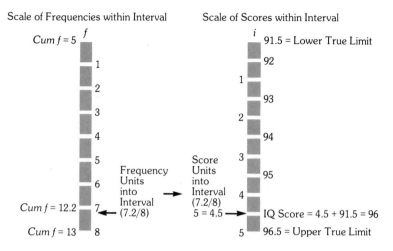

Figure 4.3.   Linear Interpolation from Scale of Frequencies to Scale of Scores

Instead of performing the linear interpolation from a scale of frequencies to a scale of scores, we may again use a general formula to compute any score corresponding to a given percentile:

$$X = X_{\ell\ell} + \left(\frac{Cum\ f - Cum\ f_{\ell\ell}}{f}\right)i \qquad (4.4)$$

where:

$X_{\ell\ell}$ = the score at the lower true limit of the interval containing $Cum\ f$,

$Cum\ f$ = the cumulative frequency of the score as determined by Formula (4.3),

$Cum\ f_{\ell\ell}$ = the cumulative frequency at the lower true limit of the interval containing $Cum\ f$,

f = the number of cases within the interval containing $Cum\ f$, and

i = the width of the interval.

To illustrate the use of Formulas (4.3) and (4.4), let us consult Table 4.1 to compute the IQ score corresponding to the 8.13th percentile. First, by using Formula (4.3), we obtain:

$$Cum\ f = \frac{(8.13)(150)}{100} = 12.2$$

Substituting the above value into Formula (4.4), we obtain a score of:

$$X = 91.5 + \left(\frac{12.2 - 5}{8}\right)5 = 96$$

Formulas (4.2) and (4.4) provide a very useful procedure for checking the accuracy of our calculations. In other words, whenever you find the percentile rank of a score, you can use Formula (4.4) to determine the score corresponding to that percentile value. Similarly, whenever you get a score corresponding to a given percentile, you can use Formula (4.2) to determine the percentile rank of that score. In this way, you check your calculations, you become familiar with the formulas, and you learn the relationships involved.

4.5 Other Fractiles: Deciles and Quartiles

A percentile (100th) is one type of *fractile*. There are other fractiles that don't partition the distribution into 100 portions of equal size—instead, they divide the distribution into different proportions. Those most frequently used are the *decile* (10th) and the *quartile* (4th).

Deciles divide a distribution into ten equal parts, each part containing 10 percent of the distribution. Therefore, the first decile, denoted as D_1, equals the tenth percentile or P_{10}; D_2 equals P_{20}; and so on. Quartiles divide a distribution into four equal parts, each part containing 25 percent of the distribution in each part. Thus, the first quartile, denoted as Q_1, equals P_{25}; Q_2 equals P_{50} (or the *median*); and Q_3 equals P_{75}.

Regardless of the type of division used, P_{22}, D_2, or Q_1 are only *points* in the distribution—not *intervals*. An individual can score *at* the twenty-second percentile, second decile, or first quartile, but not *in* the second decile or first quartile. These fractiles are merely points without dimension. An individual either scores at a given point (exactly at the first quartile, for example) or in between two points (between the fourth and fifth deciles, for example).

4.6 Caution: Know Your Reference Group

Just as a single measurement is meaningless by itself, so too are percentile, decile, and quartile ranks. Each fractile rank must always be considered in terms of the character of the *reference group*. Consider the bookish college student who obtained a percentile rank of 70 on some type of law aptitude test. Are you impressed? Well, the test was a standardized one used to assess competence in the practice of law—and all the other examinees were practicing attorneys. Now that you know this (and without passing judgment on the legal profession), isn't your impression of the student's performance different? We'll bet it is. Obviously, then, the interpretation of a fractile rank must take the character of the reference group into account.

Where We Stand

To make our data more meaningful, we obtain the percentile rank (or the percentile) either from the cumulative table, by linear interpolation between class limits, or from the ogive, using less effort and sufficient accuracy. In short, we are well on our way toward developing a type of standard score that will allow us to compare data from different distributions. Therefore, we need to sharpen the descriptive skills that we only partially developed in the last chapter. So we must introduce numerical descriptive measures of central tendency and dispersion. Although the next two chapters seem to warn against intergroup comparisons, don't be misled. The very concepts that appear to prevent such comparisons will allow us to discuss the normal curve. This curve, a theoretical model, supplies the rationale for comparing distributions that differ markedly in substance.

Terms and Symbols to Remember

Percentile Rank
Percentile
Ogive
Linear Interpolation
Fractiles (Percentiles, Deciles, and Quartiles)
Reference Group

Exercises

1. Approximate the percentile ranks of the following scores, using the ogive drawn in Figure 4.1.

    (a) 101   (f) 140
    (b) 103   (g)   97
    (c)   90   (h) 130
    (d) 146   (i) 150
    (e) 132   (j) 110

2. Approximate the scores corresponding to the following percentiles, using the ogive drawn in Figure 4.1.

    (a) 94   (f) 25
    (b) 87   (g) 23
    (c) 42   (h) 91
    (d) 60   (i) 50
    (e) 43   (j) 75

3. Compute the percentile rank of the scores in Exercise 1, using the data in Table 4.1.

4. Compute the scores corresponding to the percentiles in Exercise 2, using the data in Table 4.1.

5. Would it be proper to display the ogive as a smooth curve? Explain.

6. Refer back to Chapter 3, Exercise 1.
   (a) Prepare a cumulative percentage distribution.
   (b) Draw the ogive (cumulate *up* the cumulative distribution).
   (c) Draw the ogive (cumulate *down* the cumulative distribution).
   (d) Discuss the difference between cumulating up and down the cumulative distribution.
   (e) Approximate the percentile rank of scores of 545, 520, 650, 760, and 681.
   (f) Approximate the scores corresponding to percentiles of 20, 46, 58, 74, and 91.

7. Refer back to Chapter 3, Exercise 2.
   (a) Prepare a cumulative percentage distribution for parts (c) and (d) of Exercise 2.
   (b) Draw the ogives.
   (c) Approximate the scores corresponding to percentiles of 24, 36, 53, 70, and 90 from both ogives.

8. Compute the percentile rank of scores in Exercise 5(f).

9. Compute the scores corresponding to the percentile in Exercises 5(g) and 6(c).

10. Data on suicide rates per one million people were obtained for 100 geographic regions. Use the data from the frequency table shown below to answer the following questions.
   (a) What is the rate at the fifteenth percentile?
   (b) What is the rate at the thirty-fifth percentile?
   (c) What is the rate at the second quartile (Q_2 or median)?
   (d) What is the rate at the sixth decile?
   (e) What is the rate at the third quartile (Q_3)?

**Suicide Rates (per one million people)
for 100 Geographic Regions**

Class Interval	f	Class Interval	f
2–4	5	17–19	5
5–7	20	20–22	4
8–10	36	23–25	3
11–13	15	26–28	2
14–16	10		

The best political community is formed by citizens of the middle class. Those States are likely to be well administered in which the middle class is large, and larger if possible than both the other classes, or at any rate than either singly; for the addition of the middle class turns the scale and prevents either of the extremes from being dominant.

Aristotle

Measures of Central Tendency

5.1 Introduction: Why Measure Centrality?

In the last couple of chapters, we mainly talked about the organization of data into more compact and meaningful forms. Now it's time to define certain measures that describe important features of a set of data. One such feature relates to how the data cluster around a central value that is between the two extreme values of the variable under study. It can be very useful to know how to locate a point of central tendency, especially if you can also describe how the data vary or scatter about this point. For example, we can reduce a mass of data to a single numerical value that, in certain respects, represents the character of the entire distribution.

Most people are familiar with this concept; they call it the *average*. But this term, as used by the general public, doesn't qualify what is really intended about the central tendency of the distribution. For example, "average" may be used to convey the most popular value (the *mode*), the value in the middle (the *median*), or the sum of all the values divided by their number (the *arithmetic mean*). Each of these uses obviously represents a different aspect of central location.

The word "average" has so many different uses in everyday conversation that in a text like this, the best thing we can do is to drop it altogether from our technical vocabulary and refer, instead, to *measures of central tendency*. A measure of central tendency is an index of central location that is used in the description of a distribution. Since we can give various definitions of the center of a distribution, we can measure central tendency in a number of different ways. So this chapter's purpose is to present the three most frequently employed measures of central tendency: the *mode,* the *median,* and the *arithmetic mean.*

5.2 The Mode

Of all the measures of central tendency, the mode is certainly the easiest to compute and the simplest to interpret. *Mode* is merely the name or midpoint of the most frequent measurement class in a given set of data. In Figure 3.1, the modal class is those respondents opposing national service for women; in Figure 3.5, the mode is 109.

There are some disadvantages to the mode, however. One is that there may be more than one modal class, since a distribution doesn't necessarily have to show *unimodality* (one mode). A distribution may be *bimodal* (two modes), or perhaps even *multimodal* (more than two modes). Such cases create confusion about which class gives the mode of the distribution, as two or more values are most frequent.

Another disadvantage is that the mode is very sensitive to the size (or number) of class intervals that are used when interval data are classified; the value of the mode may be made to "jump around" considerably by changing the size of the class intervals.

Finally, the mode does not tell us certain basic things, things that we intuitively associate with centrality; that is, the central value for a distribution need not necessarily be the most typical or the most common score. For these reasons, the mode is a poor barometer of centrality—unless, of course, the data are truly nominal and we merely want to know which category has the greatest frequency. But in general, there are more precise measures of central tendency than the mode.

5.3 The Median

The *median* is another way of indicating central tendency. Median is defined in slightly different ways, depending upon whether the number of cases is odd or even, and upon whether what is to be described is raw data or a grouped distribution. For a set of raw scores, if the number of cases is odd—as in the five scores 3, 5, 6, 8, 9—then the median is the value of the $(N + 1)/2$th score, or 6. If the number of cases is even—for example, 5, 7, 8, 9—then the median is the value at the $(4 + 1)/2$th score, or the 2.5th position. This is the average of the scores at the second and third positions, or 7.5. So, for an odd *or* an even number of cases, exactly as many cases fall above as fall below the median score.

If the data are already classified, we must locate the interval within which the middle position is contained. In other words, we must determine the score at the fiftieth percentile. In computing the median, we can use the generalized procedures we discussed in Chapter 4 (Section 4.4). All we need to do is modify Formula (4.4), as follows:

$$\text{Median} = X_{\ell\ell} + \left(\frac{N/2 - Cum\, f_{\ell\ell}}{f}\right)i \tag{5.1}$$

to obtain the median for the data that appear in Table 4.1:

$$106.5 + \left(\frac{150/2 - 46}{38}\right)5 = 110.32$$

Like the concept of the mode, the concept of the median is a simple one. But unlike the mode, the median is restricted to data of the interval level. The notion that the median is appropriate with ordinal data is a common fallacy.* What we really want to know when we use the median is the value associated with the position in which half the scores are more extreme and half are less extreme. This is impossible with ordinal data, of course, since there are no values, only order of positions. In addition, the median is considerably less sensitive than the mode is to the distribution's grouping into class intervals.

Of the three measures of central tendency that we discuss here, the median is the least affected by extreme values. In each of the following sets, there are extreme scores:

199, 200, 201, 202, 203, 204, 400
8, 200, 201, 202, 203, 204, 205,

but the medians are nevertheless identical. This insensitivity makes the median particularly appropriate with distributions that are not symmetrical.

The median has still another attractive property. Suppose that the variable being studied is income, and that incomes are listed in intervals of 1,000 dollars, except for those over 50,000 dollars, which are listed simply as 50,000 dollars or more. As you can see, because of the uncertainty of the incomes in the last interval, it's not possible to compute the mean income. The *median* income, on the other hand, would not be affected by this lack of information. Therefore, in this example, it could substitute for the arithmetic mean.

---

*The appropriateness of the median with ordinal-level data is a matter of some controversy in statistics. Social scientists often use the median with ordinal-level data, maintaining that it is a "positional average" even if no actual case in the data set posseses such a value. In our opinion, this notion has limited utility.

5.4 The Arithmetic Mean

By far the most useful and familiar index of central tendency is the *arithmetic mean*—or, put more briefly, the mean. Surely everyone knows that to take the average of a set of raw scores, you simply add them all up and divide by the total number of cases. Hence, the formula:

$$\bar{X} = \frac{\Sigma X}{N} \qquad (5.2)*$$

where, you will recall, $\bar{X}$ is read as *X* bar. Note that here, *X* is *any* score and the sum is taken over *all* the *N* different observations; for example:

$$\bar{X} = \frac{70 + 50 + 80 + 90 + 92}{5} = \frac{382}{5} = 76.4$$

5.4.1 The Mean and the Grouped Distribution

It's easy enough to define and compute the mean for raw data; but when we wish to find the mean for classified data, the situation gets a little more complicated. In Section 3.2.3, we said that when a distribution is grouped into class intervals, the midpoint of each class interval is taken to represent the score of each of the cases in the interval. Thus, in an interval 5–9 with midpoint 7 and frequency 10, the sum of the scores of the ten cases falling into this interval is taken to be 10 summed 7 times, $10(7) = fMP$. Similarly, when all the scores in any interval are assumed to be the same, their sum is the frequency of that particular interval times the midpoint for that interval, or fMP. Then the sum of all the scores in the distribution is taken to be the sum of *f* times the values of *MP* over all the respective intervals. Thus the mean is found from:

$$\bar{X} = \frac{\Sigma fMP}{N} \qquad (5.4)$$

---

*A corresponding formula for ungrouped data can also be written down, if desired, as:

$$\bar{X} = \frac{\Sigma fX}{N} \qquad (5.3)$$

Note that *f* appears in the formula to remind you that each *X* should be multiplied by its corresponding frequency, prior to summing. Even when you're dealing with an array of scores, this formula is the most general—the frequency of each score is one (that is, $f = 1$). For this reason, you should regard the *f* as implied even when it is not given.

Table 5.1 illustrates the procedure involved in applying this formula. In this table, we calculate the mean age for a set of rather gruesome data. (The value of the mean given by Formula (5.4) is only an approximation of the value given by Formula (5.2). However, unless the classification is rather crude, the difference is usually so small that we can usually ignore it.)

Table 5.1. Mean Age of Murder Victims Killed by Explosives in the United States (1972)

Class Interval	f	MP	fMP
5–9	3	7	21
10–14	1	12	12
15–19	0	17	0
20–24	5	22	110
25–29	2	27	54
30–34	3	32	96
35–39	1	37	37
40–44	1	42	42
45–49	0	47	0
50–54	1	52	52
55–59	1	57	57
60–64	1	62	62
65–69	1	67	67
70–74	1	72	72
	$N = 21$		$\Sigma f MP = 682$

$$\bar{X} = \frac{\Sigma f MP}{N} = \frac{682}{21} = 32.48$$

Source: Department of Justice, *Crimes in the United States: 1972* (August 8, 1973): 118.

5.4.2 Caution: The Weighted Mean

Unlike modes and medians, means can be combined. But although the combining process is rather simple, we advise caution. Suppose, for example, that one group of ten students has a mean score of 75 on a statistics test, and that another group of ten students has a mean of 85. Since the distributions are of the same size, the weight of each mean is equal, and we can compute the grand mean as the mean of the two submeans. That is:

$$\bar{X} = (75 + 85)/2 = 80$$

But suppose that the second group's mean is based on forty cases. In this case, the two groups are not equally weighted: one mean is based on ten cases while the other is based on forty cases. Since the mean is dependent on frequency as well as on value, we must take this into consideration. So we first transpose Formula (5.2) into the equivalent statement, $\Sigma X = N\bar{X}$. This, in effect, converts the data from a distribution back into its original sum of scores.

That is:

$$X_1 + X_2 + X_3 + \ldots + X_N = N\bar{X}$$

The grand sum of scores may then be obtained by multiplying the number of cases in each distribution by its respective mean and summing—that is, $\Sigma N\bar{X}$. This grand sum is then divided by the number of cases for both groups (ΣN), which leads us to the general formula for computing the *weighted mean:*

$$\bar{X}_w = \frac{\Sigma N\bar{X}}{\Sigma N} \qquad (5.5)$$

Thus, in the present example:

$$\bar{X}_w = \frac{10(75) + 40(85)}{50} = 83$$

5.4.3 Three Characteristics of the Mean

The mean has three characteristics that set it apart from other measures of central tendency. The first is that:

> *The mean is the point in a distribution about which the summed deviations are equal to zero.*

Therefore, the mean is the point that balances all the values on either side of it. In this sense, it is like the fulcrum on a seesaw. If you have ever been on a seesaw with a somewhat heavy friend, then you know what happens—you are left hanging in the air as your friend pushes down. In order for you both to seesaw successfully, either the balance point has to be moved (that is, your friend has to move in towards the center), or you have to add more weight to your own side. In short, in order to achieve balance on a fulcrum, the values of the weights times the distances from the fulcrum must be the same on each side. Or, stated another way, the sums of the distances from the mean must be the same on each side of the mean. In symbols, $\Sigma(X - \bar{X}) = 0.^*$

---

*The algebraic proof of this statement is

$$\begin{aligned}
\Sigma(X - \bar{X}) &= \Sigma X - \Sigma \bar{X} \\
&= N\bar{X} - \Sigma \bar{X} \\
&= N\bar{X} - N\bar{X} \\
&= 0
\end{aligned}$$

In following this proof, it is important to note that (1) $\bar{X} = \Sigma X/N$ is equivalent to $\Sigma X = N\bar{X}$, and (2) $\Sigma X = N\bar{X}$ because summing the mean over all the values ($\Sigma \bar{X}$) is equivalent to multiplying $N\bar{X}$ (see summation Rule II).

This analogy leads us to the mean's second characteristic:

The mean is very sensitive to the extreme values in a distribution when these values are not balanced on both sides of it.

For example, consider these five incomes: $9,000, $11,000, $10,000, $10,000, and $50,000. Four of them are much less than the mean of this distribution, which is $18,000. Obviously, one income ($50,000) is responsible for inflating the mean. Similarly, when there are a few values that are markedly less than the rest, the mean is deflated. When such distributions are plotted (as they are in Figure 5.1), they are referred to as either *positively skewed* (as in the first case) or *negatively skewed* (as in the second case). Clearly, when such conditions prevail, the median is the better measure of central tendency. Why? Because extreme values do not have too much effect on the median.

A third, and most important, characteristic of the mean is that:

The sum of the squared deviations with respect to the mean is less than the sum of the squares of deviations about any other reference point.

We can illustrate this characteristic easily. Table 5.2 lists the sum of the squared deviations that have been taken from various reference points in a distribution. Notice that the sum of the squared deviations with respect to the mean is less than the sum of the squared deviations about any other reference point; hence $\Sigma(X - \bar{X})^2$ yields the *minimum* sum of the squared deviations. This sum is less than when it is taken from any other point. This fact suggests that the mean is the best guess of the values in the distribution using this *least squares* criterion; that is, the overall error from the mean, as a guess for all the scores, will be less

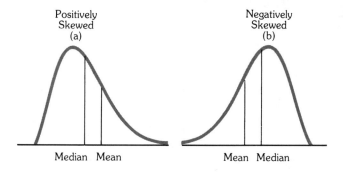

Positively Skewed (a)   Negatively Skewed (b)

Median Mean   Mean Median

Figure 5.1. Relationship Between Skewness and the Relative Position of the Mean and the Median

than if any other value in the distribution is used. This concept will take on added significance in later chapters.

Table 5.2. The Sum of the Squared Deviations Taken from Various Reference Points in a Distribution Where $\overline{X}$ = 5

X	(1) $(X - 1)^2$	(2) $(X - 2)^2$	(4) $(X - 4)^2$	(8) $(X - 8)^2$	(10) $(X - 10)^2$	$(\overline{X} = 5)$ $(X - \overline{X})^2$
1	0	1	9	49	81	16
2	1	0	4	36	64	9
4	9	4	0	16	36	1
8	49	36	16	0	4	9
10	81	64	36	4	0	25
Sum of Squared Deviations	140	105	65	105	185	60

5.4.4 Why Use the Mean?

The greatest advantage of the mean is that it gives us the most stable measure of central tendency. By "most stable" we mean that for most populations we are likely to encounter, if successive random samples were selected from a given population, the mean would vary less from these repeated samples than any other measure of central tendency. (Of course, if you have reason to suspect a skewed distribution in the underlying population, then the median might be the more "stable" measure.) You may not yet fully understand the importance of stability, but when we discuss the *central limit theorem* in Chapter 10, this benefit will become clearer. For now, remember that any statistic (in this case, the mean) is a measurement of a sample but only an estimate of the corresponding population parameter.

Another advantage of the mean is that it is used to compute other important measures. Among those are the standard deviation, the Pearson product-moment correlation coefficient, and many standard errors used in statistical inference. You can't evolve any of them unless you compute a mean first.

5.5 Caution: The Ecological Fallacy

As social scientists, we often are more interested in distributions than in individual scores; that is, we care about populations and groups rather than about samples and individuals. The indices presented in this text help us make inferences from a smaller unit of analysis to a more generalized group—or, as we sometimes say, from the particular to the general. But if we want to move in the other direction—from the general group characteristics to smaller units or individuals—we must be very careful. If we aren't, we can easily fall victim to what Robinson (1950) has called the:

Ecological Fallacy: A logical fallacy that can occur when attributing to the members of a group the properties of that group.

We can best explain this fallacy by using a couple of examples. The first involves a small Appalachian community that depends on the local coal mine as its sole source of revenue. For purposes of illustration, let's assume that each employee of the mine earns less than $8,000 per year, while the owners, who also are residents of the community, have incomes in excess of all the employees combined. Let's further assume that the mean income per family of four for the community is $20,000 per annum. Now let's consider Mr. Smith, a miner with a family of four. If we were to assume that Mr. Smith provides his family with minimum luxury, since he is from a community where the mean annual income per family of four is $20,000, we might be committing an ecological fallacy.

Let's take another example. Consider Eugene McCarthy's bid for the presidency in 1968. McCarthy, a strong critic of American involvement in Vietnam, failed to win his party's nomination. His staff, however, proclaimed a symbolic victory after the primaries because they believed that the senator made an extraordinary showing with young Democrats. Although it is true that McCarthy's main support came from election precincts with the lower mean age, the claim might have been fallacious. Why? Simply because it may have been the older voters in those "young" precincts who voted for the senator.

These examples demonstrate that we have to be fairly careful when we are drawing conclusions about a smaller unit of analysis from data based on a larger one. This doesn't mean that all such conclusions will be invalid, only that it's potentially dangerous to attribute the characteristics of a group to individual members of that group.

Where We Stand

A descriptive statistic generates a quantitative summary statement or a number that carries information relating directly to samples and indirectly to populations. Unless an entire population is surveyed, the sample may differ from the population; and since it is the sample that we are interested in, we of course want to know which statistic is least likely to deviate significantly from the one we would get if we surveyed the entire population.

This is true of statistics of all kinds. When we are only interested in centrality, the mean is the best measure. Because it varies least from sample to sample for most underlying populations it is the most stable. The mean is also related to other stable measures; in fact, it is a prerequisite for computing most of them.

The median, which is more stable than the mode (but less so than the mean), is used to pinpoint the value that cuts a distribution in half. In skewed distributions, it is the most accurate index of centrality, because it is less sensitive to extreme values.

The mode, lastly, is the fastest index to compute and the least stable, as well. It is to be used only as a preliminary estimate or as an index of the most typical (or common) value.

Terms and Symbols to Remember

Measures of Central Tendency
Mode
Unimodal
Bimodal
Multimodal
Median
Arithmetic Mean $(\bar{X})$
Weighted Mean $(\bar{X}_w)$
$\Sigma(X - \bar{X}) = 0$
$\Sigma(X - \bar{X})^2 = $ Minimal
Positively Skewed
Negatively Skewed
Least Squares
The Ecological Fallacy

Exercises

1. Calculate the mode, the median, and the mean for each of the following sets of measurements.
    (a) 3, 6, 7, 8, 9, 10, 3, 2, 1, 3
    (b) 13, 18, 18, 18, 15, 0, 0, 2, 8, 7
    (c) 6.2, 6.2, 6.5, 6.3, 6.4, 6.7, 6.7, 6.5, 6.2, 6.3, 6.3, 6.3, 6.8, 6.9, 6.2, 6.1, 6.4, 6.3, 6.0, 6.3, 6.8, 6.8, 6.8, 6.0
    (d) 102, 103, 104, 104, 105, 105, 106, 106, 105, 104, 200
    (e) 1, 2, 0, 0, 0, 3, 2, 2, 2, 2, 2, 5, 5, 5, 5, 5, 0, 0, 7
    (f) 1.375, 1.485, 1.583, 1.782, 1.625, 1.623, 1.721, 1.821
    (g) .002, .006, .007, .008, .009, .001, .001, 0, .001, 0, .002, .007, .002, .010, .012, .013
    (h) −1.6, 0, −2.0, 2.0, 2.6, 1.2, −1.1, 1.3, 1.3, 1.1
    (i) 10, 30, 46, 52, 51, 26, 51, 50, 23
    (j) 98.6, 98.6, 99.0, 98.8, 98.4, 98.8, 98.6, 98.2, 98.4

2. From the following measures of central tendency, indicate whether or not the curve is skewed; if so, what is its direction?
    (a) $\bar{X} = 100$, Median $= 50$, Mode $= 50$
    (b) $\bar{X} = 100$, Median $= 100$, Mode $= 100$
    (c) $\bar{X} = 2.5$, Median $= 3.5$
    (d) $\bar{X} = 3.5$, Median $= 1.5$
    (e) $\bar{X} = 53$, Median $= 40$, Mode $= 30$
    (f) $\bar{X} = 1.7$, Median $= 1.7$, Modes $= 1.4, 1.5, 1.8$
    (g) $\bar{X} = 63$, Median $= 21$, Mode $= 21$
    (h) $\bar{X} = 4.73$, Median $= 4.73$, Mode $= 0$
    (i) $\bar{X} = 10.00$, Median $= 10.00$, Mode $= 10.00$
    (j) $\bar{X} = 70$, Median $= 70$, Modes $= 50, 80$

3. In which of the data sets in Exercises 1 and 2 is the mean a poor measure of centrality? Explain your answer(s).

4. Explain why the mode is the only measure of central tendency that's appropriate for all levels of measurement. Why is it inappropriate to use the median with ordinally scaled data?

5. Cite some types of data for which you feel the mode would be a more appropriate measure of central location than the median or the mean.

6. Find the mode for each of the figures in Chapter 3.

7. Compute the mode, the median, and the mean for the data in Table 3.1.

8. Compute the mean and the median for the data in Table 3.3. Compare your answers with those of the above exercise. Are your answers similar? If not, why?

9. The following quiz scores were recorded for a class of students. Construct a grouped frequency table for these scores and from it calculate the mean, the median, and the mode.

```
10  6   1   7  9  18   7  11  12   9  12  13   4  15
 9  8   5  14  8   7  17   8   6  10   3   8  11  14
 6  9  12  10  5   7  10   8  13   5  11   9   8  10
```

10. In two sections of a history course the following distributions of grades are obtained:

SECTION 1		SECTION 2	
Grades	f	Grades	f
20–29	2	40–49	3
30–39	3	50–59	1
40–49	6	60–69	5
50–59	8	70–79	15
60–69	15	80–89	10
70–79	10	90–99	4
80–89	3		

(a) What is the mean and the median for Section 1?

(b) What is the mean and the median for Section 2?

(c) What is the mean and the median of the whole group?

(d) What is the mode for Section 1?

(e) What is the mode for the whole group?

11. The average (mean) age of the 100 members of the Senate is 54.5 years, and that of the 435 members of the House of Representatives is 52.6. Find the overall mean of the ages of these members of Congress.

12. For each of the following, compute the group mean.
(a) $\bar{X}_1 = 20, N_1 = 100; \bar{X}_2 = 25, N_2 = 50; \bar{X}_3 = 30, N_3 = 75$
(b) $\bar{X}_1 = 40, N_1 = 10; \bar{X}_2 = 40, N_2 = 20; \bar{X}_3 = 40, N_3 = 30$
(c) $\bar{X}_1 = 30, N_1 = 10; \bar{X}_2 = 35, N_2 = 10; \bar{X}_3 = 40, N_3 = 10$

13. Show that $\Sigma(X - \bar{X}) = 0$ for the following sets of measurements.
(a) 10, 9, 7, 8, 6, 7, 8, 9, 6
(b) 2, 3, 14, 9, 8, 4, 15, 8
(c) 1, 5, 8, 9, 7, 11, 8

14. For each of the sets of measurements in the above exercise, show that $\Sigma(X - \bar{X})^2$ is minimal.

15. Cite some type of data for which you feel the median would be a more appropriate measure of central location than the mean. Why?

16. An economist surveys a small community and finds that the median income is $15,000 and that the mean income is $27,000. What does this discrepancy tell you about the income distributions in that community?

17. Draw two different frequency polygons illustrating the median as the best measure of centrality. Do the same for the mean.

18. Explain why the following statements might turn out to be incorrect.
(a) "After studying the average suicide rates in Protestant countries, we found them to be higher than in Catholic ones. We must therefore conclude that Protestants tend to commit suicide more frequently than Catholics."
(b) "Our study revealed that the average crime rates are higher in cities having larger minority populations than in those with few minorities. And so we say that minorities commit most of the crimes in the cities we surveyed."
(c) "The records clearly show that Mr. Jones received a greater proportion of electoral support from precincts whose voters had a high mean income than in those precincts with a low median income. This suggests that Jones was more appealing to the affluent voter."

Doctor Foster went to Gloucester,
in a shower of rain.
He stepped in a puddle, up to his middle,
and never went there again.
Anonymous

Measures of Dispersion

6.1 Introduction: Why Measure Dispersion?

Suppose we are told that three different data sets all have modes, medians, and means of 50. If that's all we are told, we may conclude that the three data sets are quite similar, if not identical. But what if this conclusion is wrong? Well, let's take a look at the data we had in mind:

I: 50, 50, 50, 50, 50, 50, 50
II: 47, 48, 50, 50, 50, 52, 53
III: 10, 25, 30, 50, 50, 75, 110

Quite clearly, the sets are not identical; if we compare I with III, they are not even similar. In Set I, each score is identical to every other score; any measure of central tendency is perfectly representative of the entire set; there is no deviation from centrality, no dispersion. In Set II, there is some variation, but either the mode, the median, or the mean is still quite representative. But in Set III, most of the scores are so different from one another and from the chosen average that we must question how representative average scores are.

This set of circumstances is somewhat contrived, but not uncommon—and the obvious conclusion is that measures of central tendency do not, in themselves, fully describe a set of data. While they do help locate the central tendency of a mass of data, they tell us little about how these data are dispersed throughout the distribution. Therefore it is important to describe data by providing information about both centrality and dispersion. In short, we must define various *measures of dispersion*.

6.2 The Range

When we computed the various measures of central tendency, we located a central focal point for a data set and identified it as either the mode, the median, or the mean. Now, let us look at measures that indicate the spread of the distribution. The first measure we need to define is the *range*.

The range provides a quick reference to the spread of a distribution. It consists simply of the scale distance between the lowest and highest scores. Recall that in Section 3.2.1, as part of constructing a grouped distribution, we subtracted the lowest score from the highest score and added 1. Thus we determined the range of scores in Table 3.1 to be $(156 - 87) + 1$, or 70. With classified data, we estimate the range by first finding the difference between the midpoints of the two class intervals in which the extreme scores lie, and then adding 1.

The most striking feature of the range is also its weakness—namely, its instability. If you compute the range for 5, 6, 6, 7, 7, 8, 8, 24, 24, you will find it to be 20. Now, if we remove the scores of 24, the range is decreased fivefold. Note that not even two deviant scores were necessary to have that effect; even one would have created the same effect. The fact that the range is determined by two, and only two, values (the lowest and the highest) in any distribution makes it easy to compute and comprehend—but that also makes it unstable.

6.3 The Interquartile Range

A somewhat more useful measure of dispersion is provided by the *interquartile range*—the scale distance, including the middle 50 percent of the distribution. To find the interquartile range, we first locate the seventy-fifth percentile (or Q_3) and the twenty-fifth percentile (or Q_1), similar to the way in which we locate the median. Then we compute the difference between these percentiles. If the data are grouped, we merely modify Formula (4.4) to find the percentile with grouped data.

Another way to reduce dependence on extreme values is to use the *semi-interquartile range,* which merely halves the value for the interquartile range. Other interfractile ranges, such as those representing the middle 80 percent, 90 percent, and 95 percent, can be used—but they seldom are.

Because the *interfractile ranges* avoid the exclusive use of the two extreme scores, they create an illusion of stability. But it *is* sometimes an illusion. For example, consider the value obtained with the interquartile range. Although it provides a description of the dispersion in the central half of the distribution, it does not account for the values at the extremes. This, in effect, makes it guilty of the range's opposite defect. That is, the range is determined by the extreme scores and is insensitive to the values near the mean; however, the interquartile range is not affected by values existing beyond the first and third quartiles. On the other hand, insensitivity to extreme values can sometimes make the interquartile range the most desirable of measures, in much the same way as the median was preferred to the more stable mean. We shall see evidence of this later when we study the standard deviation.

6.4 The Mean Deviation

If we wish to make use of all scores in a distribution, common sense tells us to take the deviations of each score from some measure of centrality and then compute some kind of average of these deviations. We could use the mode or the median as the reference point, but most often we take the mean, since it is usually more stable. Suppose that we were to sum the actual deviations from the mean and divide their value by the number of deviations. But, as we already know, $\Sigma(X - \bar{X}) = 0$, suggesting that if we want to obtain a measure of dispersion about the mean, we must somehow get rid of negative signs. Two methods immediately come to mind: (1) take absolute values (remember that the absolute value merely means omitting the sign in front of a number so that all values are positive); or (2) square the differences. The first method leads to the *mean deviation,* the latter method to the *standard deviation.*

The mean deviation is defined as the sum of the absolute deviations about the mean, divided by the number of deviations. In symbols:

$$MD = \frac{\Sigma|X - \bar{X}|}{N} \tag{6.1}$$

The mean of the numbers 67, 69, 70, 71, 73 is 70. Thus, subtracting 70 from each of these numbers, omitting the signs, and then adding them up and dividing by 5 yields a mean deviation of 1.6 (that is, $(3 + 1 + 0 + 1 + 3)/5 = 1.6$). Therefore, we may conclude that on the average, the scores deviate from the mean by 1.6 units.

Although the mean deviation has a more direct interpretation than the measures we will present in the next section, it has several major drawbacks. In the first place, absolute values are not easy to manipulate algebraically. More importantly, however, the mean deviation is not easy to interpret theoretically, nor does it lead to as simple mathematical results. So for descriptive purposes, the mean deviation may seem adequate (although, as we shall see, the standard deviation can be interpreted more readily in terms of the normal curve). When we come to inferential statistics, we'll see that the standard deviation is used almost exclusively because of its theoretical manageability. This is why the mean deviation is seldom referred to in social science research.

6.5 The Standard Deviation and the Variance

The previous measures serve as theoretical building blocks for the core concept of dispersion, namely the *standard deviation*. This is defined as the square root of the sum of the squared deviations about the mean, divided by the number of deviations; or, in symbols:

$$s = \sqrt{\frac{\Sigma(X - \bar{X})^2}{N}} \qquad \textbf{(6.2)*}$$

That is, we simply take the deviation of each score from the mean, square each difference, sum the results, divide by the number of cases, and then take the square root (a divisor of a quantity that, when squared, gives the quantity). In Table 6.1, we demonstrate the computation of the standard deviation using this mean deviation method.

Table 6.1. Computation of the Standard Deviation, Using the Mean Deviation Method, from Ungrouped Data

X	$(X - \bar{X})$	$(X - \bar{X})^2$
1	−3	9
2	−2	4
2	−2	4
3	−1	1
4	0	0
5	1	1
6	2	4
9	5	25

$\Sigma X = 32$ $\Sigma(X - \bar{X})^2 = 48$
$N = 8$
$\bar{X} = 4$

$$s = \sqrt{\frac{\Sigma(X - \bar{X})^2}{N}}$$

$$= \sqrt{48/8}$$

$$= \sqrt{6}$$

$$= 2.45$$

---

*This formula describes the dispersion within a known population. As we will see, we often wish to use sample data to estimate what the population's standard deviation would be. In such cases, we will use the correction factor of $N - 1$. This modification increases the value of the sample statistic, thus making it an unbiased estimate. Of course, if N is large, this correction factor proves to be unimportant. We'll have more on this in Chapter 10 when we estimate the population mean.

The intuitive meaning of a standard deviation of 2.45 will not be apparent until later, when we use the standard deviation to obtain areas under the normal curve. But several properties of the standard deviation are apparent. For instance, the greater the scatter about the mean, the larger the standard deviation. If all the eight values in Table 6.1 were the same, the deviations would total zero, and the standard deviation would also equal zero. If we inspect the table even further, we see that a value of 25 has the greatest weight in determining the value of the standard deviation. Even though we later take the square root, we are basically giving even more weight to extreme values than we would if we were computing a mean. This suggests that in a positively skewed distribution with an extreme cluster of data to the right of the mean, the standard deviation is on one hand too large to reflect the actual dispersion to the left of the mean, and on the other hand too small to reflect the actual dispersion to the right of the mean. The same holds true for a negatively skewed distribution. In such cases, one of the interfractile ranges might be more useful as a measure of dispersion, much as the median was used instead of the mean. But for most data, the standard deviation will do nicely.

You might be wondering why we take the square root in computing a measure of dispersion. Well, one simple (but unsatisfactory) answer is: we take the square root because that is the way the standard deviation is defined. We could also point out that we want to compensate for having squared each deviation earlier. However, a much more honest answer is that we do it for practical reasons. Since in later chapters we make considerable use of the normal curve, the standard deviation, as defined, turns out to be very useful. For other purposes, we shall use the square of the standard deviation, or *variance,* which is defined as:

$$s^2 = \frac{\Sigma(X - \bar{X})^2}{N} \qquad \textbf{(6.3)}$$

The variance plays an important part later on, when it comes time to test statistical hypotheses for more than two sample means. For now, however, let's confine our attention to the standard deviation. The two concepts are so related that computational familiarity with one yields familiarity with the other. If s^2 is 25, then s is 5. Again, if the standard deviation is 3, then s^2 is 9.

6.5.1 A Short Cut for Computing the Standard Deviation

Since we know how much work computation can be, wherever possible we present a computational formula in place of the definitional one. The latter is rarely suitable for calculations, anyhow, although it does make you think about the relationships involved. The computational formula for the standard deviation is called, appropriately enough, the *machine formula,* and is written as:

$$s = \sqrt{\frac{\Sigma X^2}{N} - \bar{X}^2} \qquad\qquad (6.4)*$$

Table 6.2 shows the computational procedures for calculating the standard deviation, utilizing the machine formula. Note that the computational formula requires only one subtraction, whereas the more difficult mean deviation method demands that the mean be subtracted from each and every value. Note, too, that both methods yield identical answers, except for rounding errors.

---

*The algebraic derivation given below can demonstrate that this machine formula is equivalent to Formula (6.2). We present this derivation not to confuse you but to show you that, like any other computational formula, this one is merely a result of the expansion and combination of terms in the original formula—nothing more. So when you encounter what appears to be a strange computational procedure, rest assured that, ultimately, it refers to the original statement.

$$\frac{\Sigma(X - \bar{X})^2}{N}$$

$$\frac{1}{N} \Sigma(X - \bar{X})^2$$

$$\frac{1}{N}(\Sigma X^2 - \Sigma 2X\bar{X} + \Sigma\bar{X}^2) \qquad\text{see summation Rule VII}$$

$$\frac{1}{N}\Sigma X^2 - \frac{1}{N}\Sigma 2X\bar{X} + \frac{1}{N}\Sigma\bar{X}^2 \qquad\text{see summation Rule V}$$

$$\frac{1}{N}\Sigma X^2 - 2\bar{X}\frac{1}{N}\Sigma X + \bar{X}^2\frac{1}{N}\Sigma 1$$

$$\frac{1}{N}\Sigma X^2 - 2\bar{X}\bar{X} + \bar{X}^2\frac{1}{N}N$$

$$\frac{1}{N}\Sigma X^2 - 2\bar{X}^2 + \bar{X}^2$$

$$\frac{\Sigma X^2}{N} - \bar{X}^2$$

Thus:

$$s = \sqrt{\frac{\Sigma X^2}{N} - \bar{X}^2}$$

In following this derivation, note that if $X = 1$, then $\Sigma 1 = N$.

Table 6.2. Computation of the Standard Deviation, Using Machine Formula, from Ungrouped Data

X	X^2
1	1
2	4
2	4
3	9
4	16
5	25
6	36
9	81
$\Sigma X = 32$	$\Sigma X^2 = 176$
$N = 8$	
$\bar{X} = 4$	

$$s = \sqrt{\frac{\Sigma X^2}{N} - \bar{X}^2}$$

$$= \sqrt{\frac{176}{8} - 4^2}$$

$$= \sqrt{6}$$

$$= 2.45$$

Table 6.3. Computation of the Standard Deviation, Using Machine Formula, from Grouped Data

Class Interval	f	MP	fMP	fMP^2
5–9	3	7	21	147
10–14	1	12	12	144
15–19	0	17	0	0
20–24	5	22	110	2,420
25–29	2	27	54	1,458
30–34	3	32	96	3,072
35–39	1	37	37	1,369
40–44	1	42	42	1,764
45–49	0	47	0	0
50–54	1	52	52	2,704
55–59	1	57	57	3,249
60–64	1	62	62	3,844
65–69	1	67	67	4,489
70–74	1	72	72	5,184
	$N = 21$		$\Sigma fMP = 682$	$\Sigma fMP^2 = 29,844$

$$s = \sqrt{\frac{\Sigma fMP^2}{N} - \left(\frac{\Sigma fMP}{N}\right)^2}$$

$$= \sqrt{\frac{29,844}{21} - \left(\frac{682}{21}\right)^2}$$

$$= \sqrt{366.44}$$

$$= 19.14$$

6.5.2. Grouped Data and the Standard Deviation

The computational formula is used exclusively with grouped data. Table 6.3 shows a detailed illustration. If you examine the table, you'll see that we follow the same column structure as if we were going to compute a grouped mean (Table 5.1)—except that now we must multiply each value in the MP column by the corresponding value in the fMP column. We sum these quantities to get ΣfMP^2, which we substitute directly into the formula:

$$s = \sqrt{\frac{\Sigma fMP^2}{N} - \left(\frac{\Sigma fMP}{N}\right)^2} \qquad \textbf{(6.5)}$$

Like the computation of the grouped mean, this gives only an approximation; but nevertheless it's a rather good estimate.

6.6 The Index of Dispersion

From what we've said before, it should be clear that many interesting social science phenomena can be measured only at the nominal or ordinal level. So it helps to have a way of ascertaining dispersion in noninterval data sets. The *index of dispersion* meets this need. It is defined as the ratio of the dispersion that *does* exist to the maximum dispersion that *could* exist. Its formula is:

$$D = \frac{k(N^2 - \Sigma f^2)}{N^2(k - 1)} \qquad \textbf{(6.6)}$$

where:

$$k = \text{the number of categories,}$$
$$N = \text{the total number of cases, and}$$
$$f = \text{the frequency in each category.}$$

To understand the logic behind this qualitative measure of dispersion, let's examine Table 6.4, in which we list the different religious groups in the Sudan and the percentage of persons in each category of follower. Intuitively, we know that if maximum dispersion were present among the groups, there would be an equal percentage of followers in each of the three categories. On the other hand, if all followers were concentrated in a single category, there would be minimum dispersion. Therefore, our index value will intermediate between 0 (minimum dispersion) and 1 (maximum dispersion).

The actual computations involved in judging religious dispersion in the Sudan are worked out in detail in Table 6.5. Our computed index value shows

Table 6.4.   Religious Groups in the Sudan (1974 estimate)

Religious Group	Percentage
Islam	70% (11,200,000)
Indigenous Beliefs	25% (4,000,000)
Christianity	5% (800,000)
Totals	100% (16,000,000)

Source: Department of State, *United States Department of State Background Notes* (April 1975): 1.

Table 6.5.   Computation of the Index of Dispersion

Religious Groups	f	f^2
Islamic	70	4,900
Indigenous Beliefs	25	625
Christianity	5	25
	$N = 100$	$\Sigma f^2 = 5,550$

$$D = \frac{k(N^2 - \Sigma f^2)}{N^2(k - 1)}$$

$$= \frac{3(100^2 - 5,550)}{100^2(3 - 1)}$$

$$= \frac{13,350}{20,000}$$

$$= 0.67$$

Source: Table 6.4

Note: An examination of Formula (6.6) will show that when observed frequencies are expressed as percentages (as occurs in the present case), exactly the same results will be obtained.

us that religious diversity is two-thirds of the maximum, given three religious groups.

Dispersion, as defined here, is strictly a statistical characteristic. It is not to be confused with the sociopolitical states that characterize either social disorganization or political instability. However, it is probably true that the degree of societal unrest is related to statistical heterogeneity with regard to religion or race, for example, because this heterogeneity may be one of the factors conditioning the attitudes of the population. Thus it has been assumed that societal disorganization increases as conflict groups approach equality in power (numerical parity is an integral part of equality in power). Because of the systematic study and measurement it offers, the index of dispersion suits our purpose.

Where We Stand

The measures that we have introduced here to quantify dispersion are analogs of the measures of central tendency. The standard deviation is a kind of average of individual deviations from the mean and therefore is similar to it. It resembles the mean in two ways: it is the most stable of all measures of its kind, and it enters into a wide variety of other statistical procedures— z test, t test, and Pearson product-moment correlation coefficient, to name just a few.

The interfractile ranges avoid extremes, instead concerning themselves with the middle portion of the distribution. Although they are less stable than the standard deviation, they are preferred to it whenever the distribution is markedly skewed. Interfractile ranges are the perfect companions to the median, wherever the latter is applied properly.

Finally, the range, like the mode, is the fastest to compute; also like the mode, it is the least stable. Both measures yield rough estimates of what it is they intend to describe. And both can be used to answer questions that are directly related to their definitions. For the mode, the question is, "Which is the most typical (or common) score?" For the range, it is, "What is the scale distance between extremes of the distribution?"

Terms and Symbols to Remember

Measures of Dispersion
Range
Interquartile Range
Semi-Interquartile Range
Interfractile Ranges
Mean Deviation *(MD)*
Standard Deviation *(s)*
Machine Formula
Variance *(s^2)*
Index of Dispersion *(D)*

Exercises

1. In your own words, explain why the range, the interquartile range, and the standard deviation are analogs of the mode, the median, and the mean.

2. How does a measure of dispersion aid in describing a distribution of measurements?

3. Find the standard deviation of the following sample of measurements,

using Formula (6.2) *and* Formula (6.4). (Time yourself on both methods to see how much time you save by using the machine formula.)
13, 15, 20, 18, 16, 14, 13, 17, 12, 22

4. Find the standard deviation for each of the following samples.
   (a) 5, 3, 4, 2, 7, 3
   (b) 2, 3, 3, 4, 5, 1
   (c) 2, 2, 7, 3, 4, 5, 8, 1
   (d) 20, 15, 20, 15, 15, 25, 20, 20, 20, 30
   (e) 5, 3, 6, 36, 5, 2, 3, 4

5. Which of the samples in Exercise 4 has the largest standard deviation? Explain your answer in terms of the effect of extreme deviations on the standard deviation.

6. Find the range for the sets of measurements in Exercise 4. For which of these sets is the range a misleading measure of dispersion? Why?

7. Why does the addition or subtraction of a constant from each value in a distribution have no effect on the value of the standard deviation?

8. Calculate the range, the standard deviation, and the variance for each of the sets of measurements in Exercise 1, Chapter 5.

9. Calculate the mean deviation for the data in Exercises 3 and 4.

10. The grouped data below represent the number of questions answered incorrectly by twenty students on a fifty-question true-false examination.

Class Interval	f
10–14	1
15–19	4
20–24	3
25–29	5
30–34	4
35–39	2
40–44	1

   (a) Compute s.
   (b) Compute s^2.
   (c) Compute the range.
   (d) Compute the interquartile range.
   (e) Compute the semi-interquartile range.

11. Calculate the range, the standard deviation, and the variance for the data in Table 3.1.

12. Calculate the range, the standard deviation, and the variance for the data in Table 3.3. Compare your answers with those of the above exercise. Are your answers similar? If not, why?

13. Suppose that you obtain a set of attitude scores about gun control for a group of 1,000 respondents, and that the standard deviation of the set of scores is zero. What does this imply about the character of the group?

14. Assume that the mean of a set of ten measurements is 3, and that the sum of the squares is 100. What is the variance and the standard deviation of the set of measurements?

15. Which measures of central tendency and dispersion would be best to use for income? for religious preference? for attitudes on smoking cigarettes? for test scores? Explain your answers.

16. In a hypothetical community, 60 percent of the people are Protestant, 30 percent are Catholic, 8 percent are Jewish, and 2 percent are classified as "Other." Compute the degree of homogeneity within the community.

17. A random sample of 100 persons were interviewed to determine whether their trust in the president had changed over the past year and what direction, if any, the change had taken.

Increased trust	30
Decreased trust	60
No change	10

Use these data to compute the index of dispersion.

18. If a group of voters is composed of ten males and ten females; eight Democrats, seven Republicans, and five Independents; and fourteen whites and six blacks, is it possible to represent this diversity by means of a single index of dispersion? Explain.

CHAPTER

7

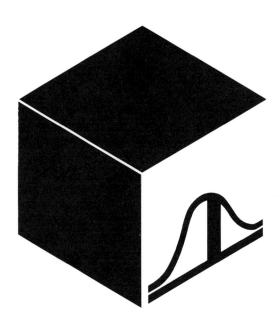

*Whenever a large sample of chaotic elements are taken
in hand and marshaled in the order of their magnitude,
an unsuspected and most beautiful form of regularity
proves to have been latent all along.*
Francis Galton

The Normal Curve

7.1 Introduction: Why Study the Normal Curve?

Jane O. Yes applies for admission to Statistics State. J.O.Y.'s high school grades were mediocre; but she put all her time into being a cheerleader, and the rest of the school's program never motivated her very much. In the appropriate space provided on the Statistics State application form, she conveys this point so convincingly that some kindly souls on the admissions committee decide to give her a chance. In their letter, they tell her to report to Anxiety Hall, where she will take aptitude tests designed to assess her mathematical and verbal abilities. If she does equally well on both tests, and if she does better than 75 percent of the rest of the applicants, she will be allowed to enter the hallowed halls of Statistics State.

Along with a large number of other hopefuls, Jane takes the tests. She scores 30 on the mathematics test and 60 on the verbal. While Jane suns in the Caribbean, college officials compare the mean and standard deviation for both the math test ($\bar{X} = 20$ and $s = 5$) and the verbal test ($\bar{X} = 40$ and $s = 10$) with Jane's grades. She receives her letter of acceptance.

Perhaps you're wondering how Jane could have done equally well on both parts of the test, when from the looks of it she is twice as good with words as she is with numbers. Also, note that the test distributions have different standard deviations—as we mentioned in Chapter 6, this is apt to produce dramatically different interpretations for scores above and below the mean. So we may well ask, what type of statistical manipulation was performed on Jane's test scores to make them equal and at the same time above 75 percent of the scores in their respective distributions?

In the case of Jane's test scores, we assume (quite reasonably, with such data) that they can be described by a frequently encountered distribution in statistical thinking, the *standard normal distribution* (or *normal curve*). The properties of this theoretical probability distribution are defined in such a way that we can know the proportion of cases in a distribution falling between any two scores. Then we can use the properties of the normal curve to compare Jane's position on one variable (verbal ability) with her position on the other variable (mathematical aptitude). This might not be clear to you right now, but it will be in a short while.

When we discussed fractile ranks earlier, we noted that they have a particular limitation: they cannot be used to make comparisons among distributions of a different character. However, as we will see, the normal curve can be used for just this purpose. How this is accomplished has to do with the concept of standard deviation. We hope that by explaining this achievement, we'll justify our interest in a measure that until now may have seemed to prevent such intergroup comparisons. However, once you have mastered the material contained in this chapter, not only will you have the means to compare data from different distributions but you'll also become sophisticated enough to start learning about inferential statistics.

7.2 The Normal Curve: A Model of the Real World

Are different distributions similar in shape? Surprising as it may sound, Florence Nightingale (1820–1910) is one person who thought so. Throughout her entire adult life, she believed that statistics could unearth the similarity among different social phenomena and thus reveal the equality present in the character of God. For her, the study of statistics had religious overtones. Although we cannot comment on the latter concept, we can declare that there is an uncanny similarity among most phenomena in terms of their frequency distributions.

In the nineteenth century, the Belgian mathematician Adolphe Quètelet (1796–1874) set out to find similarities in the universe. He suspected that man's physical, as well as social, traits could be described in terms of a curve that rose to a rounded peak in the middle and tapered off symmetrically at both tails. After collecting data on the heights and chest expansions of French and Scottish soldiers, respectively, Quètelet had his proof. Most Frenchmen, for example, fell within the five- to six-foot range; far fewer of them were either very short (less than five feet) or very tall (more than six feet). When plotted, each set of observations conformed with the bell-shaped normal curve drawn in Figure 7.1.

Since those early days of Quètelet's studies, countless researchers in innumerable investigations have found that a variety of physical and social

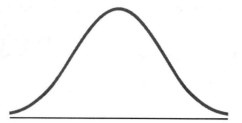

Figure 7.1. The Normal Curve

phenomena, ranging from the ridiculous to the sublime, follow the symmetry of this remarkable unimodal distribution. Phenomena such as height, chest expansion (both male and female), IQ, and even the wear of the carpet in a hallway are just a few of the phenomena that were found to distribute themselves normally. We leave the more ridiculous investigations to your vivid imagination. However, the point is clear: the mathematical world of the normal curve does not differ radically from the real world in which we live. Therefore, we use the normal curve as a standard by which to describe and compare a countless variety of data sets.

7.3 The Characteristics of the Normal Curve

The normal distribution contains an infinite number of cases, is unimodal and symmetrical, and is unbounded at either end. The mean, the median, and the mode all coincide to divide the distribution into two equal parts. The graphic version of this distribution is a smooth curve, rising to a peak in the middle, with a slope that tapers off without ever touching the base line (see Figure 7.2). Starting at its peak, the curve falls more and more rapidly to the *point of inflection*—that is, the point at which the curve changes from concave downward to concave upward—and then gradually levels off, extending indefinitely in either direction. This point of inflection is exactly one standard-deviation distance from the zero (mean) origin.

If you look closely at Figure 7.2, you'll see that the distance from the mean in units of the standard deviation may be converted into a percentage of the total area under the curve. Such distances are convenient measures of relative position that allow us to find the percentage of area between any two standard deviations under the curve. If we travel one standard-deviation unit above the mean, we leave behind 34.13 percent of the cases; if we continue on to a point two standard-deviation units above the mean, we leave behind not 68.26 percent of the cases but rather 47.72 percent, because of the steady decline in frequency as we move away from the mean. Since the curve is symmetrical, 68.26 percent of the cases lie in an interval extending from one standard-deviation unit above the mean to one such unit below the mean. If we move two standard deviations above and below the mean, we find 95.44 percent of the cases. And finally, the area between the mean and three standard deviations both above and below the mean, six in all, includes 99.74 percent (or almost all) of the cases.

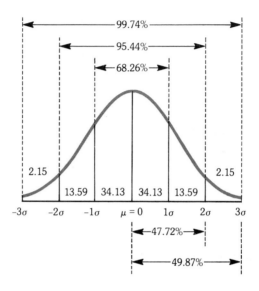

Figure 7.2. Areas Between Selected Points Under the Normal Curve

Note: The Greek letters μ and σ represent the population mean and standard deviation, respectively.

7.4 The Normal Curve and Standard Scores

The description of the normal curve given in Section 7.3 is solely an abstract one. It is presented in terms of the standard deviation measured from the mean as an origin. Therefore, any unit of measure is independent not only of diverse measurement systems but also of the concrete values themselves. So it makes no difference whether we are dealing with ages in years or centuries, with weights in ounces or pounds, or with varying grades on an examination. It also will make no difference whether we are dealing with peaches, pears, or apples.

Invariably, data come to us as raw scores and give every indication of being noncomparable. In that case, can we readily compare Jane's test scores by eye? We can solve the problem of comparability of measurement units by converting raw scores into standard deviation units of measure, which *are* comparable. In other words, we express raw deviations from the respective means as multiples of their standard deviations. For this reason, we call such measures *standard scores*. When and if these standard scores are distributed normally, they are referred to as *normal scores* and are conventionally symbolized as *z*. Thus:

$$z = \frac{X - \bar{X}}{s} \tag{7.1}$$

If we're careful, we can convert to *z* scores relatively easily; but before we turn to their computation, let's make sure that we understand the translation of Jane's test scores to standard scores. Consulting Figure 7.3, we see that Jane's math score of 30 coincides with a standard score of 2.00. This illustrative graph also demonstrates that Jane's math score of 30 is statistically identical with her verbal score of 60—that is, both lie two standard deviations above the mean. If we now refer to Figure 7.2, we find that 47.72 percent of the total area lies between the mean and two standard deviation units above the mean. Since 50 percent of the area also falls below the mean of this symmetrical distribution, we may conclude that in both instances Jane scored much higher than the required 75 percent—97.72, to be exact. Now do you see why she was accepted to Statistics State? It should be equally evident why the *z* measure establishes the relative position of a score in a normal array and thereby renders corresponding scores in two or more normal arrays comparable.

Just to make our point clear, we confined the preceding discussion of area under the normal curve to selected points. But in fact, it's possible to determine the percentage of area between any two points by consulting the tabled values of the area under the normal curve in the *Normal Table* (Table B in Appendix III). The left-hand column (Column A) represents the deviation from the mean, expressed in standard deviation units. By referring to the body of the table, we can determine the percentage (or proportion) of total area between a given

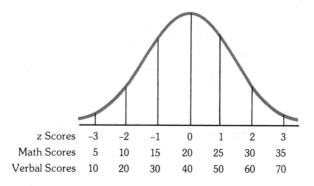

z Scores	-3	-2	-1	0	1	2	3
Math Scores	5	10	15	20	25	30	35
Verbal Scores	10	20	30	40	50	60	70

Figure 7.3.　Math Scores and Verbal Scores as z Scores

score and the mean from Column B, and the area beyond a given score from Column C. (Keep Table B handy—perhaps insert a bookmark there. We'll be using this table repeatedly throughout the remainder of this chapter.) Therefore, if Jane had scored 26 on the math test, her z score would be:

$$z = \frac{26 - 20}{5} = 1.2$$

Referring to Column A and reading across under Column B, we find that .3849 or 38.49 percent* of the area lies between her score and the mean. Since 50 percent of the area falls below the mean in a symmetrical distribution, we can conclude that 38.49 + 50 or 88.49 percent of all the area falls below a score of 26. Or we can translate this score into a percentile rank of 88.49.

If, in contrast, Jane had scored 14 on the same test, her z score would now be:

$$z = \frac{14 - 20}{5} = -1.2$$

Since the normal curve is symmetrical, only areas corresponding to the positive z values are given in Table B. Negative z values have precisely the same percentages as their positive counterparts. The area between the mean and a z of −1.2 is also 38.49 percent. We now must subtract 38.49 percent from 50 percent or else find the value directly from Column C, which specifies the proportion of scores lower than a z score of −1.2. In either case, Jane's percentile rank is 11.51. Figure 7.4 further clarifies the relationships among z scores, raw scores, and percentile ranks for the math test given by Statistics State.

7.5 Some Problems Illustrating the Use of the Normal Table

So that you can become skillful at using the standard normal table, we ask you to verify the following problems. The variables used are the tests given by Statistics State. You will recall that the math test had $\bar{X} = 20$ and $s = 5$, and the verbal test had $\bar{X} = 40$ and $s = 10$.

Problem I   Mary, Jane's friend, obtained a math score of 24. What percentage of the scores fall between her score and the mean? Also, what is her percentile rank?

---

*To express the areas under the normal curve as percentages of area, we either multiply by 100 or move the decimal point two places to the right.

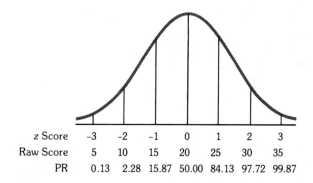

z Score	-3	-2	-1	0	1	2	3
Raw Score	5	10	15	20	25	30	35
PR	0.13	2.28	15.87	50.00	84.13	97.72	99.87

Figure 7.4. Relationship Among z Scores, Raw Scores, and Percentile Ranks of a Normally Distributed Variable, with X̄ = 20 and s = 5

From the start, it's wise to construct a diagram that illustrates the problem. This will help you clarify where z scores are in relation to the mean, and decide whether to add or to subtract areas from each other or from .5000. Thus, in the present problem, the diagram would look like the one in Figure 7.5.

To find the standard score equivalent or z score for a score of 24, we subtract the distribution mean from 24 and divide by 5. Thus:

$$z = \frac{24 - 20}{5} = 0.80$$

Looking up 0.80 in Column A and reading across under Column B of the normal table (Table B), we find that 22.81 percent of the area falls between the mean and 0.8 standard deviations above the mean. Thus, 22.81 percent of the area falls between Mary's score and the mean; therefore, her percentile rank is 50 + 22.81, or 72.81.

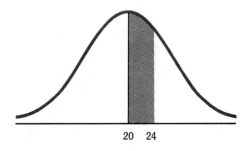

20   24

Figure 7.5. Diagram of Normal Curve, with Shaded Portion Representing the Area Between Mean and Score of 24

Problem II   On the verbal portion of the test, Mary obtains a score of 29. What is her percentile rank (Figure 7.6)?

To find the value of z corresponding to a score of 29, we subtract the distribution mean from 29 and divide by 10. Thus:

$$z = \frac{29 - 40}{10} = -1.10$$

The minus sign indicates that the score is below the mean. Looking up 1.10 in Column A and reading across under Column C, we find that 13.57 percent of the cases fall below Mary's score; therefore, her percentile rank is 13.57.

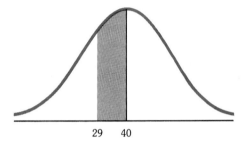

29   40

Figure 7.6.   Diagram of Normal Curve, with Shaded Portion Representing the Area Between Mean and Score of 29

Problem III   Bruce, Jane's high school sweetheart, obtained a score of 49 on the verbal test. How many applicants had scores over 49? Assume that 1,000 applicants took the test.

After drawing the diagram (Figure 7.7), we compute the standard score equivalent for a score of 49 to be:

$$z = \frac{49 - 40}{10} = 0.90$$

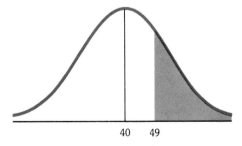

40   49

Figure 7.7.   Diagram of Normal Curve, with Shaded Portion Representing the Area Beyond a Score of 49

In Column C, we note that 18.41 percent of the area is between a z of 0.90 and the extreme. Thus, a shade over 184 applicants did better than Bruce.

Problem IV   What percentage of the students have scores between Jane's math score of 60 and Bruce's math score of 49?

We can't obtain the area in question in Figure 7.8 directly from Table B. Instead, we must calculate it by first finding the percentage of scores between the mean and Bruce's score of 49, and then subtracting this value from the percentage of scores between the mean and Jane's score of 60. The z scores corresponding to 49 and 60 are 0.90 and 2.00, respectively. By subtracting the percentage between 0.90 and the mean, 31.59 percent, from the percentage between 2.00 and the mean, 47.72 percent, we obtain the 16.13 percent that lies between Jane's and Bruce's scores.

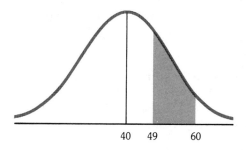

40   49   60

Figure 7.8. Diagram of Normal Curve, with Shaded Portion Representing the Area Between the Scores of 49 and 60

7.6 Caution: When Distributions Are Not Normal

In social science research, as in other kinds of research, some phenomena simply do not conform to the theoretical notion of the normal curve. Many distributions are highly skewed; others are multimodal; and still others are symmetrical but not bell-shaped. Our point is that if the original distribution is not normal, then the distribution of standard score equivalents will not be normal. In other words, transformation to z scores will not convert a non-normal distribution into a normal one.

When we have good reason to expect radical departures from normality (as often happens in some laboratory experiments), we should not use the normal curve as a model. Nevertheless, it has its advantages: as a statistical model in fitting nature's variation, the normal curve is still unrivaled in the scope of its application, even though it lacks the universal application to physical and social data that Quètelet attributed to it.

Where We Stand

How, you may ask, do we know the useful properties of the standard normal distribution? By empirical means—that is, by observation? If this were the particular data-accumulation process we used, we would need an infinite amount of data. Then we would have to place them into class intervals so small that they virtually would not exist. Then, and only then, would we obtain a smooth polygon like the normal curve. But who has the money, time, and energy to collect this amount of data? No, it is probability theory that shows the powerful properties of this remarkable theoretical distribution.

As we will see, the normal curve takes on a new perspective when viewed from the context of probability theory. And within this context, we can view all statements made about the area between two points of the curve as probability statements. Keep this in mind throughout the next two chapters, when we explore probability theory and statistical inference. By the time we return to discussing the normal curve, you'll really understand its important role in testing statistical hypotheses.

Terms and Symbols to Remember

Standard Normal Distribution (Normal Curve)
Standard Score (z Score or Normal Score)
Normal Table

Exercises

1. Discuss the nature and properties of the normal curve.

2. Is there more than one normal distribution? Explain.

3. Find the percentage of area under the normal curve between the mean and the following z scores.
   (a) +2.67   (f) −1.87
   (b) +2.08   (g) +2.60
   (c) −1.63   (h) −1.09
   (d) +0.28   (i) +3.04
   (e) −0.28   (j) −0.52

4. Find the percentage of area under the normal curve that lies:
   (a) To the left of $z = 1.52$
   (b) To the right of $z = -1.65$
   (c) To the right of $z = 2.02$
   (d) To the left of $z = -1.14$
   (e) To the left of $z = 1.14$

(f)   Between $z = 1.15$ and $z = 1.85$
(g)   Between $z = -1.05$ and $z = 1.45$
(h)   Between $z = -0.60$ and $z = -1.80$
(i)   To the right of $z = 0.60$
(j)   To the right of $z = -0.60$

5.   Given a normal distribution with a mean of 68.2 and a standard deviation of 10.8, find the standard score equivalents for the following scores.

(a)   65.0   (f)   70.2
(b)   43.0   (g)   80.8
(c)   29.6   (h)   69.0
(d)   30.4   (i)   73.3
(e)   20.1   (j)   46.0

6.   In a normal distribution with a mean of 80 and a standard deviation of 15:
(a)   What is the score at the first decile?
(b)   What is the score at the third decile?
(c)   What is the score at the first quartile?
(d)   What is the score at the second quartile?
(e)   What is the score at the third quartile?
(f)   What is the score at the sixty-third percentile?
(g)   What is the score at the seventieth percentile?
(h)   What is the score at the eighty-second percentile?
(i)   What is the score at the ninety-first percentile?
(j)   What is the score at the ninety-second percentile?

7.   A large set of IQs approximates a normal curve with a mean of 100 and a standard deviation of 10. Find:
(a)   The percentage of these IQs that can be expected to lie on the interval from 89 to 109
(b)   The score at the ninetieth percentile
(c)   The score at the mode
(d)   The score at the median
(e)   The score at the mean

8.   At Statistics State, the mean IQ is 125, with a standard deviation of 10. At Verbal College, the mean IQ is 115, with a standard deviation of 15. Assuming normal distributions, what percentage of the Verbal College students are below 65 percent of the Statistics State students?

9.   Suppose that a set of final examination grades in Introductory Statistics has a mean of 70 and a standard deviation of 15, and suppose that it approximates a normal curve. If the highest 10 percent of the students are to get As, which score would be the lowest A? If the lowest 10 percent of the students are to fail, which score would be the highest F?

10. A high school senior takes the College Entrance Examination Board (CEEB) test and receives a score of 1,050. Assume that the scores are normally distributed, the mean score is 1,000, and the standard deviation is 100.
    (a) What is the student's z score?
    (b) What is the student's percentile rank?
    (c) What percentage of students performed better?

11. Two students in different classes take final examinations in economics; each receives a score of 90. Suppose that the mean and standard deviation in one class were 80 and 8, respectively, but that in the second class the mean and the standard deviation were 82 and 6. Assuming a normal distribution for both classes, which student did better, relative to the other students of his/her class? What else might you say about making such a comparison?

12. Assume that students in a college are given point scores based on grades, service, and attendance. Assume, further, that the distribution of these point scores approximates a normal curve whose mean and standard deviations are 100 and 8, respectively. How many students from this college of 2,500 would be given honors certificates if the college awarded honors certificates to all students with more than 118 points?

13. A personnel manager of a company uses a test of ability to screen applicants for employment. Assuming that the distribution of test results is normally distributed with a mean test score of 100 and standard deviation of 5, answer the following questions.
    (a) If the personnel manager hires every applicant with a score of 110 or more, what proportion of the applicants will be hired? (The score of 110 will be called the *breaking point*.)
    (b) What breaking point should the personnel manager use if he wants to hire the top 30 percent of the applicants?
    (c) What scores should he use as the breaking point if he wants to hire all applicants except the top 10 percent and the bottom 10 percent?
    (d) Last year, the personnel manager hired fifty applicants by taking every applicant with a score of 105 or more. This year, he wants to hire seventy-five employees. What score should he use as a breaking point if he assumes that there will be the same number of applicants this year as there were last year? What score should he use as the breaking point if he assumes that there will be twice as many applicants this year as there were last year?

14. In what way might the normal curve be a bridge between descriptive and inferential statistics?

The theory of probabilities is nothing more than good
sense confirmed by calculations.
Marquis de Laplace

Probability

8.1 Introduction: Why Study Probability Theory?

So far, we've been concerned with accurately and economically describing different aspects of selective data sets. As we have seen, there are certain statistical techniques that enable us to meet these objectives. But when we consider inferential statistics, our concern shifts from describing samples to reaching some conclusions about populations.

Suppose that we have a coin of unknown origin, and that we wish to know whether it is an unbiased or a biased one. That is, we wish to know whether it will produce half heads and half tails or some other proportion. We can't toss the coin an infinite number of times, so we obtain a sample of only ten tosses; the results are nine heads and one tail. Is the coin biased? We can't answer conclusively, but we can make a reasonable guess. How? By determining the probability that a sample with this type of split, or even a more extreme one, would occur if the coin were, in fact, unbiased. If the probability is so low in getting a result this extreme with an unbiased coin, then it's reasonable to conclude that the coin is likely to be biased. If, on the other hand, the probability associated with getting this split is so high, then it would be reasonable to conclude that the coin is indeed unbiased. In this example, we have suggested how probability theory is used to test an hypothesis about a population parameter, in order to reach some conclusion about it. Probability theory thus gives us both the techniques and the rationale for making reasonable guesses of inferences about populations, based on the data from samples.

In this chapter, we will present some of the elementary principles associated with probability theory. Although the following material is selective rather than comprehensive, nevertheless it will help us understand the various concepts of inferential statistics that will appear in later chapters.

8.2 Probability Defined

A *probability* is basically a special type of proportion. In descriptive statistics, we can define a simple proportion as the frequency in any given distribution, divided by the total number of cases. Similarly, we can define a probability as the proportion of favorable events to the total number of possible outcomes. Stated symbolically:

$$p(A) = \frac{\text{Outcomes Favoring Event } A}{\text{Total Number of Outcomes}} \qquad (8.1)$$

where $p(A)$ is read as "the probability of A."

More specifically, a probability represents the proportion of times an event will occur, given an extremely large (technically, an infinite) number of trials. Here, we are using the concept of probability to refer not to single events, such as one toss of a coin, but to a large number of events or to what happens in the "long run."* To understand this, let's consider the probability involved with tossing a coin. If you were asked to give the probability of getting a head in one toss of an unbiased coin, you would most likely say, "one half" —which is, of course, impossible for one toss. The true answer is either zero or one. Yet "one half" comes to mind, probably because you are accustomed to thinking in terms of what would happen over a very large number of trials.

So, even though we toss a coin only once, we know that the probability of getting a head is ½.† Even though we roll the die only once, we know that the probability of it coming up six is $1/6$. We make these probability statements because we can identify the total number of possible outcomes and can enumerate the number of favorable events. This means that we can logically de-

---

*We can also approach probability from the standpoint of the single event, which presents at least as many conceptual difficulties as the one we already used. If you're interested in philosophical and mathematical arguments about the relative merits of the two approachs, you can read about them in Nagel (1939).

†Probabilities are not always expressed as ratios. They may be expressed as percentages, odds, or chances. These other forms are consistent with our definition, because they may all be translated into basic fraction or ratio form. For example, the 50 percent probability of a head may be divided by 100 percent to get the fraction ½.

termine the probability beforehand; in other words, we can compute the a priori probability.

In certain situations, we do not have sufficient information about the population to anticipate the relevant probabilities. Consider the probability associated with the cure rate for a new serum. How do we go about specifying such a probability? The answer partially revolves around the distinction made by statisticians between *deductive,* or *a priori probability* (made before observation) and *empirical,* or *a posteriori probability* (made after observation).

Deductive probability allows us to know all the possible outcomes in advance, and to assign probabilities to them. Thus, we know that there are two sides to a fair coin, fifty-two cards in a standard deck of playing cards, and six sides to an unloaded die. If we assume that each outcome is equally likely to occur, we can specify the probability of each.

When we cannot logically determine the probabilities beforehand, such as the probabilities associated with the cure rate for the serum we mentioned, we must use the concept of empirical probability. To get an empirical probability distribution, we must observe patiently and record dutifully the frequency of each event to obtain the total number of outcomes. Once we have arrived at an estimate of favorable events (usually after making a very large number of observations), we must use deductive probability theory to answer this type of question: What is the probability that, in a random sample of 500 patients, 50 percent will be cured with this serum?

8.3 The Ideal Couple: Probability Theory and Games of Chance

When discussing probabilities, it is a time-honored tradition to refer to coins, dice, or cards, and we have already begun our discussion in this way. Historically, probability has always been linked closely with games of chance. It is said that the pharaohs of ancient Egypt had the inner walls of the great pyramids inscribed with probabilities associated with the games of kings. However, the first written work on the subject of probability appeared in 1477, in a commentary on Dante's *Divine Comedy,* where mention was made of the probability of various outcomes in the throw of three dice. Given the connection between probability theory and games of chance, it is fitting that history should credit Girolamo Cardano (1501–1576), an able mathematician and an avid gambler, as the father of probability theory. Appropriately enough, one of his major works, roughly translated as the *Book on Games of Chance,* is essentially a gambler's manual.

The science of probability developed rapidly in the seventeenth century, because of the pioneering efforts of Blaise Pascal (1623–1662), Jacques Bernoulli (1654–1705), and Abraham de Moivre (1667–1754). These noted mathematicians were also fascinated by games of chance, and they were employed by wealthy gamblers, particularly the French nobility, who were trying to win at dice and cards. At first, the mathematicians' research was unsystematic, often leading to erroneous results; we do not know whether heavy losses at the gaming tables were suffered as a consequence. Nevertheless, during the second half of the seventeenth century, probability theory developed into a respectable scholarly pursuit.

Although our concern is not really with the plight of the gambler, we maintain this tradition of discussing gambling for two reasons. First, the probabilities associated with games of chance provide excellent teaching devices for the student of probability and social science statistics. Second, games of chance offer clear-cut examples for the derivation of deductive probabilities. For these reasons, except where some other type of example appears more suitable, throughout this chapter and the next we will continue to illustrate with coins, dice, or cards.

8.4 Four Formal Characteristics of Probabilities

So that we can understand probability better, we distinguish among four formal characteristics of probabilities. The first two help us define probabilities; the second two are rules for computing them. Pay close attention to the logic underlying these characteristics—we will be calling on them in later discussions.

8.4.1 Probabilities Vary from Zero to One

The first formal characteristic can be stated as follows:

> *A probability cannot be negative or exceed one; namely, it must be a number on the interval zero to one.*

Thus for any given event, say A:

$$0 \leq p(A) \leq 1.00$$

in which the symbol $\leq$ is read as "less than or equal to."

This characteristic is consistent with our earlier definition of probability, since a proportion is neither negative in value nor greater than one. Therefore, any value between zero and one expresses the probability associated with the event—which, of course, reflects the degree to which an event is expected (or not expected) to occur. At one extreme, then, where $p(A) = 0$, event A cannot possibly occur. At the other extreme, where $p(A) = 1.00$, event A is certain to occur. For example, the probability of drawing a red ace of spades from a standard deck of fifty-two playing cards is zero, whereas the probability of drawing a card from the same deck with some type of marking on it is one.

8.4.2 The Probabilities in a Distribution Sum to One

The second formal characteristic can be stated as follows:

Since all the outcomes of an experiment must exhaust the possibilities, all the probabilities in a distribution must sum to one.

In symbols:

$$\Sigma p(A) = 1.00$$

You can see the reasonableness of this characteristic quickly enough. If something is going to occur, like drawing a card with some type of marking on it, then the sum of the alternatives must add up to certainty (that is, $\Sigma p(A) = 1.00$) that you will draw a card with some sort of marking on it.

Here's another illustration of this characteristic. Consider the individual probabilities associated with the six sides of an unloaded die. Since each side of the die has a probability of $\frac{1}{6}$ of coming up, the sum of all the probabilities in the distribution of outcomes equals one.

8.4.3 The Addition Rule

The third formal characteristic, the addition rule, helps us compute probabilities. It is:

If A and B are events, the probability of obtaining either of them is equal to the probability of A plus the probability of B minus the probability of their joint occurrence.

In symbols:

$$p(A \text{ or } B) = p(A) + p(B) - p(A \text{ and } B) \qquad \textbf{(8.2)}$$

The $p(A \text{ or } B)$ is obtained by first adding the individual probabilities associated with both A and B and then subtracting out the probability of obtaining both A and B simultaneously. The reason for subtracting $p(A \text{ and } B)$ is that the probability of this joint occurrence has been computed twice, once in $p(A)$ and again in $p(B)$.

As a concrete example, suppose that we draw one card from a well-shuffled deck of fifty-two playing cards. What is the probability that this card is either a king or a club? The probability is:

$$4/52 + 13/52 - 1/52 = 4/13$$

We can verify this result intuitively by noting that either $p(KING)$ or $p(CLUB)$ could be obtained if we draw any club or one of the remaining kings. Had we simply added $p(KING)$ and $p(CLUB)$, the king of clubs would have been considered twice. Since we won't always find it so easy to obtain this quantity, in the next section we will consider a general rule for computing joint occurrence. In any case, when outcomes are all *mutually exclusive* (that is, if both events *cannot* occur simultaneously)—such as the probability of getting a king or queen—then obviously there is no overlap, and the last term in the above formula drops out. When this happens, Formula (8.2) takes the simplified form:

$$p(A \text{ or } B) = p(A) + p(B) \text{ if A and B are mutually exclusive} \quad \textbf{(8.3)}$$

It's easy to generalize this special rule of addition so that it applies to more than two mutually exclusive events. If we have three mutually exclusive events (A, B, and C), the probability that one of them will occur is:

$$p(A \text{ or } B \text{ or } C) = p(A) + p(B) + p(C)$$

Thus, the probability that a card drawn at random would be either an ace, a king, or a queen is:

$$4/52 + 4/52 + 4/52 = 3/13$$

8.4.4 The Multiplication Rule

The fourth formal characteristic also helps us to compute probabilities. Because it is so simple, we shall first take up a special case of the multiplication rule. The rule states:

> **If A and B are two independent events, the probability of obtaining A and B is the product of the two probabilities.**

In symbols:

$$p(A \text{ and } B) = p(A)p(B) \text{ if A and B are independent} \quad \textbf{(8.4)}$$

The term "independent" means that the probability that A will occur remains constant regardless of whether or not B has occurred, and vice versa. That is, even if we know that one of the events has happened, this knowledge does not help us one iota to predict the occurrence of the other event; in other words, one event does not affect the outcome of the other event. Events like tossing coins or spinning a roulette wheel all have a fixed number of outcomes. These results do not change from trial to trial. Thus, the probability of obtaining two heads in two tosses of a fair coin are independent of each other, and the probability is (½)(½) or ¼.

Notice that what we are denying is the celebrated

Monte Carlo Fallacy: The unanimous belief that after a long run of successes a failure is inevitable, and vice versa.

Many gamblers believe that if a coin comes up heads many times in a row, a tail is long overdue and therefore has a higher probability of occurrence than one half. This reasoning, of course, is fallacious. The very calculation that makes a long run of heads improbable is based on the assumption that each toss of the coin is independent of every other; therefore the probability of heads is the same in each instance, no matter what the results of the preceding tosses have been. Unlike a human, a coin has neither memory nor a will of its own.

This simplified version of the multiplication rule may be generalized to incorporate any number of events, as long as they are independent of one another. Thus for three events *(A, B,* and *C)*, we write:

$$p(A \text{ and } B \text{ and } C) = p(A)p(B)p(C)$$

To understand how this formula is applied, consider the probability of obtaining a head, a six, and a king in one toss of a coin, one roll of a die, and one draw from a deck of cards. Since the events are all independent of each other, our answer is:

$$(1/2)(1/6)(4/52) = 1/156$$

When events are not independent, we must know the probability of obtaining the joint occurrence of both events. To acquire this probability value, we apply a more general form of the multiplication rule:

If A and B are two events, the probability of obtaining both A and B is the product of the probability of obtaining A times the conditional probability of obtaining B given that A has occurred.

In symbols:

$$p(A \text{ and } B) = p(A)p(B/A) \qquad \qquad \textbf{(8.5)}^*$$

---

*We selected the symbols *A* and *B* for their clarity. Therefore, no time order is implied in the way they occur. Thus:

$$p(A \text{ and } B) = p(A)p(B/A) = p(B)p(A/B)$$

where *p(B/A)* represents conditional probability and is read as "the probability of *B*, given that *A* has occurred."

The term *conditional probability* takes into account that the probability is subject to a condition. That is, one event's occurrence is influenced by the other event's occurrence. The probability that a person will become an octogenarian is increased if that person has already reached the age of seventy-five. The probability that a person will contract lung cancer is greatly increased if that person is a heavy cigarette-smoker. In both examples, the events are not independent of each other. However, obtaining a head on one toss of the fair coin does not affect the probability of obtaining one on the next toss. Whenever the latter exists, *p(B/A)* = *p(B)*, and the general form of the multiplication rule reduces to Formula (8.4).

To understand the use of conditional probability, let's consider the probability of drawing at random two aces in succession from a standard deck of fifty-two playing cards. The probability of drawing the first ace, of course, is 4/52. But getting one ace (without replacing it) obviously affects the probability of drawing another ace on the next draw. Since only fifty-one cards remain (including three aces), the probability of the next ace is now reduced to 3/51. We can then obtain the probability of drawing two successive aces by multiplying the unconditional probability 4/52 by the conditional probability 3/51, which equals 1/221. If we had replaced the first ace, it could not affect the second draw and the answer therefore would be (4/52)(4/52), or 1/169. This indicates that sampling with replacement assures independence, and that its alternative, sampling without replacement, forces us to compute conditional probability.

8.5 The Concept of Randomness

We have based our explanation of probability theory on the assumption that the events we have described have been selected randomly. Indeed, from time to time we have said outright that an event has been selected "at random." Now we will explain precisely what this term means since, when we use sample data, it will be very important to make inferences about populations.

Events are said to be random if one event has no predictable effect on the next event. We have already said that if we know one toss of a coin, one roll of a die, or one selection of a card (assuming replacement), this knowledge will not help us to predict future outcomes. Thus, the first characteristic of random events is *independence*. Only if independence is achieved can events be said to be truly random. The second characteristic of randomness is that when the sample is taken from a population, each event must have an *equally likely chance* of being selected. However, if the selection process favors certain events or certain combinations of events, we cannot justifiably claim randomness. In this case, any such procedure is said to be *biased*. The dangers of generalizing to the population from biased sources should be apparent from our earlier example of the 1936 *Literary Digest* poll (Section 1.4). Unless the condition of randomness is met, we may never know to what population we should generalize our results. And nonrandom samples play havoc with the rules of probability, as well as with the concepts relative to inferential statistics that we will discuss in the following chapters.

The most popular sampling technique used to guarantee randomness is the *simple random sample* (or *random sample* for short). We get this in much the same way we would if we were drawing numbers in a bingo game or cards from a well-shuffled deck. In this way, each event (or combination of events) has an equally likely chance of being selected. At other times a *stratified sample* is used. This divides the population into homogeneous strata, such as upper, middle, and lower class. Then a random sample is taken from each stratum. The *area* (or *cluster*) *sample* is still another method of random sampling. Here, the population is subdivided into smaller geographic segments, and all (or a subsample) of the individuals are chosen at random.*

Each of these sampling techniques faithfully maintains the two conditions of randomness. The researcher may select any one of them, depending on such important considerations as its relation to the study at hand, its cost, or its ability to produce the wanted results.

Where We Stand

We opened this chapter with an adage attributed to the Marquis de Laplace (French mathematician Pierre Simon, 1749–1827): "The theory of probabilities is nothing more than good sense confirmed by calculations." We hope that you can see the concept probability in this way, by now. Probability concepts can be formalized into some basic rules, but they still have more to do with logical reasoning than with anything else. Although we are all prone to commit the well-known Monte Carlo fallacy, a player who witnesses twenty successive red calls in twenty spins of the roulette wheel would do well to bet red on the twenty-first spin of the roulette wheel just as a matter of intelligent strategy, assuming that either the house, the croupier, or the wheel is dishonest. In class, in research, and in everyday life, you would do well to realize what the concept of probability entails and to warn all those who would misuse it against doing so.

---

*We must confess that we have greatly oversimplified our presentation of these sampling techniques. But a detailed discussion of the fine points of sample designs is beyond the scope of this text. If you are interested in finding out more, Kish (1965) offers a good discussion of sample designs and the practical problems facing the researcher in designing a sample.

By now, you probably realize that you need to know and understand the concepts covered here in order to shift from descriptive to inferential statistics. In the next chapter, we'll see how probability theory is the basis for the whole inferential process.

Terms and Symbols to Remember

Probability *(p(A))*
Deductive (or A Priori) Probability
Empirical (or A Posteriori) Probability
Addition Rule *(p(A or B))*
Mutually Exclusive
Multiplication Rule *(p(A and B))*
Independence
Monte Carlo Fallacy
Conditional Probability *(p(B/A))*
Sampling with Replacement
Sampling without Replacement
Randomness
Simple Random Sample (or Random Sample)
Stratified Sample
Area Sample (or Cluster Sample)

Exercises

1. What is the difference between *a priori* and *a posteriori* probabilities?

2. A card is drawn at random from a standard deck of fifty-two playing cards. Determine the probability that:
   (a) It will be an ace
   (b) It will be the ace of diamonds
   (c) It will be a club or a picture card
   (d) It will be a club or an ace
   (e) It will be a king or red card

3. In a single roll of two dice, determine the probability that:
   (a) A 6 will appear
   (b) A 7 or 11 will appear
   (c) Two of the same number will appear
   (d) An odd number will appear
   (e) An even number or 7 will appear

4. Which of the following pairs of events are mutually exclusive?
   (a) Drawing an ace or a red card from a standard deck of fifty-two playing cards
   (b) Being over thirty and being female

    (c)   Being a college student and owning a car

    (d)   Rolling a five or an 8 with a pair of dice

    (e)   Getting heads in two successive tosses of a coin

5.   Which of the following pairs of events are independent?

    (a)   Smoking cigarettes and contracting lung cancer

    (b)   Drinking alcohol and having an automobile accident

    (c)   Being a Democrat and being female

    (d)   Being a Democrat and owning a car

    (e)   Being male and having brown hair

6.   Determine the probability of obtaining:

    (a)   A tail, a 5, and an ace in one toss of a coin, one roll of a die, and one draw from a deck of fifty-two playing cards

    (b)   A tail, a 3, and a picture card in one toss of a coin, one roll of a die, and one draw from a deck of fifty-two playing cards

    (c)   A tail or a head, a 5 or a 7, and an ace or a 4 in one toss of a coin, one roll of a die, and one draw from a deck of fifty-two playing cards

    (d)   A tail, a 3 or a 5, and a picture card or an ace in one toss of a coin, one roll of a die, and one draw from a deck of fifty-two playing cards

7.   Enumerate all the possible outcomes of a coin tossed four times. Determine the probability of obtaining:

    (a)   Four heads

    (b)   Three heads and one tail

    (c)   Three tails and one head

    (d)   At least two tails

    (e)   At least two heads

8.   If A and B are independent events, $p(A) = 0.30$, and $p(B) = 0.50$, compute each of the following:

    (a) $p(A \text{ or } B)$

    (b) $p(A \text{ and } B)$

    (c) $p(A/B)$

    (d) $p(B/A)$

9.   Assume that the above events are mutually exclusive. Find the following probabilities:

    (a) $p(A \text{ or } B)$

    (b) $p(A \text{ and } B)$

    (c) $p(A/B)$

    (d) $p(B/A)$

10.   Four cards are drawn at random (without replacement) from a standard deck of fifty-two playing cards. Determine the probability that:

    (a)   All three will be spades

    (b)   All three will be picture cards

    (c)   All three will be aces

(d) None of the cards will be aces

(e) None of the cards will be clubs

11. Calculate the probabilities in Exercise 10 if each card is replaced after it is drawn.

12. An urn contains ten red marbles, five white marbles, and five blue marbles. Suppose that one marble is selected at random from the urn. Determine the probability that it will be:

(a) Blue

(b) Blue or red

(c) Red

(d) Red or white

(e) White

13. Determine the probability that a score selected at random from a normally distributed population with a mean of 100 and a standard deviation of 15 will be:

(a) Less than 85

(b) In the 80s

(c) Between 92 and 105

(d) Greater than 85 or less than 73

(e) Greater than 106

(f) Either less than 69 or between 102 and 109

(g) Greater than 110 and between 80 and 83

(h) In the 90s

(i) Less than 85 or greater than 117

(j) Between 65 and 70 or between 106 and 121

14. Discuss the concept of random sampling.

15. Which of the following selection techniques will result in random samples? Explain.

(a) Population: 300 students attending an introductory class in music. Sampling technique: interviewing every fifth student in the class

(b) Population: shoppers in a supermarket. Sampling technique: interviewing every shopper with a child

(c) Population: all the residents of New York City. Sampling technique: interviewing every other person passing by the Empire State building

By a small sample we may judge of the whole piece.
Miguel de Cervantes

Statistical Inference: An Introduction

9.1 Introduction: Why Sample?

As we've said throughout this book, the social scientist is usually concerned with the entire *population*. At times, populations may be small—for example, all the participants in an athletic contest, or all the members of a standing committee of Congress—but more often, however, they are so large that they defy exhaustive study—for example, the entire adult population of the United States. Populations may be of infinite dimensions, such as all infants who have ever been or will ever be born. In these cases, it is impossible to study all the members of the population: the population, as defined, is unlimited. No matter what resources we use or how much time we spend, we could never observe the population as a whole—at any point in time, the population would be only partially completed. The collection process would take literally an eternity. In addition, some populations are structured in such a way that portions of them are physically inaccessible. Consider the case of the survey attempting to query all smokers on their attitudes towards low-tar cigarettes. Not all the smokers can be reached to participate. Some of them may be incarcerated in prisons or mental institutions and therefore have only limited contact with the outside world. Others may be inaccessible to persistent survey takers for other reasons; for example, the president of the United States is not easy to reach in person.

Generally, most of the populations that scientific investigations concern themselves with are either quite large, indefinite in number, or partly inaccessible. This means that they cannot be studied in detail. Therefore, if we want to get information about the population, we are forced to *sample*. And why do we want to observe this sample? To use the data we obtain to draw inferences about the population. So we aren't really concerned with descriptive statistics generated from sample data, per se. We want to make inferences about the population. When pollsters like Gallup and Harris ask a nationwide sample of 1,500 voters how they intend to vote in a forthcoming election, they don't so much care about the behavior of these particular respondents as about predicting how the entire voting population will cast their ballots. Descriptive statistics give pollsters (as well as social scientists) a factual basis upon which to make the transition from samples to populations.

This chapter introduces the concept of statistical inference. Now that you have some knowledge of probability theory, let's take a look at the important concept of a *sampling distribution*.

9.2 The Concept of the Sampling Distribution

In practice, we make inferences about the parameters of a population from statistics that are calculated from a sample of observations drawn at random from this population. (If you need to, review the explanation of these concepts in Section 1.3.) If we continued to draw random samples of the same size from the population, we would find some variation among the values for the sample statistics. This type of thinking lies behind the concept of a:

> *Sampling Distribution: A theoretical probability distribution of all the values of a sample statistic that would occur if we were to draw all possible random samples of a fixed size from a population.*

Suppose that we draw successive random samples from a large population. We draw one sample of 100 subjects and measure them on some behavioral act; then we replace the sample, draw another of equal size, and measure the subjects again; and so on. Now suppose that we repeat this procedure until we have exhausted all the unique random samples of 100 in the population, computed the mean for each, and constructed a frequency polygon of these means. The distribution of the means of all these samples would constitute a sampling distribution of means.

Of course, the social scientist who's engaged in research doesn't obtain a sampling distribution in quite this way. In actual practice, the researcher draws a single sample, computes a statistic, and uses his or her knowledge of the sampling distribution to generalize to the corresponding population parameter.

Every statistic, whether it is a mean, a standard deviation, or something else, has a sampling distribution. If the statistic is to help us make inferences from samples to populations, we must know its sampling distribution. Why? Because whenever we make an inference about a population from a sample, we want to know if the estimate is reasonable. It's logical for us to ask questions like: Is our sample statistic identical with the corresponding population parameter? And if not, then how much sampling error do we have? By providing a set of "expected results" by which to judge our estimates, the appropriate sampling distribution will give us the information we need to answer these questions. But some questions still remain: What does the sampling distribution of a statistic look like? How can we ever come to know the form of the distribution, and from that know the expected results?

Each statistical technique has a corresponding sample distribution(s). These sampling distributions are usually presented in tabular form, as they are in Appendix III of this text. For example, whenever large samples are used, the normal curve, which is a theoretical probability distribution, is the appropriate sampling distribution for statistics like the mean and/or standard deviation. But we use the sampling distribution known as the *binomial distribution** whenever: we encounter simple dichotomies, such as whether an individual possesses a certain attribute, or whether an experiment has been a success or failure; we can assume a probability of successes equal to one half; trials are independent of each other; and the number of trials is relatively small.

Although many statistical tests are more suitable than those that use the binomial, we think it's important to demonstrate this distribution, primarily because of its simplicity. When we use the binomial, we can explain the steps that are involved in making inferences with a discrete two-category variable. In the process, we can better understand the general procedures used in *all* statistical tests of significance. Keep in mind, however, that such tests are simply statistical techniques that we use to establish the probability of obtaining certain results. In this way, we are using probability theory to determine whether or not the sample statistic we obtain is a reasonable estimate of the corresponding population parameter.

9.3 The Binomial Distribution

Imagine the following scenario. You want to postpone reading the remainder of this chapter so that you can chat with a friend on the telephone. But since you're feeling a bit guilty about neglecting your work, you decide to absolve yourself of any guilt by tossing a coin: "Heads, I'll study; tails, I'll call." It's heads, and reluctantly you gear up to study. But then your little brother enters the room and turns on the TV. Angered by the distraction, you offer to toss a

---

*The binomial and normal distributions bear a striking resemblance to each other. In fact, the binomial distribution will resemble the normal curve as the sample size increases without limit. Because the discrete binomial converges on the continuous normal curve, we will present the *central limit theorem* in the next chapter. If you are interested in learning more about the mathematics behind both distributions, consult Alder and Roessler (1972: 91–117).

coin: "Heads, I'll leave; tails you'll leave." He agrees. You toss the coin you tossed before, and again you obtain a head. Frustratedly, you toss the coin again insisting, "Heads, I lose." It's heads once more. You toss the coin seven more times and obtain six more heads, making a total of nine heads out of ten tosses. Can the coin be expected to turn up heads more often than tails? How could you find out?

One thing is clear. There is no way you can come to know the true proportion of heads and tails for this coin. Even if you started tossing the coin right this very second and continued ad infinitum (assuming a long life, an infinite amount of patience, and a remarkably durable coin), you would not exhaust the population of possible outcomes. The true proportion of heads and tails is unknowable, because the population is infinite in number.

So in order to determine whether the coin is biased, we will assume that the nine heads-one tail (9H-1T) split indicates the behavior of the coin. On the basis of this result, are we then justified in concluding that the coin is biased? Or is it reasonable to expect such a result from an unbiased coin? Before we can answer these questions, we have to look at the sampling distribution of all possible outcomes. The intuitive construction of this sampling distribution is the next topic.

9.3.1 Enumerating the Binomial Distribution

Assume that we have an unbiased coin in our possession, and that there are only two possible outcomes for each toss: heads or tails. Assume further that the coin will not come up both heads and tails simultaneously; that it will not stand on edge; and that it will not become worn unevenly in the process of being tossed and thereby favor a particular outcome.

If we toss this hypothetical coin, we can determine the number of different possible outcomes of heads and tails on a given number of occasions and the probability of the occurrence of each by applying the multiplication rule. If we toss the coin twice, for example, four outcomes are possible (HH, HT, TH, TT), and the probability of each is $(\frac{1}{2})^2$. That is, the probability of each is $(\frac{1}{2})(\frac{1}{2})$, or $\frac{1}{4}$. Similarly, with five tosses, there are thirty-two possible outcomes, each with a probability of $(\frac{1}{2})^5$ or $\frac{1}{32}$.

Now, suppose that for the five-trial case we count the thirty-two possible outcomes of heads and tails. For each outcome, we then count the number of heads, ranging from zero to five. We find that five heads occur in only one of the thirty-two possible outcomes, but there are five outcomes in which one head occurs (HTTTT, THTTT, TTHTT, TTTHT, TTTTH). Similarly, there are ten outcomes in which two heads occur, and so on. A complete frequency distribution of these various number of heads, from five to zero, is shown in Table 9.1. The bottom row of the table contains the probability of occurrence of each number of heads. This list of probabilities is known as a *binomial probability distribution*.

Table 9.1.   Theoretical Frequency Distribution of Possible Outcomes Obtained by Tossing an Unbiased Coin Five Times

	5H	4H	3H	2H	1H	0H
	HHHHH	HHHHT	HHHTT	HHTTT	HTTTT	TTTTT
		HHHTH	HHTHT	HTHTT	THTTT	
		HHTHH	HTHHT	HTTHT	TTHTT	
		HTHHH	THHHT	HTTTH	TTTHT	
		THHHH	HHTTH	THHTT	TTTTH	
			HTHTH	THTHT		
			THHTH	THTTH		
			HTTHH	TTHHT		
			THTHH	TTHTH		
			TTHHH	TTTHH		
$f/$	1	5	10	10	5	1
$p/$	0.03125	0.15625	0.31250	0.31250	0.15625	0.03125

We obtained the binomial probabilities in the table by using the addition rule for mutually exclusive events. That is, each of the individual outcomes of heads and tails has a probability of $\frac{1}{32}$, or 0.03125. Thus, the probability of obtaining five heads is $\frac{1}{32}$, since five heads occur only once among the thirty-two outcomes. Four heads occur in five of the outcomes; adding 0.03125 together five times gives 0.15625, and so on.

Clearly, this method of enumeration allows us to discover all the possible outcomes for an unbiased coin that's tossed five times. Our purpose, however, is to obtain the sampling distribution for an unbiased coin tossed ten times. Here, since simple enumeration is too cumbersome, we resort to the formula computation described in Section 9.3.2.

9.3.2 A Formula for Computing the Binomial Distribution

An alternative method of obtaining the sampling distribution for a coin tossed ten times is given by expanding the binomial: $(p + q)^N$. Formula (9.1) presents the binomial expansion in its most general form:

$$(p + q)^N = p^N + \frac{N}{1} p^{N-1} q + \frac{N(N-1)}{(1)(2)} p^{N-2} q^2$$

$$+ \frac{N(N-1)(N-2)}{(1)(2)(3)} p^{N-3} q^3 + \ldots + q^N \qquad \textbf{(9.1)}$$

where:

p = the probability of obtaining the favored event,

q = the probability of obtaining the nonfavored event, and

N = the number of independent trials.

Looks pretty complicated, doesn't it? But it isn't. Let's illustrate its use with the five-trial case (Table 9.2). Note. $p = \frac{1}{2}$ (the probability of obtaining a head), $q = \frac{1}{2}$ (the probability of obtaining a tail), and $N = 5$ (the number of tosses).

We'd guess that after you've studied the substitutions and mathematics presented in the table, your anxiety has subsided. Let's look at the numerators for each term in the expansion. The numerator for the first term on the right-hand side of the equation, $(\frac{1}{2})^5$, is 1, and the remaining numerators, also referred to as the *coefficients of the binomial,* are 5, 10, 10, 5, 1. Notice that they correspond exactly to the frequencies of each outcome in Table 9.1, and that the terms themselves coincide with the probabilities listed in the table. For example, the probability of five heads is given by the first term, $(\frac{1}{2})^5$ or 0.03125; using the addition rule, we obtained this same figure earlier.

9.4 Testing Statistical Hypotheses: Null Hypothesis and Research Hypothesis

Once you've investigated the binomial distribution, you'll most likely have a hunch about the nature of the coin in our present experiment. In general, a statistical hypothesis is merely a calculated hunch. Of course, you would like to prove your hunch correct! Social scientists often are more interested in finding differences than in determining that differences don't exist. Differences among groups, whether they're expected on theoretical or empirical grounds, often provide the basis upon which a study is conducted. However, since it is a much more clear-cut process to reject a statistical hypothesis than it is to accept it, what usually happens is that you frame a statistical hypothesis contrary to the one you are hoping to prove—which, in the present case, is that the coin is biased. Such an hypothesis is known as a *null hypothesis* (H_0). The word "null" comes from the Latin root meaning none, which is precisely what the hypothesis assumes: that any difference between the sample statistic and the population parameter (assumed by the null hypothesis) is to be regarded as a chance occurrence resulting from a sampling error alone. The *alternative* or *research hypothesis* (H_1) is merely an alternative hunch about the value(s) of the parameter that may seem more plausible.

The advantage of using the null hypothesis in this way is that we can use it as a basis for selecting a specific sampling distribution—that is, the sampling distribution that would be found if the null hypothesis were, in fact, true. We then use this sampling distribution to judge whether the sample data warrant rejection of the null hypothesis in favor of the research hypothesis. In the present context, these two hypotheses read as follows:

$$H_0\text{: } p = \frac{1}{2} \quad \textit{the coin is unbiased}$$
$$H_1\text{: } p > \frac{1}{2} \quad \textit{the coin is biased}$$

Table 9.2. Illustration of the Binomial Expansion when $p = q = \frac{1}{2}$ and $N = 5$

5H		4H		3H		2H		1H		0H
$(p+q)^5 = p^5$	$+$	$\dfrac{N}{1}\,p^4q$	$+$	$\dfrac{N(N-1)}{(1)(2)}\,p^3q^2$	$+$	$\dfrac{N(N-1)(N-2)}{(1)(2)(3)}\,p^2q^3$	$+$	$\dfrac{N(N-1)(N-2)(N-3)}{(1)(2)(3)(4)}\,pq^4$	$+$	q^5
$(\frac{1}{2}+\frac{1}{2})^5 = (\frac{1}{2})^5$	$+$	$5(\frac{1}{2})^4(\frac{1}{2})$	$+$	$10(\frac{1}{2})^3(\frac{1}{2})^2$	$+$	$10(\frac{1}{2})^2(\frac{1}{2})^3$	$+$	$5(\frac{1}{2})(\frac{1}{2})^4$	$+$	$(\frac{1}{2})^5$
$1.00 = 1/32$	$+$	$5/32$	$+$	$10/32$	$+$	$10/32$	$+$	$5/32$	$+$	$1/32$
$1.00 = 0.03125$	$+$	0.15625	$+$	0.31250	$+$	0.31250	$+$	0.15625	$+$	0.03125

The research hypothesis can be either *directional* (as in the present example) or *nondirectional*. When it only asserts that the population parameter is different from the one hypothesized, it is referred to as a nondirectional or *two-tailed* research hypothesis (for example, $p \neq q \neq \frac{1}{2}$). But if the research hypothesis is directional or *one-tailed,* not only do we assert that the population parameter is different from the one hypothesized, but we also state the direction of that difference (for example, $p > \frac{1}{2}$). This distinction will become clear when we discuss the procedures used to determine the critical region for the present coin experiment.

9.5 Testing Statistical Hypotheses: The Critical Region

Now that we have formulated the necessary hypotheses, we are ready to find out about the honesty of the coin in our current example. Let's begin with a question: Would we suspect the honesty of the coin if it had produced an equal number of heads and tails? Probably not. After all, this is what we would expect from an honest coin. The same would probably be true if we had gotten six heads (or six tails). If we expand the binomial to accommodate ten trials, as we did in Table 9.3, we see that the 60–40 (or 40–60) split is quite common and is fully expected to occur almost as frequently as the 50–50 split. So the crucial question is: At what point can we say that a departure is no longer a chance occurrence? Is the cutoff at 70–30, 80–20, or 90–10?

Table 9.3.   Theoretical Frequency Distribution of Number of Heads Obtained by Tossing an Unbiased Coin Ten Times

Heads	f	p
0	1	0.001
1	10	0.010
2	45	0.044
3	120	0.117
4	210	0.205
5	252	0.246
6	210	0.205
7	120	0.117
8	45	0.044
9	10	0.010
10	1	0.001

The answer to these questions reveals the probabilistic orientation of statistical inference. That is, any decision we make concerning the coin's honesty is *probabilistic* rather than *absolute*. If we knew for certain, we would not have to sample, and inferential statistics would not be necessary. We sample so that we can draw inferences. Why do we want to? Because, for one reason or another, our population values are unknowable. If they become known, we do not need statistical inference.

In the social sciences, two cutoff points, 0.01 and 0.05, generally are used as a basis for inferring the operation of nonchance factors. These are commonly called the 0.01 (or 1%) and 0.05 (or 5%) *significance levels*. The level of signficance helps us determine the *critical region*—those values of the test statistic for which the null hypothesis will be rejected. The size of this region is determined by direction and by willingness to commit Type I or Type II errors (we discuss these matters in the following paragraphs). We call the size of the critical region (for example, 0.01 and 0.05) *alpha* (α). Alpha corresponds to the percentage of all samples in the sampling distribution that would show values sufficiently deviating from the hypothesized parameters so as to lead us to reject the null hypothesis. This means that if we obtain a result falling within this region, we would reject the null hypothesis since such a value is so unlikely as to not occur by chance alone.

In our current coin experiment, let's assume that we wish to use the 0.01 significance level to evaluate the 9H–1T split. The crucial question becomes: Can we expect this occurrence on the basis of chance alone? That depends on how much risk we are willing to take. Using the 0.01 significance level is like saying that our outcome could occur theoretically by chance factors in fewer than 1 percent of the cases. If the theoretical probability distribution says that the occurrence could happen by chance alone in more than 1 percent of the cases, then we cannot reject the null hypothesis and must conclude that the coin is honest.

In contrast, the 0.05 significance level imposes a less restrictive criterion for rejecting the null hypothesis about the coin. When we use this level, we assume that the theoretical probability distribution is telling us to expect the 9H–1T split to occur in at least 5 percent of the cases by chance alone, if the coin is unbiased. Anything less than this theoretically expected result seems to indicate that our outcome is unusual and is probably caused by nonchance factors. In other words, Table 9.3 tells us that the 9H–1T split should theoretically occur only 1 percent (0.010) of the time. The 0.05 level tells us that if this outcome were not the result of bias, theoretically it would have to occur in at least 5 percent of outcomes by chance alone. Clearly, this is not the case; so we must reject the null hypothesis.

If, by this time, you feel that no more discussion about the critical region is needed, excellent—you have really mastered the concept. However, the construction of the critical region is so important that we have to examine it from just one more important angle.

Suppose, for example, that the critical region consisted of only the probability of nine heads. But the choice of such a critical region would seldom make sense theoretically, since ordinarily we would be even more hesitant about accepting the null hypothesis if ten heads were to come up; yet this alternative would not belong to the critical region. What this suggests is that we almost always want to use at least an entire tail of the distribution. It isn't the probability of getting nine heads that we're interested in, but the probability of getting nine heads or something even *more* unusual.

But why not also include zero and one head in the critical region, since these alternatives are just as unlikely as nine and ten heads? Often, we can't predict ahead of time the direction in which the unusual results may occur. Although we suspect that the coin is dishonest, we may have no hint as to whether it is biased in favor of heads or tails. In such a case, we'd want to play safe and use both tails of the sampling distribution. If we used a critical region consisting only of nine and ten heads, and if we obtained only one head, we would find ourselves not rejecting the null hypothesis, even though it might be incorrect. In short, our interest in two directions of the sampling distribution necessitates a *two-tailed* rather than a *one-tailed test* of the null hypothesis.

Don't let the formulation of a null hypothesis and the designation of one- and two-tailed tests worry you. All you need to know is that the research hypothesis dictates both the form of the null hypothesis and whether we have a one- or a two-tailed test. You'll become more skillful at hypothesis formulation as we continue to discuss appropriate statistical tests of significance for particular population parameters. For now, let's assume that the coin is biased in favor of heads and set the critical region to include the probability of an event as unusual as either nine or ten heads. Table 9.3 reveals that the one-tailed probability of an event as unusual as nine heads out of ten tosses is:

$$0.010 + 0.001 = 0.011 \text{ or } 1.1\%$$

Using the 0.01 significance level, we cannot reject the null hypothesis that $p = \frac{1}{2}$, because 0.011 falls outside the critical region. But if we use the broader 0.05 level, we can reject the null hypothesis and support the research hypothesis that the coin is indeed biased in favor of heads.

9.6 Testing Statistical Hypotheses: Two Types of Errors

When we rejected the null hypothesis at the 0.05 alpha level, we said that if the null hypothesis is tenable, theoretically the 9H-1T split should occur by chance alone 5 percent of the time or more. However, our application of the binomial distribution to our coin experiment tells us that our obtained result could happen theoretically by chance alone less than 5 percent of the time (see Table 9.3). This result deviates sufficiently from the hypothesized population parameter to lead us to conclude that the null hypothesis was faulty. Could we have made a mistake? We hope not; but everyone is fallible. In fact, if we set the critical region at 0.05, about 5 percent of the time we'd mistakenly reject the null hypothesis. If the null hypothesis were true, and if we repeated our experiment over and over again, we would get outcomes that differed this much or more from the expected outcome, 5 times out of 100. So when we reject the null hypothesis on a given occasion, we risk labeling a true hypothesis as a false one.

Now it's time to talk about *Type* I and *Type* II errors. A Type I error is the rejection of the null hypothesis when, in fact, it is true. The probability of a Type I error is denoted by α. It would seem that in order to avoid Type I errors, we should set the rejection level as low as possible; for example, at 0.01 or 0.001. In the latter case, we would only risk a Type I error about 1 time in every 1,000. However, the lower we set the α level, the greater is our chance of committing a Type II error.

A Type II error occurs when we accept the null hypothesis when it is, in fact, false. Beta (β) is the probability of making a Type II error. Obviously, we lower the risk of a Type II error in exactly the opposite way that we'd use in trying to minimize a Type I error—that is, by raising the alpha level.

The fact that changing the size of the critical region has opposite effects on the likelihood of Type I and Type II errors does not mean that the probability of the two errors are exact mathematical complements of each other. In other words, if the critical region is 0.05, the probability of a Type I error is 0.05 when we reject an hypothesis, but is greater than 5 percent for a Type II error when we fail to reject an hypothesis. We can see the greater likelihood of making a Type II error if we assume a value for p of 0.5 and find it to be 0.4. Upon testing the null hypothesis we conclude that, at a certain alpha level, the hypothesis is reasonable. But perhaps we could have set the null hypothesis at 0.40 or 0.45, among other values; and if we had, perhaps we also would have concluded that the hypothesis was tenable. In short, many null hypotheses are acceptable for any given critical region. When we retain a null hypothesis, we're basically saying that it is one of a number of acceptable hypotheses. For this reason, we should say nothing stronger than that we failed to reject the null hypothesis, and we should try not to speak in terms of accepting it.

At this point you may ask, "How is it possible to tell when we are making a Type I or Type II error?" Unfortunately, we cannot tell. Furthermore, there is no way of adjusting the critical region to reduce the probability of one type of error without increasing the probability of the other. The critical region set by most researchers depends on the relative costliness of the two types of errors in the study being done.

Consider the plight of public health authorities who are trying to decide whether or not to attempt mass innoculations of a new vaccine. Feeling reluctant to damage their credibility in the future, the authorities may not want to risk putting their stamp of approval on an ineffective serum. So they decide to risk erring in the direction of failure to claim a result rather than in the direction of claiming a result when it is wrong. The alpha level is set rather low, with the possible consequence that the results of the experimentation would lead public health authorities to refrain from promoting the vaccine. On the other hand, the failure to use an effective vaccine could cause considerable damage to the impact of public health policies, not to mention the social cost of the thousands of premature deaths that might have been prevented.

This illustration should make it clear that there is nothing sacred or absolute about the selection of one alternative over the other. In the final analysis, the choice must be based upon various subjective factors that have to do with how the risks of incorrect decisions should be balanced. Only the researcher can judge whether saving the life of one individual is more or less serious than approving an ineffectual vaccine and running the risk of damaging the credibility of future claims. The researcher is responsible for establishing the degree to which one error is considered worse than the other.

As a rule, however, scientists prefer to be conservative about rejecting null hypotheses, believing that any decision that would allow a Type I error to occur more than 5 percent of the time is unacceptable. But setting the alpha level lower than 1 percent will often result in a Type II error. This is why, in most scientific research, the alpha ranges between a low of 0.01 and a high of 0.05.

9.7 An Application: The Sign Test

This chapter's main goal has been to help you understand the process of inference. We discussed the binomial in terms of tossing coins, primarily because it made it even easier to derive this sampling distribution. It also has social science applications, one of which is the *sign test,* used to study change of all kinds.

Suppose, for example, that we want to find out whether higher education makes people less prejudiced toward minorities. We test the subjects both before and after the experiment and are able to determine whether or not prejudice has been reduced. Let us indicate a success (symbolized as a plus sign $(+)$) as a person who has less prejudice after attending college, and a failure (symbolized as a minus sign $(-)$) as a person in whom prejudice is actually increased. Assuming that our measurement device is sensitive enough to pick up even the slightest changes and thereby to avoid cases that show no change whatsoever, we should be able to judge the impact of education on prejudice. In order to maintain independence, we should use random sampling of the population we wish to generalize about. This will help minimize mutual influence (which might effect their prejudice) among participants.'

Since we cannot establish directly whether or not the reduction of prejudice is a result of the college experience, we set up the null hypothesis that no systematic change has occurred through exposure to higher education. If this is true, then if the entire population from which the sample was drawn went through similar experiences, we would expect to find the same proportion of plus signs as minus signs. Using p as the proportion of successes, we have $H_0 : p = \frac{1}{2}$; and the problem reduces to a basic binomial situation. Now all we need to do is specify the direction of the research hypothesis, which is $H_1 : p >$

½, and establish a critical region, say 0.05. If the experiment shows that in seven cases prejudice has been reduced, and in three cases it has increased, we do not reject the hypothesis that the experiment has no effect. With a little imagination, this kind of test can be used in many situations in social science.

9.8 Caution: On Making Decisions Before Testing Statistical Hypotheses

All decisions concerning the formulation and rejection of statistical hypotheses should be made before the test. In other words, note all the outcomes that you expect to reject and all those you expect to accept. If you don't do this before the test, you can easily be tempted to retain an hypothesis by simply changing the rules as you go along.

Here's an analogy: You are going to use a coin to decide whether to go out or to stay home and study for tomorrow's statistics quiz. You decide, "Heads, I'll go; tails, I'll study." If the coin happens to come up heads, you breathe a sigh of relief and leave. If the coin comes up tails, you immediately decide on the best two out of three trials, and continue to toss the coin. Thus, you always have your way (unless, of course, you have been mistakenly tossing a two-tailed coin, or happen to always obtain tails—a most fortunate but unexpected event).

Where We Stand

Not all tests of significance are alike; most will involve distributions other than the binomial. However, several processes are involved in all hypothesis testing.

Step I: State the hypothesis to be tested (that is, the null hypothesis), together with the research hypothesis (or hypotheses) that state a specific value or range of values. That is, determine the alternative decision (or decisions) that can be made.

Step II: Specify the appropriate sampling distribution to be used. Remember that every sample statistic, whether it is a proportion, a mean, or whatever, has a sampling distribution. If you're testing a sample proportion as an estimate of a population parameter, use the binomial. If you're testing an hypothesis about a population using a sample mean, either use the normal or Student t distributions.

Step III: Select a critical region. This entails making two decisions: first, decide whether the critical region is to be in one or two tails of the distribution. As a rule of thumb, both tails are to be included in the critical region (two-tailed test) whenever the research hypothesis asserts that the population parameter is different from the one hypothesized. When the research hypothesis is directional and therefore asserts the direction of that difference, the tail in question is used to determine the critical region (one-tailed test). Second, decide how much of a Type I or Type II error you are willing to make. This decision is obviously going to be influenced by how important it is to reject the null hypothesis and be incorrect, as opposed to being more conservative by making the alpha level as low as possible. Remember, however, that in the final analysis *you* must decide which error is worse.

Step IV: Compute the actual test statistic (such as a proportion) and compare it with the cutoff value (such as the probability) corresponding to the alpha value. On the basis of this comparison, decide to either reject or not reject the null hypothesis. Generally speaking, if the computed test statistic equals or exceeds the tabled value associated with the selected critical region, reject the null hypothesis, even though you know full well that you might be making a Type I error with a probability equal to that of your selected significance level.

Terms and Symbols to Remember

Population
Sample
Sampling Distribution
Binomial Distribution
Binomial Expansion ($(p + q)^N$)
Coefficients of the Binomial
Null Hypothesis (H_0)
Research (or Alternative) Hypothesis (H_1)
Directional (or One-Tailed) Hypothesis
Nondirectional (or Two-Tailed) Hypothesis
Significance Level
Critical Region
Alpha (α) Level
One-Tailed Test
Two-Tailed Test
Type I Error (α Error)
Type II Error (β Error)
Sign Test

Exercises

1. Explain, in your own words, why it is necessary to sample.
2. Explain the importance of the concept of a sampling distribution.

3. Construct a binomial sampling distribution when $N = 6, p = q = \frac{1}{2}$, and test the following hypotheses at both the 0.01 and 0.05 levels of significance. (Assume that a sample produces five tails in six tosses.)
    (a) $H_0: p = q = \frac{1}{2}; H_1: p \neq q \neq \frac{1}{2}$
    (b) $H_0: p = \frac{1}{2}; H_1: p < \frac{1}{2}$
    (c) $H_0: q = \frac{1}{2}; H_1: q > \frac{1}{2}$
    (d) $H_0: q = \frac{1}{2}; H_1: q < \frac{1}{2}$
    (e) $H_0: p = \frac{1}{2}; H_1: p > \frac{1}{2}$

4. In a ten-item true or false test:
    (a) What is the probability that a student will obtain seven correct answers by chance alone?
    (b) What is the probability that a student will obtain either eight correct or incorrect answers by chance alone?
    (c) What is the probability that a student will get half the answers correct?
    (d) What is the probability that a student will get all ten answers correct by chance alone?
    (e) If four incorrect answers constitute a failing grade, what is the probability that a student will fail?

5. Construct a frequency polygon for the data displayed in Table 9.3. Note that it begins to approximate the shape of the normal curve (see footnote on p. 119).

6. Identify whether each of the following hypotheses is the null or research hypothesis.
    (a) Females are more in favor of gun control than males.
    (b) There is no difference between male and female attitudes toward capital punishment.
    (c) Blue-collar workers make less money than professionals.
    (d) Female college students are no more alienated than male college students.
    (e) American politicians care less about their constituents than their Canadian counterparts.
    (f) Americans spend the same amount of money in Italy as do tourists from other countries.
    (g) There is a difference between the median income of college professors and the median income of dentists.
    (h) Children of college graduates do better in college than children whose parents never attended college.

7. For Exercise 6, state whether the research hypothesis is one- or two-tailed.

8. For Exercise 6, state whether the null hypothesis is one- or two-tailed.

9. Match the null hypothesis with its alternative or research hypothesis.
   (a) The proportion of blue-collar workers in Canada is equal to 0.40.
   (b) The population mean in education is less than twelve years.
   (c) The population mean in age is equal to fifty-two years.
   (d) The population mean in education is equal to twelve years.
   (e) The proportion of blue-collar workers in Canada is not equal to 0.40.
   (f) The population mean in age is greater than fifty-two years.

10. Is the null hypothesis in a one-tailed test different from the null hypothesis in a two-tailed test? If so, give an example.

11. Is the research hypothesis in a one-tailed test different from the research hypothesis in a two-tailed test? If so, give an example.

12. Discuss the considerations involved in deciding on whether or not to use a one- or two-tailed test of significance.

13. Give an illustration in which a Type I error would be worse than a Type II error.

14. Give an illustration in which a Type II error would be worse than a Type I error.

15. A nuclear physicist is testing the following two hypotheses:
    (a) The lead walls surrounding a nuclear plant are safe.
    (b) The lead walls are too thin to be safe.
    With respect to the decision on each of these hypotheses, which error (Type I or Type II) would have the most dangerous implications?

16. Select any journal in your discipline that emphasizes statistical studies, and do the following:
    (a) Identify the null and research hypothesis (or hypotheses) used in three different articles.
    (b) Identify the sampling distribution(s) used.
    (c) Specify the level of significance used and indicate whether the statistical test used was one- or two-tailed.

17. Discuss the concept of randomness.

18. Discuss how you would go about drawing a random sample of:
    (a) Registered Democrats in Texas
    (b) The adult population of the United States
    (c) Students in a class
    (d) Factory workers in Detroit
    (e) Television viewers throughout a state

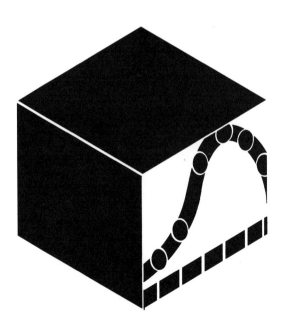

O world invisible, we view thee,
O world intangible, we touch thee,
O world unknowable, we know thee.
Francis Thompson

Statistical Inference: One Sample

10.1 Introduction: Why Another Sampling Distribution?

In the last chapter, we "flipped out"—that is, we tossed a hypothetical coin to illustrate the binomial. Thus, by simple enumeration and application of the multiplication rule, we could see exactly how probability theory is used to obtain the sampling distribution for a discrete, two-category variable. In this chapter, we'll take a look at tests of hypotheses about population means. This means that we have to introduce a sampling distribution that's appropriate to continuously distributed data. But we will not discuss in detail how probability theory is used to obtain such a sampling distribution. Mathematical derivations are available, of course, but most of them involve more probability and mathematics that you may know right now. In fact, from this point on, the mathematical considerations involved become so complex that, despite the desirability of knowing what lies behind every statement, you will have to take certain statements on faith alone. With this in mind, let's discuss the fascinating *central limit theorem*.

10.2 The Central Limit Theorem

If we sample sets of observations, and if we compute a mean for each set, we can describe the sampling distribution of these sample means by a normal distribution. The theorem that allows us to make that statement is called the:

> *Central Limit Theorem: If repeated random samples of a fixed size are drawn from any population having a mean of μ and a standard deviation of σ, then as the size of the sample increases, the sampling distribution of sample means approaches normality, with a mean of μ and a standard error of $\sigma/\sqrt{N}$ (abbreviated as $\sigma_{\bar{x}}$).*

This remarkable theorem is among the most important in all of inferential statistics. In brief, it says that no matter how unusual a distribution we start with, provided the sample is sufficiently large, we can use a sampling distribution that is approximately normal. Since the sampling distribution, and not the population, is used to test hypotheses, this means that whenever the sample is large we can completely relax the assumption about normality of the population and still use the known characteristics of the normal curve to test hypotheses.

But perhaps you need to be convinced that the central limit theorem makes sense empirically. The best way to accomplish this, and at the same time to persuade yourself that the *standard error of the mean* (the standard deviation of the sample means) is really $\sigma_{\bar{x}}$, is to draw random samples from a known population, compute the sample means, and compare the result obtained with $\sigma_{\bar{x}}$. Still, you might wonder: "Why should the sampling distribution become normal if the population isn't?" To answer this question, we are going to look at a population that is far from normal to see what happens when we begin to increase the sample size.*

Imagine that we are experienced crapshooters, hooked on the dubious delights of dancing dominoes (otherwise known as dice). Imagine further that we have a pair of unloaded dice for which the probabilities of getting each of the six sides are $\frac{1}{6}$. The probability distribution drawn in Figure 10.1 for one roll of a single die is then rectangular, since each outcome is equally likely. To be exact, of course, this distribution should be discrete and not continuous, as the diagram implies. At any rate, the distribution is in marked contrast to a normally distributed one, in which extreme values are less likely than those close to the mean.

Considering such a distribution as our population of all possible dice rolls, let's develop the sampling distribution of the means of samples for rolling two dice. This means that we are going to roll two dice, sum up the face values, and divide by 2 to get the sample mean. Being experts at craps, we know that these sums can range from 2 to 12. We also know that there are (6)(6) or thirty-six possible outcomes that will produce these eleven different sums. Thus the probability associated with a particular mean depends on the number of sums capable of producing that mean. Plotting these eleven sample means and their corresponding probabilities of occurrence gives us the triangular probability distribution of Figure 10.2.

---

*This example has been adapted from Blalock (1972: 181–184).

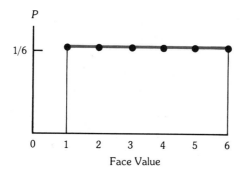

Figure 10.1.   Probability Distribution of the Face Values for a Perfect Die

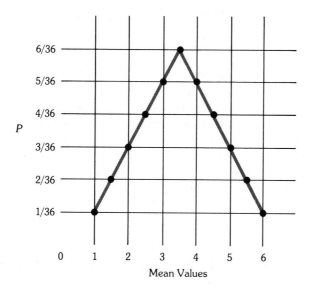

Figure 10.2.   Sampling Distribution of Means for Two Dice ($N = 2$)

What happens if we increase the sample size to three? This will naturally produce (6)(6)(6), or 216 possible outcomes that yield sixteen different means. When plotted, this probability distribution of sample means looks like the one in Figure 10.3. Now the distribution is beginning to approximate the normal curve, even though the sample size is still relatively small.

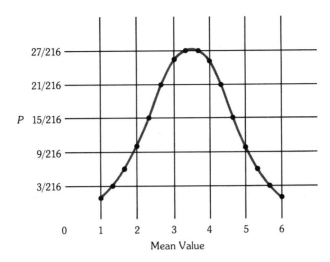

Figure 10.3. Sampling Distribution of Means for Three Dice ($N = 3$)

While this example does not constitute a formal proof of the central limit theorem, it does demonstrate how a sampling distribution of means rapidly approaches normality, even when the parent population is not distributed normally. It also demonstrates that the mean of the sampling distribution is equal to the population mean. Even with samples of only thirty cases, the shape of the sampling distribution is so close to normality that the error involved is trivial.

Now we are ready to demonstrate the use of the central limit theorem by putting it to work, in much the same way as we used the binomial distribution to test hypotheses.

10.3 Testing Statistical Hypotheses: The *z* Distribution

Suppose that you're planning to open a McDonald's restaurant. One of the first things you do is consult the latest census tract. Suppose further that you're planning to base your decision on age, because burgers, fries, and thick shakes appeal much more to the young. Checking the census you find that the mean age of the south area of town is thirty years, with a standard deviation of eight years. Your immediate reaction is optimistic. Now all you need to do, you think gleefully, is visit the area and pick the exact site. Once you're there, however, you are surprised by the extraordinarily large number of senior citizens. You wonder whether the age composition of the area has changed. Must you wait until the next census to find out? Of course not. What you should do is draw a random sample of people from the area and ask their ages. Assume, for illustrative purposes, that a sample of 100 residents produces a mean age of thirty-two years. Is this statistic indicative of a significant change since the last census? To find out, let's perform a single sample test of significance.

But first, ask yourself whether you feel as if you understand the four steps involved in performing all tests of significance. If you do, then go directly to the four steps common to all tests of significance that follow. If you feel the need to refresh your memory, go back and reread the last "Where We Stand" section, and proceed with the test stated in formal terms.

Hypotheses   The first step, as always, is to state the hypotheses as briefly as possible. In the present problem they are:

$$H_0: \mu = 30$$
$$H_1: \mu > 30$$

Sampling Distribution   The sampling distribution that we will use to solve this problem has already been derived for us. Since we know that the sampling distribution of sample means is normal or nearly so, we can utilize the Normal Table (Table B). From now on, in fact, sampling distributions will always appear in the form of tables in Appendix III. So remember that whenever we are using a table to conduct a statistical test of significance, we are using a sampling distribution as well.

Critical Region   Since a lot of money is needed to open a McDonald's, before you rejected the null hypothesis, most likely you'd want to be very sure of your decision. So let's set alpha at 0.01 and use a one-tailed test of significance, since the research hypothesis is directional. Referring to Column C (Table B), we find that the area beyond a z of 2.33 is approximately 0.01. This means that if the result of our test is equal to or greater than $z_{0.01}$ or 2.33, then we should reject the null hypothesis. Figure 10.4 depicts this critical region of rejection.

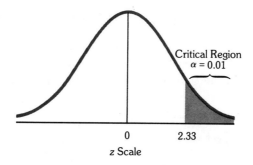

Figure 10.4.   Critical Region for Rejection of the Null Hypothesis when $\alpha = 0.01$, One-Tailed Test

Test and Decision   We know that any normally distributed variable can be transformed into the normally distributed z scale. We also know that we can establish probability values in terms of the relationships between z scores and areas under the normal curve. So if we assume, as we have, that the form of the distribution of sample means is normal, we can establish probability values in terms of the relationships between z scores and areas under the normal curve. But before we can obtain the z value and compare it with the z value associated with alpha, we must modify Formula (7.1) for use with the sampling distribution of sample means. Therefore, $\bar{X}$ replaces X, μ replaces $\bar{X}$, and $\sigma_{\bar{x}}$ (that is, $\sigma/\sqrt{N}$) replaces s in Formula (7.1) for z. Thus, we obtain the z *ratio:*

$$z = \frac{\bar{X} - \mu}{\sigma/\sqrt{N}} \qquad (10.1)$$

where:

$$\bar{X} = \text{the sample mean,}$$
$$\mu = \text{the population mean, and}$$
$$\sigma/\sqrt{N} = \text{the standard error.}$$

In our present problem, we obtain the following z value:

$$z = \frac{32 - 30}{8/\sqrt{100}} = 2.5$$

Since this obtained z score is greater than the z value associated with an alpha of 0.01 (that is, $2.5 > z_{0.01}$), we may reject the null hypothesis and must tentatively conclude that the mean age of the neighborhood is not now the same as it was in the past.*

10.4 Testing Statistical Hypotheses: Unknown Variance

In general, we are confronted with the problem of not knowing the population variance. But we could estimate this population parameter by replacing σ with $\hat{s}$† in the formula for the standard error, thus enabling us to continue with z as

---

*A positive z value merely indicates that the mean of the sample is greater than that of the population. A negative value, on the other hand, indicates the opposite.

†We will use a circumflex (i.e., $\hat{s}$) to indicate an unbiased estimate of a parameter.

our test statistic and with the normal curve as the model for our sampling distribution. In actual practice, however, this replacement only works well with large samples. With small samples, it turns out to consistently underestimate the population standard deviation; therefore, it constitutes *bias*. Bias in statistics means that there is a systematic error inherent in the sampling procedure, thus giving us a consistent distortion of the population parameter(s) we are trying to estimate. To offset this bias, we use the *Student's t distribution*.

10.4.1 The Student's *t* Distributions

Early in this century, William S. Gosset (1876–1937) made one of the first discoveries—and a rather startling one it was—concerning modern sampling theory. Gosset, an Oxford scholar in mathematics, was employed by the famous Guinness Brewery in Ireland to apply his knowledge to the quality-control aspect of production. His job was to take rather small samples from large brewery vats. After these samples had been analyzed, decisions were then made about the production of the entire vat. Gosset knew that, when estimating a population mean, using small samples in testing procedures led to considerable error. So drawing on his knowledge of mathematical statistics, he derived a *t ratio*, based on the *z* ratio. In it, he replaced σ with *s* and $\sqrt{N}$ with $\sqrt{N-1}$ in the formula for the standard error. It wasn't long before Gosset realized the ramifications of his discovery, but since he was an employee of a thriving enterprise he was forbidden to give away any trade secrets. Eventually, a compromise was reached: he was allowed to publish his discovery under the rather obvious pseudonym of "Student."

Some time later, Gosset developed a sampling distribution—rather, a family of distributions—that permit hypotheses to be tested with small samples drawn from normally distributed populations when σ is known. These distributions, referred to variously as the *t distributions* or *Student's t*, are based on the observation that as the size of a sample gets smaller, the effect of the unbiased estimate of the population standard deviation becomes more important. Thus, the larger the sample, the more the *t* distribution resembles the normal curve. With small samples, however, the *t* distribution provides a better estimate of the standard error of the mean. To help you see the contrast between a *t* distribution and the normal curve, we have reproduced the two curves, based on a sample of eight measurements, in Figure 10.5.

The expression under the radical sign is the denominator of Gosset's standard error, $N-1$, called the *degrees of freedom* (abbreviated *df*). Degrees of freedom is a rather subtle concept; it refers to information restrictions placed on the data by the population parameters. Some values will be constrained or not allowed to vary because of information that's already known. For example, if we know the mean of a particular data set and if, in the process of computing

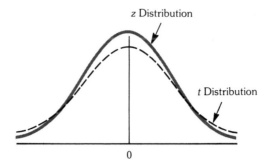

Figure 10.5.   Standard Normal Distribution and *t* Distribution ($N = 8$)

the standard deviation, we subtract it from all the individual scores, in effect we have constrained the value of the last individual score, since we know the mean and all the other scores. The last individual score must be a certain value—sometimes we say that it's *determined*. Similarly, if we have three numbers on which we place the restriction that the sum must equal 50, two of the numbers can take on any value, but the third is fixed. Thus, if two values are 10 and 15, respectively, the third has to be 25 to make the sum add up to 50. In general, the degrees of freedom is equal to the number of values N minus the number of parameters being tested.

Thus, the *t* ratio is defined in much the same way as the *z* ratio, except that the standard error has been estimated. Therefore:

$$t = \frac{\bar{X} - \mu}{s/\sqrt{N-1}} \tag{10.2}*$$

---

*The algebraic proof below demonstrates how we arrive at this estimate of $\sigma_{\bar{x}}$:

$$s^2 = \frac{\Sigma(X - \bar{X})^2}{N}$$

$$\left(\frac{N}{N-1}\right)\left(\frac{\Sigma(X - \bar{X})^2}{N}\right)$$

$$\hat{s}^2 = \frac{N\Sigma(X - \bar{X})^2}{N(N-1)}$$

$$= \sqrt{\frac{s^2}{N-1}}$$

$$\hat{s} = \frac{s}{\sqrt{N-1}} = s_{\bar{x}}$$

In following this proof, note that if we multiply the sample variance s^2 by $N/(N-1)$, the result is an unbiased estimate of σ^2 (that is, $\hat{s}^2$).

In contrast to our use of the normal curve, the tabled values of t (Table C) are critical values—that is, those values that bound the critical region corresponding to the various alpha levels. So when we use the table for the distribution of t, first we need to locate the appropriate number of degrees of freedom down the side of the table; then we need to find the chosen alpha level across the top. The value corresponding to the intersection of that row and column represents the t value required for significance. As with the z ratio, if our t value equals or exceeds the tabled value, we reject the null hypothesis.

10.4.2 The t Test: An Illustration

Let's take a familiar example: the miles per gallon you get from your gas guzzler. Suppose that you randomly select five people in your neighborhood who also own the same model, and that you compute the average miles per gallon to be ten miles, with a standard deviation of two miles. But suppose that the auto maker proudly declares that the car averages thirteen miles per gallon. Is it reasonable to assume that this sample is representative of a population in which $\mu = 13$?

Hypotheses   In the present problem, the two hypotheses read as follows:

$$H_0: \mu = 13$$
$$H_1: \mu \neq 13$$

Sampling Distribution   The sampling distribution is the Student's t (Table C), since we are going to assume that the population is distributed normally and that the population standard deviation is unknown.

Critical Region   If we were to set the alpha level at 0.05, the critical region for rejecting the null hypothesis would consist of all values of t greater than or equal to ($\geq$) 2.78 and $t \leq -2.78$. Is this an adequate critical region? Sure. Remember, the choice of an alpha level is mostly arbitrary, whereas the position of the critical region is not. Figure 10.6 shows the position of the critical region as dictated by our nondirectional (two-tailed) research hypothesis.

Test and Decision   The value of t corresponding to $\bar{X} = 13$ is:

$$t = \frac{\bar{X} - \mu}{s/\sqrt{N-1}} = \frac{10 - 13}{2/\sqrt{5-1}} = \frac{-3}{2/\sqrt{4}} = -3.00$$

Since the computed t falls within the critical region (that is, $-3.00 < t_{0.05}$), we reject the null hypothesis. We must tentatively conclude that the average number of miles per gallon for our sample of five cars is not representative of the population of cars. (Please pardon our pessimism but we are always wary of any claims made by auto makers, faith healers, and salesmen in general.)

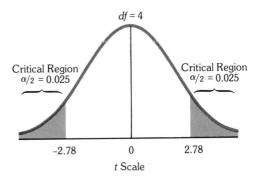

Figure 10.6. **Critical Region for Rejection of the Null Hypothesis when** $\alpha = 0.05$, **Two-Tailed Test**

10.5 Caution: To *t* or Not to *t*? That Is the Question

It sounds as if we use the *t* ratio whenever we don't know the population standard deviation. But this is not necessarily so. When *N* is larger than 30, there is virtually no difference between the *t* and *z* distributions. This is easily verified by comparing the *t* and *z* values needed for significance for samples larger than 30. Once more, it is also true that larger samples eliminate the difference between $\sqrt{N}$ and $\sqrt{N-1}$. Thus we can conclude that when *N* is larger than 30, we can use the *z* distribution with the ratio:

$$\frac{\bar{X} - \mu}{\hat{s}/\sqrt{N}} \qquad (10.3)$$

as our test statistic, with little resulting error. When the *N* is less than or equal to 30 and the population standard deviation is unknown, $\sqrt{N-1}$ corrects the bias that results from using the sample standard deviation as our parameter estimator. The *t* ratio thus is the preferred alternative, even though we must assume that the population is distributed normally. In Chapter 16, we will discuss some statistical tests of significance—alternatives to the *z* and *t* tests—that don't make the rather strict assumption of normality or require intervally scaled data. For some data sets, these alternative tests might be more appropriate.

10.6 Testing Statistical Hypotheses: Unknown Parameters

Perhaps you are about to ask, "Can we use a single sample to test an hypothesis about a population mean if the value of the population mean is unknown?" Unfortunately, we cannot, since we cannot specify any value for the population mean being tested. But we can specify a range of values within which we believe the population mean lies. So instead of making a *point estimate* (which is based on a single sample value), we will make an *interval estimate* (which is based on a range of values). The interval within which we consider an hypothesis defensible is referred to as a *confidence interval,* and the limits defining the area are referred to as *confidence limits.* We use the term "confidence" because probabilities tell us how much confidence we should place in our estimate. So even though point estimates are easier to understand, interval estimates are more desirable: they indicate the accuracy of our estimate in terms of the alpha level we choose to set.

A confidence interval can be derived algebraically from either the z or t ratio for the population mean. Which formula we should use again depends on the size of our sample. Thus the upper and lower limits for a confidence interval based on samples larger than 30 are:

$$\mu = \bar{X} \pm z(\hat{s}/\sqrt{N}) \qquad \qquad \textbf{(10.4)}$$

Similarly, the limits surrounding the confidence interval for samples less than or equal to 30 are:

$$\mu = \bar{X} \pm t(s/\sqrt{N - 1}) \qquad \qquad \textbf{(10.5)}$$

As an illustration, consider the hourly wages for factory workers. Suppose that a random sample of 100 employee records is examined, and that the sample mean and standard deviation are \$10.00 and \$1.50, respectively. What are the limits of the interval in which we can be 95 percent sure that the population mean is contained? Using Formula (10.4), we find that the upper 95 percent confidence limit is:

$$\bar{X} + z(\hat{s}/\sqrt{N})$$
$$10 + 1.96(1.5/\sqrt{100})$$
$$10.29$$

Similarly, we find that the lower confidence limit is:

$$\bar{X} - z(\hat{s}/\sqrt{N})$$
$$10 - 1.96(1.5/\sqrt{100})$$
$$9.71$$

Now that we have established the 95 percent confidence limits as \$9.71 and \$10.29, we can conclude that the interval defined by these limits encompasses the population mean 95 percent of the time. (See Figure 10.7 for a schematic representation for the confidence interval of this example. Stating

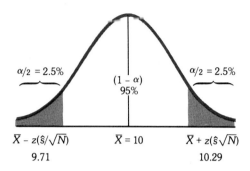

Figure 10.7. The 95 Percent Confidence Interval When $\bar{X} = 10$ and $s = 1.5$, Two-Tailed Test

that the confidence interval includes the population mean can refer to a number of different confidence intervals, as depicted in Figure 10.8.)

However, just because we have established the interval within which the population mean falls does not mean we should conclude that the sample mean represents the true population mean—in other words, that our sample mean is accurate. In still other words, we cannot claim that the chances are 95 in 100 that the population is $10.00. Our statements are valid only with respect to the interval, not with respect to any particular value of the sample mean. So when we state that the 95 percent confidence interval for the mean hourly range of the population of factory workers is $9.71 to $10.29, what we're really saying is that if we took all possible random samples of a fixed size from the population, 95 percent of the intervals constructed from these samples would cover the true value of the population mean. Figure 10.8 illustrates this; most of the intervals cover the vertical line, and hence include the population mean μ.

We construct the confidence interval for the population mean for samples less than or equal to 30 in the same way, except that we must: (1) use Formula (10.5), and (2) look up the t value (the critical value) in accordance with the degrees of freedom. Confidence intervals of 90 percent and 99 percent are used, of course, but not as widely as the 95 percent confidence interval. This is because most researchers in the social sciences want to avoid criteria that are either too lenient or too strict.

Where We Stand

We have learned how to use the z ratio and the standard normal curve to test hypotheses about population parameters. We have also learned how to use the t ratio and the t distributions to test hypotheses about population parameters when the variances are unknown and our sample is relatively small. Since the population parameters are almost never available, we must use samples to

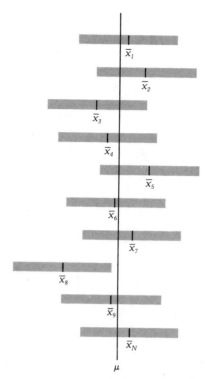

Figure 10.8. Distribution of Different 95 Percent Confidence Intervals for μ

estimate these parameters. Within particular confidence levels, interval estimation allows us to specify a range of values within which the true parameters can be expected to fall.

Often we want to compare two groups to see whether or not they could have come from the same population. This means that we have to acquire some new tools to explain the differences or similarities that we have observed between two samples. The next chapter will deal with this matter.

Terms and Symbols to Remember

Central Limit Theorem
Standard Error of the Mean ($\sigma_{\bar{x}}$)
z Ratio
t Distributions (or Student's t)
Estimated Standard Error of the Mean ($s_{\bar{x}}$)
Degrees of Freedom (df)
t Ratio

Point Estimate
Interval Estimate
Confidence Interval
Confidence Limits

Exercises

1. Explain, in your own words, why the sampling distribution of the means is the equivalent of the distribution of sampling errors.

2. Explain, in your own words, what happens to the distribution of sample means when you increase the number of samples.

3. On the Stanford-Binet intelligence test, the mean is 100, with a standard deviation of 15. Twenty school children are tested, with a resulting mean of 107 and a standard deviation of 14. Does this group have a mean IQ significantly higher than the norm group? Use the 0.05 significance level.

4. A random sample of fifty voters in a western state produces a mean family income of $14,000, with a standard deviation of $2,000. If $\mu = \$13,800$ and $\sigma = \$2,000$, can we assume that the sample mean is different from the population mean? Use $\alpha = 0.05$. Can we assume that the sample mean exceeds the population mean? Use $\alpha = 0.01$.

5. An institute devoted to the study of extrasensory perception (ESP) gives a group of volunteers a test that, as they know from thousands of similar tests, yields a $\mu = 30$. The present group of fifteen obtains $\bar{X} = 33$ and $s = 3.5$. Should the institute assume that the group has superior ESP powers? Use the 0.05 significance level and the 0.01 significance level. Are the results the same?

6. A large state university claims that its faculty is rather young, and bases this assertion on the fact that the national mean age for a university or college faculty is forty-eight, whereas the mean age of their 800-member faculty is forty-six, with $s = 5.00$. Test the validity of their assumption at the 0.01 level of significance.

7. Suppose that we wish to find out whether the members of the arts division of the state university mentioned in Exercise 6 are younger than the members of the other divisions. Assume that a sample of nineteen produces a mean age of forty-five, with a standard deviation of 5.00. Is the result indicative of a younger arts faculty? Use $\alpha = 0.05$ to test the null hypothesis.

8. How do the t distributions differ from the z distribution? Are they ever the same? Explain.

9. Discuss the difference between point and interval estimations.

10. It is true that "the wider the confidence interval, the greater the degree of reliability." Explain.

11. It is true that "the greater the degree of precision, the less reliable the confidence interval." Explain.

12. Given that $\bar{X} = 50$ and $s = 5$ for $N = 29$, use the t distribution to find:
    (a) The 95 percent confidence limits for μ.
    (b) The 99 percent confidence limits for μ.

13. Given the $\bar{X} = 50$ and $s = 5$ for $N = 250$, use the z distribution to find:
    (a) The 95 percent confidence limits for μ.
    (b) The 99 percent confidence limits for μ.
    (c) Compare these results with Exercise 12, and explain why they are dissimilar.

14. Using the data in Exercise 4, find the interval within which we can be 95 percent confident that the true population mean probably falls.

15. Using the data in Exercise 5, find the interval within which we can be 99 percent confident that the true population falls.

16. A simple random sample of 1,600 households shows the average rental to be $110, with a standard deviation of $30. If we want to estimate the average rental for the population with 95 percent certainty, what estimate would we use?

17. Another survey of a city of 100,000 households is conducted. Suppose that we estimate the standard deviation to be $30. Suppose, further, that we want to estimate the average income for the population with 95 percent certainty, and that we want the confidence interval to be ± $1.50. What size sample should we use? (Hint: Assume that the sample will be over 30.)

18. Suppose that still another survey is conducted in a city of 100,000 households. Suppose that the average rental per month for the 1,600 residents sampled is $110, that the standard deviation is $30, and that the average income for the population is given as $\bar{X} = 110 \pm 1.50$. What level of significance can be associated with the confidence interval?

Like, but oh how different!
William Wordsworth

Statistical Inference: Two Samples

11.1 Introduction: Why Test the Difference Between Means?

Does the recidivism rate of juvenile offenders who are provided with job skills differ from those who are not? Are the IQ scores of black children lower or higher than those of white children? What is the effect of one teaching method over another?

All these questions have at least one thing in common: they deal with differences in some trait or characteristic between two samples of elements. Thus far we have confined our discussion of hypothesis testing to the one sample case. But social scientists are interested in testing to see whether two independent samples (typically, nonquantitative categories such as black versus white, male versus female, or Democrat versus Republican), on which some interval level trait (such as income) has been measured, could have been drawn from the same population. If the data indicate otherwise, these researchers conclude that the two groups differ with respect to the trait in question. Testing the difference between two sample means is a popular way to assess population similarity or dissimilarity.

In this chapter, we add the difference of means test to our growing pool of inferential skills. From our discussion in Section 9.2, you may have guessed that there is a sampling distribution for testing the difference between two population means. And this probability distribution is presented in the following section.

11.2 The Central Limit Theorem, Revisited

In Section 10.2 we said that from the basic result of the central limit theorem, we learn that any variable that's sampled repeatedly and randomly tends to be distributed normally, the larger the size of the sample is. Now we are going to extend the theorem to get the sampling distribution for the difference between means. We do this by drawing pairs of samples, finding the differences between the means of each pair, and obtaining a distribution of these differences. This new distribution (referred to as the *sampling distribution of the difference of means*) is also distributed normally, but with a mean ($\mu_1 - \mu_2$) equal to zero and a standard deviation ($\sigma_{\bar{x}_1 - \bar{x}_2}$, referred to as the *standard error of the difference of means*) equal to:

$$\sqrt{\frac{\sigma_1^2}{N_1} + \frac{\sigma_2^2}{N_2}}$$

You should have no trouble figuring out why we now have a sampling distribution of the difference of means rather than a sampling distribution of means. Nor should you have any trouble explaining why $\mu_1 - \mu_2 = 0$. But our earlier discussion may not have made it clear why we add two standard errors of the mean to get the standard error of the difference of means. However, if we remember that the standard error ($\sigma_{\bar{x}}$) is just the square root of the variance of that error ($\sigma_{\bar{x}}^2$), we realize that whenever we combine squared terms we also must weigh them by the square of their coefficients. Why? Because we are taking one unit of variance from the first population and subtracting from it one unit of variance from the second population. Thus:

$$(1)^2 \frac{\sigma_1^2}{N_1} + (-1)^2 \frac{\sigma_2^2}{N_2} = \frac{\sigma_1^2}{N_1} + \frac{\sigma_2^2}{N_2}$$

The standard error of the difference of means is therefore:

$$\sqrt{\frac{\sigma_1^2}{N_1} + \frac{\sigma_2^2}{N_2}}$$

Now we are ready to construct a test for the difference of population means using *z* scores. You'll understand the transition from the test for one sample case using *z* scores to the test for a difference between means better if you keep in mind the relationship of μ and $\mu_1 - \mu_2$ and $\sigma_{\bar{x}}$ and $\sigma_{\bar{x}_1 - \bar{x}_2}$ to the central limit theorem.

11.3 Testing the Difference Between Means: The z Test

Remember that *z* is computed by taking the difference between the obtained sample value and the mean of the population, and then dividing by the standard error. Now, however, we are concerned with the difference between sample means $\bar{X}_1 - \bar{X}_2$. Since the mean of the population is now $\mu_1 - \mu_2$, we obtain the following expression for:

$$z = \frac{(\bar{X}_1 - \bar{X}_2) - (\mu_1 - \mu_2)}{\sigma_{\bar{x}_1 - \bar{x}_2}}$$

Since we want to see whether or not there is a difference between two types of populations, our null hypothesis will be that there is *no* difference; or, in symbolic form, $\mu_1 = \mu_2 = 0$. Thus $\mu_1 - \mu_2$ in the above expression drops out, and our formula becomes:

$$z = \frac{\bar{X}_1 - \bar{X}_2}{\sigma_{\bar{x}_1 - \bar{x}_2}} \qquad \textbf{(11.1)}$$

where:

$$\sigma_{\bar{x}_1 - \bar{x}_2} = \sqrt{\frac{\sigma_1^2}{N_1} + \frac{\sigma_2^2}{N_2}}$$

The resemblance between the numerator in Formula (11.1) and the one used in Formula (10.1) is purely coincidental; it occurs because the difference between population means dropped out under the null hypothesis. Therefore it would be incorrect to conclude that the μ in Formula (10.1) has simply been replaced by the sample mean of the second population. Actually, the expression $(\bar{X}_1 - \bar{X}_2)$ has replaced $\bar{X}$; $(\mu_1 - \mu_2)$ has replaced μ (but dropped out, assuming that the null hypothesis $\mu_1 = \mu_2 = 0$ is correct); and $\sigma_{\bar{x}_1 - \bar{x}_2}$ has replaced $\sigma_{\bar{x}}$.

When we know the population standard deviation or when our sample is large enough (that is, greater than 30), we can use z scores in exactly the same way we did for the one sample case. Specifically, we would compare our computed value to the tabled z value and decide to reject the null hypothesis if our z equals or exceeds that associated with the chosen critical region.

Unfortunately, neither of these requirements is met very often. For one thing, if we knew the standard deviations of the two populations, we'd probably be in a position to know the mean, as well. In that case, we wouldn't need to hypothesize about their differences; to obtain the exact difference, we'd need only to subtract one from the other.

Now let's assume that the z test still applies because our sample size is sufficiently large. For social science research, however, it's unwise to make the latter assumption; the bulk of the research done in many fields (psychology and sociology, for example) uses very small samples on which to make decisions about differences.

We will use the t distribution to test a difference between means. But this should not trouble us, because the t distribution provides a better test when the sample is small. Besides, we can always revert back to z scores in later discussions, if necessary.

11.4 Testing the Difference Between Means: The t Test

We use the z ratio when we know the population standard deviations or when the samples are at least equal to or greater than 30. Since this seldom occurs, we must estimate the standard error of interest; that is, $\sigma_{\bar{x}_1 - \bar{x}_2}$. Thus, if we draw at random one sample from one population and a second sample from another population, the standard error of the difference of means may be estimated by:

$$s_{\bar{x}_1 - \bar{x}_2} = \sqrt{\frac{s_1^2}{N_1} + \frac{s_2^2}{N_2}}$$

In its present state, this formula represents a biased estimate of the standard error of the difference of means. However, when both samples are equal to or greater than 30, the bias is eliminated. Thus, with large samples, the above formula provides an estimate for the standard error in Formula (11.1). To obtain an unbiased estimate for smaller samples (30 or less), we divide both variances by $N - 1$. Thus:

$$s_{\bar{x}_1 - \bar{x}_2} = \sqrt{\frac{s_1^2}{N_1 - 1} + \frac{s_2^2}{N_2 - 1}}$$

To use the t distribution, we simply exchange this unbiased standard error for the standard error in Formula (11.1). Thus we get the t ratio:

$$t = \frac{\bar{X}_1 - \bar{X}_2}{\sqrt{\dfrac{s_1^2}{N_1 - 1} + \dfrac{s_2^2}{N_2 - 1}}} \qquad (11.2)$$

We use this formula when we assume the inequality of variances (*heteroscedasticity*) among samples. However, if we assume that the null hypothesis $\mu_1 = \mu_2 = 0$ is valid and that the variances of the two populations are $\sigma_1^2 = \sigma_2^2$, we can pool the two sample variances (that is, take a weighted average of the variances) to estimate the standard error:

$$s_{\bar{x}_1 - \bar{x}_2} = \sqrt{\left(\frac{N_1 s_1^2 + N_2 s_2^2}{N_1 + N_2 - 2}\right)\left(\frac{1}{N_1} + \frac{1}{N_2}\right)}$$

Thus:

$$t = \frac{\bar{X}_1 - \bar{X}_2}{\sqrt{\left(\dfrac{N_1 s_1^2 + N_2 s_2^2}{N_1 + N_2 - 2}\right)\left(\dfrac{1}{N_1} + \dfrac{1}{N_2}\right)}} \qquad (11.3)^*$$

11.4.1 The Degrees of Freedom for *t*

The *t* distribution is a family of distributions. Therefore, we must find in Table C the value based on the number of degrees of freedom. In contrast to the one sample case, here we have $N - 1$ values that can vary in each sample. Therefore, the degrees of freedom for implementation with Formula (11.3) is determined as follows:

$$df = (N_1 - 1) + (N_2 - 1)$$
$$= N_1 + N_2 - 2$$

When using Formula (11.2), we encounter a difficulty in selecting degrees of freedom. Suppose that both samples differ in their variances and that the first sample is unusually small. Then it would be rather misleading to use $N_1 + N_2 - 2$ as the degrees of freedom, since our sample statistic is a poor estimate of the corresponding population parameter and since the value of $s_1^2/(N_1 - 1)$ would most likely be larger than that of $s_2^2/(N_2 - 1)$. This situation can easily occur if the values of both sample standard deviations do not differ greatly. In these situations, the relative sizes of the two fractions will be determined primarily by their denominators. Therefore, we recommend that you use the following formula to obtain an approximation of the correct degrees of freedom:

$$df = \frac{\left(\dfrac{s_1^2}{N_1 - 1} + \dfrac{s_2^2}{N_2 - 1}\right)^2}{\left(\dfrac{s_1^2}{N_1 - 1}\right)^2\left(\dfrac{1}{N_1 + 1}\right) + \left(\dfrac{s_2^2}{N_2 - 1}\right)^2\left(\dfrac{1}{N_2 + 1}\right)} - 2$$

---

*Exercise 9(c) at the end of this chapter asks for the derivation of formulas for interval estimates of the difference between two means for small and large samples. Also see the "Answers to Selected Exercises" section under Chapter 11, Answer 9(c).

11.4.2 The *t* Test: An Illustration

To illustrate the *t* test, let's examine whether or not ascorbic acid has any effect on reducing the incidence of common colds. Twenty subjects (the *experimental group*) are selected randomly and given a daily dosage of vitamin C. Twenty more subjects (the *control group*) are also selected at random and given a placebo. (The experimental group always receives a treatment, whereas the control group does not.) After a specified period of time has passed, we find that the mean number of colds for the experimental group is 6.4, with a standard deviation of 1.7. For the control group, the mean is 7.2, with a standard deviation of 2.0. Is there a significant difference between the means? To find out, once again we set up the problem at hand, using the four steps common to all tests of significance.

Hypotheses   To test the difference between means, we set up our hypotheses as follows:

$$H_0: \mu_1 = \mu_2$$
$$H_1: \mu_1 < \mu_2$$

Sampling Distribution   The sampling distribution is the Student *t* distribution (Table C) with $df = N_1 + N_2 - 2$, or $20 + 20 - 2 = 38$

Critical Region   Let's set the alpha level at 0.01. Since our research hypothesis is directional, we'll use a one-tailed test of the null hypothesis. Therefore, the critical region for rejection, using an alpha level of 0.01 and 38 *df*, is $t \leqslant -2.423$. Since the jump in degrees of freedom down the side of Table C is from 30 to 40, we use the degrees of freedom appropriate for 40. (You can use interpolation if you want a somewhat more accurate method.)

Test and Decision   Since the population variances are assumed to be equal, we use Formula (11.3) to obtain the *t* value:

$$t = \frac{\bar{X}_1 - \bar{X}_2}{\sqrt{\left(\dfrac{N_1 s_1^2 + N_2 s_2^2}{N_1 + N_2 - 2}\right)\left(\dfrac{1}{N_1} + \dfrac{1}{N_2}\right)}}$$

$$= \frac{6.4 - 7.2}{\sqrt{\left(\dfrac{20(1.7)^2 + 20(2)^2}{20 + 20 - 2}\right)\left(\dfrac{1}{20} + \dfrac{1}{20}\right)}}$$

$$= \frac{-0.8}{\sqrt{.363}}$$

$$= \frac{-0.8}{.602}$$

$$= -1.33$$

Since the computed t value falls outside the critical region (that is, $-1.33 > -2.423$), we do not reject the null hypothesis. Instead, we conclude that there is insufficient evidence to show that vitamin C reduces the occurrence of common colds. (Sorry, Dr. Pauling.)

11.4.3 To Pool or Not to Pool? That Is the Question

As you've guessed by now, we tested the difference between means under the assumption of the equality of variances *(homoscedasticity)* between samples. But you may think that the formula to use in this instance is (11.2). We agree with you; but if we get samples of equal size from each population, we avoid the problem of assuming homoscedasticity. Why? Because the unpooled standard error simplifies to:

$$\sqrt{\frac{s_1^2}{N_1 - 1} + \frac{s_2^2}{N_2 - 1}} = \sqrt{\frac{s_1^2 + s_2^2}{N - 1}} \quad \text{if } N_1 = N_2$$

Similarly, the pooled estimate of the two variances also simplifies to:

$$\sqrt{\left(\frac{N_1 s_1^2 + N_2 s_2^2}{N_1 + N_2 - 2}\right) \left(\frac{1}{N_1} + \frac{1}{N_2}\right)} = \sqrt{\frac{s_1^2 + s_2^2}{N - 1}} \quad \text{if } N_1 = N_2$$

Thus, Formulas (11.2) and (11.3) are algebraically identical, except with respect to the estimate of the degrees of freedom. Therefore, we recommend Formula (11.3).

11.5 Testing the Equality of Variances: The F Test

Often, we may discover that data from one group is distributed more widely than from another group. In such instances, we need to consider the possibility that we are sampling two different distributions. Are these distributions different in their means or variances?

To determine whether two variances differ significantly from one another, we consult the F *distribution*. The "F" refers to its creator, the late Sir Ronald A. Fisher (1890–1962), one of the most prolific writers in the field of statistics. (In fact, it was Fisher who applied the t ratio to test the difference of means, something that Gosset did not do.) The F *ratio* tests the equality of variances in a test of difference between means. In order to obtain this ratio, we compute the ratio of the two sample estimates of the population variances (always using the larger one as the numerator to assure a value greater than one). Formally:

$$F = \frac{\hat{s}^2}{\hat{s}^2} \qquad\qquad (11.4)$$

To facilitate the computation of Formula (11.4), let's rewrite it to include the degrees of freedom and magnitudes appropriate for each of the two sample estimates of the population variances. Thus:

$$F = \frac{\hat{s}^2/(N-1)}{\hat{s}^2/(N-1)} \quad \begin{array}{l} \textit{larger variance} \\ \textit{smaller variance} \end{array}$$

After computing the F ratio, we consult the F table (see Table D). Since both variances in the F ratio estimate a value, we have two different degrees of freedom, corresponding to the two sample variances. Looking at Table D, we note probabilities for the 0.05 and 0.01 levels. We list only these levels. Why? Because it would take an entire book to give all the possible alpha levels that could be used with t or z. At any rate, we find the critical F *value* by locating the degrees of freedom of the smaller variance down the side of the table, and the degrees of freedom of the larger variance across the top. The value corresponding to the intersection of that row and column represents the F value required for rejecting the null hypothesis of no difference in variances.

In the preceding example, the variance for our experimental group is $(1.7)^2/19$ or 0.152, whereas the variance for our control group is $(2.0)^2/19$ or 0.211. Our computed F ratio is therefore:

$$F = \frac{0.211}{0.152} = 1.39$$

Referring to Table D, under 19 and 19 degrees of freedom,* we find that an F ratio of approximately 3.00 or larger is required for significance at the 0.01 level. Since $1.39 < 3.00$, we conclude that it is reasonable to assume that both samples were drawn from a population with the same variances.

---

*Since the jump in degrees of freedom along the top of Table D is from 16 to 20, we use the degrees of freedom appropriate for 20. As we said before, it is more accurate to interpolate.

What if we find a significant difference in variances? Would we be more inclined to reject the null hypothesis of no difference between means? Probably not. If anything, the probability of rejection would be lowered, and so, in turn, would the probability of committing a Type I error. Now do you see why we avoided Formula (11.2) and the hard work of preparing the approximation to the correct degrees of freedom? Why, then, should we concern ourselves with the possible difference among the variances of two samples?

Sometimes, for reasons that may not be very clear, the variance of the experimental group is significantly larger than that of the control group, thus indicating that there are more extreme scores at both ends of the experimental group's frequency distribution. In these instances, analysis of variances can unearth what is commonly referred to as the "dual effect" of the experimental conditions. We may find this kind of effect, for example, when we're testing the effects of the outflow of adrenalin during physical performance. Surely, we've all heard of the frantic mother who lifted a Volkswagen to free her trapped child. Well, the increasing outflow of adrenalin may make some individuals likely to do better than expected; but it may make others "freak out." So the astute researcher will want to study the variance within the experimental group further, whenever there is suspicion of a dual effect.

11.6 Testing the Difference Between Means: Matched Samples

So far we have considered only the comparison of samples when each sample is drawn randomly and independently from the population. But rather than testing the difference between means for two independent samples, suppose that we test one group on two occasions—in other words, suppose that we perform a *before and after experimental design*. (This kind of test is like checking to see whether a group of inmates has the same opinion about the correctional facility after spending a year participating in a work-release program.) Now, suppose that we want to determine the effect of some drug on learning ability. We select individuals to participate in both the experimental and the control groups that are matched on some variable known to be associated with our dependent variable (learning ability). Now we have what is commonly known as a *matched experimental design*. Experiments of this nature assume that all other factors are equal, or do not affect the variable under investigation.

Both designs statistically help to remove the effects of individual differences (variance) among subjects and thus improve our ability to estimate the effect of the experimental variable on subjects participating in the experiment. Realistically, the before and after design is more helpful. In the first place, the actual mechanics of getting subjects to participate are usually simpler because all we need is a single random sample. More importantly, however, there is no problem of obtaining comparable groups. The subjects serving in the different conditions are ideally matched; they are, after all, the same people. We are most likely to be our own best match. Both designs produce matched samples and are most welcome when the study at hand lends itself to such matched sample analysis.

To test the difference between means for matched samples, we compute a test statistic based on t. The value is the ratio of the mean difference $(\bar{D})$ to the standard error of the mean difference $(s_{\bar{D}})$. In symbols:

$$t = \frac{\bar{D}}{\sqrt{\dfrac{s_D^2}{N-1}}} \qquad (11.5)^*$$

where:

$\bar{D}$ = the mean difference across pairs having matched scores ($\Sigma D/N$ where D equals the difference between a pair of scores), and

s_D^2 = the variance of the differences ($\Sigma D^2/N - \bar{D}^2$).

The degrees of freedom for this variation of the t test is $(N-1)$, where N is the number of pairs having matched scores.

Consider this example as a test of the difference between means for matched samples. A well-known pharmaceutical house decides to test the effects of a new drug on reduction of cigarette smoking. Ten subjects are selected at random, and before the drug is administered each subject records the number of cigarettes consumed over the last twenty-four hours. One week after the specified drug dosage has been administered, each subject again records the number of cigarettes consumed for the same time period (see Table 11.1). Is the drug effective in reducing cigarette consumption? To find out, we set up the problem in formal statistical terms.

---

*Actually, the formula for t with matched samples is:

$$t = \frac{\bar{X}_1 - \bar{X}_2}{\sqrt{s_1^2 + s_2^2 - 2rs_1s_2}} \qquad (11.6)$$

This formula requires a knowledge of the *Pearson product-moment correlation coefficient* (r). But since we won't be covering this concept until Chapter 13, we use the *direct-difference method* (Formula (11.5)) to compute t for matched (or correlated) samples, instead. After familiarizing yourself with the concept correlation, you may wish to use the above formula to see whether it gives the same value as that generated by Formula (11.5). (It does, of course.)

Table 11.1.   An Illustration of the Before and After Experimental Design

Subject	(Before Drug) X_1	(After Drug) X_2	Difference $(X_2 - X_1)$ D	(Difference)² $(X_2 - X_1)^2$ D^2
1	15	13	-2	4
2	22	15	-7	49
3	35	30	-5	25
4	9	10	1	1
5	16	15	-1	1
6	42	37	-5	25
7	21	25	4	16
8	32	28	-4	16
9	42	39	-3	9
10	11	10	-1	1
			$\Sigma D = -23$	$\Sigma D^2 = 147$

Hypotheses   To test the difference of means, we set up our hypotheses as follows:

$$H_0: \mu_1 = \mu_2$$
$$H_1: \mu_1 > \mu_2$$

Sampling Distribution   The sampling distribution is the Student's t distributions (Table C) with $df = N - 1$, or $10 - 1 = 9$.

Critical Region   Let's set the alpha region at 0.01. Since our research hypothesis is directional, we'll use a one-tailed test of the null hypothesis. Therefore, the critical region for rejection, using an alpha level of 0.01 and 9 df, is $t \leq -2.821$.

Test and Decision   Utilizing Table 11.1, we have $\Sigma D = -23$ and $\Sigma D^2 = 147$; thus:

$$\bar{D} = \Sigma D/N = -23/10 = -2.3$$

and:

$$s_{\bar{D}} = \sqrt{\frac{s_D^2}{N-1}} = \sqrt{\frac{\Sigma D^2/N - \bar{D}^2}{N-1}} = \sqrt{\frac{147/10 - (-2.30)^2}{10-1}}$$
$$= \sqrt{1.05} = 1.02$$

Plugging the above values into Formula (11.5), we obtain a t value:

$$t = \frac{\bar{D}}{\sqrt{\dfrac{s_D^2}{N-1}}}$$

$$= \frac{-2.30}{1.02}$$

$$= -2.25$$

Since the t value falls outside the critical region (that is, $-2.25 > -2.821$), we do not reject the null hypothesis. We conclude that the drug does not help in reducing cigarette consumption. (Had we established a significance level of 0.05, we would have decided to reject the null hypothesis. Here we see the importance of specifying alpha in advance of the test of significance.)

11.7 Testing the Difference Between Means: Sandler's A Test

Thanks to Sandler (1955), we now have an extremely simple procedure for finding probability values in all situations for which the t ratio for matched samples is appropriate. The Sandler A ratio is derived from t, thereby making the two tests comparable. To compute A, we merely divide the sum of the square of the differences by the square of the sum of the differences. In symbols:

$$A = \frac{\Sigma D^2}{(\Sigma D)^2} \tag{11.7}$$

Let's illustrate the computation of A from our previous example. Remember that $\Sigma D^2 = 147$ and $\Sigma D = -23$. Thus:

$$A = \frac{147}{(-23)^2}$$

$$= 0.278$$

Referring to the table of critical A values (Table E) under 9 degrees of freedom, we find that an $A \leq 0.213$ is required for significance at the 0.01 level. Since our computed value of A is greater than the critical value, we cannot reject the null hypothesis. For one thing, this is the same conclusion we reached when using the t ratio; for another, Sandler's A test requires much less computation and therefore is most preferred whenever matched samples are used.

Where We Stand

Most social science research focuses upon establishing relationships between variables. This contrasts with the point and interval estimates in which finding the parameters of a single variable is the primary concern. Whenever we make comparisons between samples, we have the most simple kind of problem that relates two variables together. Up until this chapter, we were primarily concerned with *univariate analysis* (that is, with one variable at a time). In this chapter, we explored how a dichotomous variable can be related to another variable. Now we can compare a dichotomous variable like sex, for example, to a host of interesting independent variables. Similarly, we can make comparisons between an experimental and a control group into which some variable has been introduced. To extend our analysis beyond these relationships, we must broaden our knowledge base to allow for the testing of the differences among two or more samples. Appropriately enough, the simultaneous treatment of several sample means, called a *one-way analysis of variance,* is our next topic.

Terms and Symbols to Remember

Central Limit Theorem
Sampling Distribution of the Difference of Means
Standard Error of the Difference of Means ($\sigma_{\bar{x}_1 - \bar{x}_2}$)
The z Ratio (Test of Difference between Means)
The t Ratio (Test of Difference between Means)
Experimental Group
Control Group
Homoscedasticity
Heteroscedasticity
F Distribution
F Ratio
F Value
Before and After Experimental Design
Matched Experimental Design
Matched Samples
Direct-Difference Method
Sandler's A Test
Univariate Analysis

Exercises

1. For a difference of means test, specify the conditions under which you would employ:
    (a) A one-tailed test of significance.
    (b) A two-tailed test of significance.
    (c) The z distribution.
    (d) The t distribution.
    (e) The A distribution.

2. If we find a significant difference between means at the 0.05 level of significance, should we then conclude (yes or no) that:
    (a) This difference is significant at the 5 percent significance level?
    (b) This difference is significant at the 0.01 level of significance?
    (c) This difference is not significant at the 0.001 level of significance?
    (d) This difference is significant at the 0.10 level of significance?
    (e) This difference between means reflects the true population difference? Explain.

3. Below are the wages per week for two independently drawn random samples of white-collar workers for two corporations. Test whether or not there is a significant difference between the wages for the two corporations.
    (a) Corporation A: $200, $187, $190, $226, $270, $175, $165, $190, $202
    (b) Corporation B: $160, $210, $200, $170, $163, $187, $145

4. A survey is conducted in two large cities to determine the average length of time it takes to get to work. Given the following data, can we conclude that there is a difference between the two cities with 99 percent certainty (that is, 0.01 level of significance)?

City A	City B
$\bar{X}$ = 25 minutes	$\bar{X}$ = 20 minutes
s = 2 minutes	s = 4 minutes
N = 300	N = 400

5. A psychologist wants to test the hypothesis that distribution of power in the family makes a difference in the number of days a year a child is absent from school. Families are divided into two groups—those where the mother is dominant and those where the father is dominant. Does the data listed below allow us to accept the null hypothesis as the 0.01 level of significance?

Mother Dominant	Father Dominant
$\bar{X} = 10$	$\bar{X} = 9$
$s = 3$	$s = 2$
$N = 15$	$N = 15$

6. Do the above exercise without assuming homoscedasticity.

7. Given that:
$$\bar{X}_1 = 60 \qquad \Sigma(X - \bar{X})^2 = 1{,}550 \qquad N = 14$$
$$\bar{X}_2 = 55 \qquad \Sigma(X - \bar{X})^2 = 1{,}980 \qquad N = 24$$
   (a) Set up the proper hypotheses.
   (b) Test the null hypothesis at the 0.01 level of significance.

8. Given two random samples of sizes 8 and 10 from two independent normal populations with $\bar{X}_1 = 25$, $\bar{X}_2 = 28$, $s_1 = 6$, $s_2 = 5$:
   (a) Test by means of the t distribution the hypothesis that $\mu_1 = \mu_2$, assuming $\sigma_1 = \sigma_2$.
   (b) Test by means of the t distribution the hypothesis that $\mu_1 = \mu_2$, assuming $\sigma_1 \neq \sigma_2$.

9. A college administrator wishes to estimate the difference in high school averages for students attending two ivy league colleges. Two independent random samples selected for each college produce the following hypothetical data:

Yale	Harvard
$\bar{X} = 93.1$	$\bar{X} = 90.1$
$s = 1.1$	$s = 2.1$
$N = 10$	$N = 12$

   (a) Construct a 95 percent confidence interval for the difference in population means. (*Hint:* Use the same logic as was used to derive Formulas (10.4) and (10.5).)
   (b) Change $N = 10$ to $N = 50$ and $N = 12$ to $N = 51$, and construct a 99 percent confidence interval for the difference in population means.
   (c) Write general formulas for constructing confidence intervals for both small and large samples.

10. A physician wishes to test the null hypothesis that patient recovery, as measured in days, using a bland diet to treat ulcers, is less than or equal to the duration of recovery using a regular diet and medication. If the hypothesis is rejected, the physician will adopt the latter treatment on

future patients; if the sample results are not significant, the physician will continue to recommend the bland diet. A random sample of twenty patients is selected from those treated with the bland diet, and another group of eighteen patients are given drugs and left to eat whatever they want. The following occurs:

Bland Diet	Regular Diet and Medication
$\bar{X} = 30$	$\bar{X} = 28$
$s = 2.8$	$s = 2.0$

(a) Determine whether or not this test is one- or two-tailed.
(b) Which treatment will the physician rely on in the future? Use the 0.01 level of significance.

11. From a pool of statistics majors, ten pairs who are matched on mathematical ability are selected. One subject in each pair is allowed to use a calculator throughout the course in mathematical statistics (the treatment group); the other subject is assigned the control (no calculator). The results of the experiment appear below, as measured by the score on the final examination. Use the t test for matched samples and the 0.01 level of significance to test the null hypothesis.

Subject	Experimental	Control
A	80	70
B	80	60
C	70	75
D	75	80
E	65	63
F	90	95
G	95	70
H	70	55
I	50	59
J	85	65

12. A group of test animals are given a test to determine reaction time, as measured in seconds. Then a state of anxiety is induced and the same test is readministered. Using the data shown below, determine the following:
(a) Is there a significant difference between the two sets of scores?
(b) Did a state of anxiety lower the reaction times of the test animals?

Animals	Pre-Test	Anxiety-Induced Test
A	25	30
B	24	26
C	28	26
D	30	31
E	31	30
F	24	26
G	25	27

13. Explain why the design in the above exercise might be faulty. How could you improve upon it?

14. An electronics manufacturer wishes to study the effect of music on the productivity of assembly-line workers. The manufacturer selects fifteen workers and measures their productivity on a day when there is no piped-in music. Later he measures the productivity of the same fifteen workers on a day when they are exposed to piped-in music. From the data listed below, determine the following.
  (a) Is there evidence to suggest that listening to music raises productivity?
  (b) Is there evidence of heterogeneity of variance? If so, why?

Worker	No Music	Music	Worker	No Music	Music
A	50	55	I	20	14
B	39	40	J	51	49
C	38	40	K	50	58
D	37	35	L	38	41
E	43	44	M	50	36
F	45	45	N	46	46
G	36	37	O	38	37
H	39	45			

15. Compute the A ratio for the data in Exercises 11, 12, and 14. Are your answers similar, using the A test? If they are, then why?

E Pluribus Unum (One Out of Many)
Anonymous

Statistical Inference: Multiple Samples

12.1 Introduction: Why Make Multigroup Comparisons?

We just reviewed statistical techniques that allow us to test the difference between two groups. But do all variables conveniently order themselves into two groups—for example, an experimental group and a control group? Obviously not. Besides, if we observed the difference between blacks and whites, would that constitute a complete test of racial differences? Or if we tested only the differences between Protestants and Catholics without including Jews and other religious groups, could we say that we were really testing *all* differences among the denominations? Nominally measured variables (such as political party, education, and a host of other widely used nominal variables) often include more than two categories. Therefore, if we, as social scientists, restrict our observations to only two groups, we oversimplify the phenomena that we are obliged to investigate.

So, what do we do when we have three or more groups? Maybe you're wondering why multigroup comparisons should pose a problem at all. Why not simply repeat the *t* test, taking two means at a time, until we have made all possible comparisons? For example, if we categorize our subjects into four groups (A, B, C, D), conceivably we could conduct six separate *t* tests (A with B, C, and D; B with C and D; and C with D) to find out which pairs of means differ significantly. It can, indeed, be boring to perform that many tests, but tedium isn't the main reason why this procedure is unacceptable. There are two other more important reasons.

First, let's assume that we have conducted a study involving the calculation of a hundred or so separate t tests. Would we be shocked if, say, five of the computed t values proved to be significant at the 0.05 alpha level? Of course not. With that many comparisons, we should be shocked if at least five comparisons that are "significant" didn't come about by chance alone. In a hundred such comparisons, five should show significance at the 0.05 level and one should show significance at the 0.01 level, even if there are no real differences at all. Therefore, multiple t tests increase the likelihood of making an incorrect decision about the null hypothesis; in other words, they increase our chances of making Type I errors. The second reason involves the more advanced research design, in which two or more independent variables can interact with one another. *Interaction* can occur when two or more identifiable categories within an independent variable affect the dependent variable in different ways, or when two or more independent variables affect each other and thus have a confounding effect on the dependent variable as well. Since the t test does not "partial out" these interaction effects, we need a technique that will produce results that are free from this bias.

The procedure known as *analysis of variance* (sometimes abbreviated ANOVA) lets us simultaneously compare more than two sample means. In its most elementary form, analysis of variance is used to compare the several groups of a single independent variable, while holding the Type I error at a constant level. A more advanced two-way design is available to partial out the interaction effect of one independent variable on the other, to test for the significance of each effect, and to estimate the proportion of overall variation caused by each effect. However, we will talk largely about the simple one-way design.

12.2 Variance Estimation and Sums of Squares

A comparison of variances between and within each of the categorical groups is central to the concept of analysis of variance. That is, we wish to know whether the between-group variance is large or small relative to the within-group variance. The larger it is, the more likely it is that the groups were drawn from different populations. This means that the groups differ with respect to the dependent variable in question. Referring to Table 12.1, suppose that we wish to know whether the differences in the cumulative grade-point averages for the groups differ in the underlying population. In other words, is there a meaningful difference among the groups?

Since we're dealing with samples and do not know σ^2, we must estimate it. The null hypothesis is that the samples are part of the same population (that is, $\mu_1 = \mu_2 = \mu_3$). Just as we did in the last chapter, we must estimate the population parameters from our knowledge of the samples. In analysis of var-

Table 12.1. Scores of Three Groups of Subjects

	HISTORY		CHEMISTRY		SOCIOLOGY		TOTAL	
	X_1	X_1^2	X_2	X_2^2	X_3	X_3^2		
	3.1	9.61	2.7	7.29	2.8	7.84	$\Sigma X =$ 8.6	24.74
	3.2	10.24	2.4	5.76	2.8	7.84	8.4	23.84
Cumulative	3.4	11.56	2.3	5.29	2.9	8.41	8.6	25.26
Grade-Point	3.5	12.25	3.4	11.56	3.3	10.89	10.2	34.70
Averages	3.3	10.89	2.2	4.84	3.7	13.69	9.2	29.42
	$\Sigma X = 16.5$ $\Sigma X^2 = 54.55$		$\Sigma X = 13.0$ $\Sigma X^2 = 34.74$		$\Sigma X = 15.5$ $\Sigma X^2 = 48.67$		$\Sigma X = 45.0$ $\Sigma X^2 = 137.96$	
	$N = 5$ $\bar{X} = 3.3$		$N = 5$ $\bar{X} = 2.6$		$N = 5$ $\bar{X} = 3.1$		$N_t = 15$ $\bar{X}_t = 3$	

iance, as the name suggests, we're analyzing the variance among samples so that we may make some decision about the null hypothesis. To do so, we must use the pooled estimates of the variances for each sample. The first step in obtaining *variance estimators* is to compute the *sums of squares* for each sample (categorical group), which serve as the numerators for our variance estimators. We partition the sums of squares into two components: *within-group sums of squares* and *between-group sums of squares*.

12.2.1 The Within-Group Sum of Squares

The first thing we want to look at is how the scores within each group differ from their respective means. That is, we want to know the amount of deviation within each group. The within-group sum of squares for each group tells us this.* To compute, we use a double sigma ($\Sigma\Sigma$), one to indicate the sum within each group $\left(\sum_{i=1}^{N_j} \right)$ and one to indicate the sum across the groups $\left(\sum_{j=1}^{k} \right)$. Symbolically, the formula for the within-group sum of squares (SS_w) is written as:

$$SS_w = \sum_{j=1}^{k} \sum_{i=1}^{N_j} (X_{ij} - \bar{X}_j)^2 \tag{12.1}$$

where:

$$X_{ij} = \text{the individual score, and}$$
$$\bar{X}_j = \text{the group mean.}$$

Using the data in Table 12.1, we look at Formula (12.1), which directs us to subtract the group mean from an individual score in a group and square the difference. Then we sum the squared differences for the group. We repeat the procedure for each group and add each group total to each other to obtain the within-group sum of squares:

$$(3.1 - 3.3)^2 + (3.2 - 3.3)^2 + (3.4 - 3.3)^2 + (3.5 - 3.3)^2 + (3.3 - 3.3)^2$$
$$+ (2.7 - 2.6)^2 + (2.4 - 2.6)^2 + (2.3 - 2.6)^2 + (3.4 - 2.6)^2 + (2.2 - 2.6)^2$$
$$+ (2.8 - 3.1)^2 + (2.8 - 3.1)^2 + (2.9 - 3.1)^2 + (3.3 - 3.1)^2 + (3.7 - 3.1)^2$$
$$= 1.66$$

---

*Note that the sum of squares is defined as $\Sigma(X - \bar{X})^2$. To review the concept, refer back to Section 5.4.3.

Although the theoretical Formula (12.1) intuitively conveys the meaning of within-group sum of squares, we suggest that you use the alternative machine formula:

$$SS_w = \Sigma \left[\Sigma X_j^2 - \frac{(\Sigma X_j)^2}{N} \right]$$

(12.2)

where:

X_j = the individual scores within a group.

For the data in Table 12.1, the within-group sum of squares is:

$SS_w = [54.55 - (16.5)^2/5] + [34.74 - (13.0)^2/5] + [48.67 - (15.5)^2/5]$
$= 0.10 + 0.94 + 0.62$
$= 1.66$

12.2.2 The Between-Group Sum of Squares

Now we're ready to obtain the *between-group sum of squares*. To do so, we subtract the overall group mean from the mean of each group and square the result. Then we multiply this value by the number of cases in each group and sum across the groups. Thus, we have the theoretical formula for the between-group sum of squares (SS_b):

$$SS_b = \sum_{j=1}^{k} N_j(\bar{X}_j - \bar{X}_t)^2$$

(12.3)*

where:

N_j = the number of cases in each group,
$\bar{X}_j$ = the mean of each group, and
$\bar{X}_t$ = the overall group mean.

As with the within-group sum of squares, there is a way to avoid the tedious computations required by the theoretical formula. Therefore, we again give an alternative machine formula:

$$SS_b = \Sigma \left[\frac{(\Sigma X_j)^2}{N_j} \right] - \frac{(\Sigma \Sigma X_{ij})^2}{N_t}$$

(12.4)

---

*If the groups are of unequal size, use the weighted mean given in Formula (5.5).

where:

$$N_t = \text{the total number of cases for all groups.}$$

For the data in Table 12.1, the between-group sum of squares is:

$$
\begin{aligned}
SS_b &= [(16.5)^2/5 + (13.0)^2/5 + (15.5)^2/5] - (45.0)^2/15 \\
&= [54.45 + 33.80 + 48.05] - 135.00 \\
&= 1.30
\end{aligned}
$$

12.2.3 The Total Sum of Squares

The *total sum of squares* (SS_t) is equal to the within-group sum of squares plus the between-group sum of squares. In other words:

$$SS_t = SS_w + SS_b$$

Thus, the total sum of squares is 1.66 + 1.30, or 2.96.

We can obtain the total sum of squares directly by subtracting the overall mean from each score and squaring the difference:

$$SS_t = \sum_{j=1}^{k} \sum_{i=1}^{N_j} (X_{ij} - \bar{X}_t)^2 \tag{12.5}$$

Here, too, a machine formula is available:

$$SS_t = \sum X_{ij}^2 - \frac{(\sum X_{ij})^2}{N_t} \tag{12.6}$$

For the data in Table 12.1, the total sum of squares is:

$$
\begin{aligned}
SS_t &= 137.96 - (45.0)^2/15 \\
&= 2.96
\end{aligned}
$$

It is clear that we can determine SS_t indirectly, by adding together SS_w and SS_b, or directly, using either Formula (12.5) or (12.6). It is equally clear that we can determine SS_w if we know SS_b and SS_t, or SS_b if we know SS_w and SS_t. Thus:

$$
\begin{aligned}
SS_w &= SS_t - SS_b \\
SS_b &= SS_t - SS_w
\end{aligned}
$$

In general, it is easier to compute SS_t directly first, and then to compute either SS_w or SS_b to obtain SS_b or SS_w. On the other hand, using all three formulas serves as a check on computational errors.

12.2.4 The Variance Estimates

The computed sums of squares give us numerators for our variance estimates. To compute these estimates, we must divide the sums of squares by their appropriate degrees of freedom. The degrees of freedom (see Section 10.4.1) is the result of the assumptions we're required to make about the quantities we are trying to estimate. As a general rule, the more parameters we're trying to estimate, the more constraints we're placing upon our estimated values. As we know from earlier discussion, $df = N - 1$ for the estimate of a population parameter. This constitutes our total degrees of freedom for a one-way analysis of variance. However, we partitioned our sums of squares into "between" and "within" components, which constitute the numerators for the between and within variance estimates, respectively. After all, we computed the between-group and within-group sum of squares to get variance estimators. Therefore we must complete this process of estimation by using the appropriate degrees of freedom as denominators for our respective between-group and within-group sum of squares. This yields the between and within variance estimators, which together constitute the total variance estimate.

To arrive at the between-group variance, we divide the between-group sum of squares by $k - 1$, where k is the *number of groups*. Since each group mean is treated as an observation, we are estimating one parameter; so $k - 1$ is the appropriate degrees of freedom. Likewise, we estimate the within-group variance by the within-group sum of squares, divided by its appropriate degrees of freedom—which in this case is $N - k$. This tells us that the number of groups within which our observations *(N)* are categorized places further restrictions on the components of our estimate. The two components of estimated variance add up to the total variance estimate. Just as the total sum of squares is the sum of the within-group and between-group sum of squares, so too is the total degrees of freedom $(N - 1) = (k - 1) + (N - k)$. Now we have computed both our numerators and denominators for the two components of variance estimators.

As we've said earlier in this section, we're interested in determining whether the between-group variance $(\hat{s}_b^2)$ is larger or smaller than the within-group variance $(\hat{s}_w^2)$. This can be summarized by the *F ratio*, which is defined as:

$$F = \frac{\hat{s}_b^2}{\hat{s}_w^2}$$

(12.7)

where:

$$\hat{s}_b^2 = SS_b/(k-1), \text{ and}$$
$$\hat{s}_w^2 = SS_w/(N-k).$$

The F ratio tells us whether we can reject our null hypothesis that the groups come from the same underlying population.

12.3 Caution: Some Assumptions Underlying Analysis of Variance

If a simple one-way analysis of variance is to be legitimate, we must be able to assume that data on the independent variable are measured at the interval level, and that they are distributed normally in the underlying population. In addition, the samples must be selected randomly and drawn independently.

We must also be able to assume that the variances (which, when summed together, make up $\hat{s}_w^2$) are homogeneous. As with the t ratio, there is the Bartlett (1937) test for determining whether more than two variances are homogeneous; but this test is much too advanced for our purposes.*

Obviously, the above conditions might keep us from using analysis of variance. In Chapter 16, we will discuss a workable alternative—but for the present, let's just assume that we meet each condition, and proceed at once to a worked example.

12.4 Analysis of Variance: An Illustration

Here's one problem that lends itself to multigroup comparisons: checking to see whether social science majors graduate with higher averages than humanities majors or natural science majors do, or whether natural science majors have consistently higher averages than all other majors, and so forth. Although many factors enter into solving this problem, we can start the investigation with a one-way analysis of variance.

Suppose that we take a random sample for each of three broad areas of academic concentration, as in Table 12.1. Suppose further that we obtain the cumulative grade-point averages of five students in each area, as in Table 12.1. Now let's use these data to see whether they come from the same population (that is, whether that major has no effect on a student's average).

---

*However, if you are interested in getting your feet wet, you can check with Edwards (1958) and Snedecor and Cochran (1967) for applications of this test.

Hypotheses   To test the difference between means for our three groups, we set up the hypotheses as follows:

$$H_0: \mu_1 = \mu_2 = \mu_3$$
$$H_1: \mu_1 \neq \mu_2 \neq \mu_3$$

Sampling Distribution   The sampling distribution for an analysis of variance is the *F distribution* (Table D). Incidentally, since we will be dealing with squared values, all values of *F* are necessarily positive. Thus the *F* distribution is a positively skewed type of bell-shaped curve, which, from left to right, reads zero to infinity (see Figure 12.1). This tells us that tests of significance that use this distribution are invariably one-tailed.

Assume an alpha of 0.05. To obtain the corresponding critical *F* value we consult Table D, using the between-group degrees of freedom $(k - 1)$ to locate the proper column across the top, and the within-group degrees of freedom $(N - k)$ to locate the proper row down the side. The intersection of row 2 (i.e., 3 − 1) and column 12 (i.e., 15 − 3) contains the critical *F* value of 3.88.

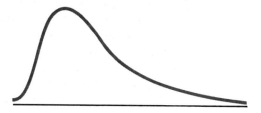

Figure 12.1.   The F Distribution

Test and Decision   To perform an *F* test, we must compute the between- and within-group variances. Since we've already computed all the information necessary for these estimates, we should place these values in a summary table, such as the one shown in Table 12.2. We suggest that you, too, place all the relevant information for an analysis-of-variance problem in this form; it will condense all your computations and keep you from losing them.

Table 12.2.   Summary Table for Analysis-of-Variance Problems

Source of Variance	Sum of Squares	Degrees of Freedom	Variance Estimate	F
Between-Group	1.30	2 $(k - 1)$	0.65 $(\hat{s}_b^2)$	4.64
Within-Group	1.66	12 $(N - k)$	0.14 $(\hat{s}_w^2)$	
Total	2.96	14 $(N - 1)$		

In order for us to reject the null hypothesis at the 0.05 level with 2 and 12 degrees of freedom, our computed F value must equal or exceed 3.88. Since our $F = 0.65/0.14$ or 4.64 exceeds the value, we can reject the null hypothesis and accept its alternatives. Specifically, we may conclude that the three groups differ with respect to cumulative grade-point average.

12.5 Interpreting F: The Honestly Significant Difference Test

A significant F ratio tells us that the three or more group means are not all estimates of a common population. Should we stop at this point? Don't we at least partially want to determine whether one of three groups is, say, superior to the other two? In short, a significant F ratio should tell us: "Something's here — start looking."

As we explained earlier in this chapter, we'd have to work really hard to obtain a t ratio for each comparison, and we'd be more likely to make Type I errors, as well. Fortunately, both these obstacles are easy to overcome. Statisticians have thought up a number of statistical tests for making multiple comparisons after a significant F ratio is obtained. These tests show us where the significant mean differences lie, and also consider the likelihood that Type I errors will increase as the number of means being compared increases. And since those tests will obtain fewer significant differences than the t ratio will, a significant difference is more likely to surface in a multiple group comparison of three means than in a multiple group comparison of four or five means.

As we know, there are quite a few methods for testing a significant F ratio. But we will give only one of them here — Tukey's (1953) method, often referred to as the *HSD* test (for "honestly significant difference"). We use the *HSD* test only when the overall F ratio is significant and the groups are of equal size. (A test for unequal samples appears in Hays (1973).) With these warnings in mind, we judge a difference between two means as "honestly significant," at a given alpha level, if it equals or exceeds the *HSD*. The formula for *HSD* is:

$$HSD = q\alpha \sqrt{\frac{\hat{s}_w^2}{N}} \qquad (12.8)$$

where:

$q\alpha$ = the tabled Studentized range value (Table F) for a given alpha level for the degrees of freedom $(k - 1)$ associated with the within-group sum of squares,

$\hat{s}_w^2$ = the within-group variance estimate, and

N = the number of cases in each group.

Let's illustrate the application of this test by returning to the present example, in which the three groups of majors were found to differ with respect to cumulative grade-point average.

Hypotheses   To make pairwise comparisons among the three groups, we must involve the following hypotheses:

$$H_0: \mu_1 = \mu_2 \text{ and } H_1: \mu_1 \neq \mu_2$$
$$H_0: \mu_1 = \mu_3 \text{ and } H_1: \mu_1 \neq \mu_3$$
$$H_0: \mu_2 = \mu_3 \text{ and } H_1: \mu_2 \neq \mu_3$$

Sampling Distribution   The sampling distribution for the *HSD* test is the percentage points of the Studentized range (Table F).

Critical Region   Referring to Table *F* under $df = 12$ (degrees of freedom for $\hat{s}_w^2$), $k = 3$ (number of means) at $\alpha = 0.05$, we find $q_{0.05} = 3.77$. We enter this value, along with the values of $\hat{s}_w^2$ and *N*, into the *HSD* formula:

$$HSD = 3.77\sqrt{0.14/5} = 0.631$$

Now we have computed the critical value that the difference between a pair of means must equal or exceed in order to be significant at the 0.05 alpha level.

Test and Decision   To compare all possible pairs of means (assuming that we wish to do so), we first must find the absolute difference between each pair of means. These values are usually presented in matrix form, as in Table 12.3. Comparing *HSD* against each value in the matrix, we note that the difference between $\bar{X}_1$ and $\bar{X}_2$ exceeds our critical value of 0.631. We therefore may conclude that this difference is statistically significant at the 0.05 alpha level.

Table 12.3.   Absolute Differences Between Pairs of Means

	$\bar{X}_1$	$\bar{X}_2$	$\bar{X}_3$
$\bar{X}_1$ (3.3)	—	0.70	0.20
$\bar{X}_2$ (2.6)	—	—	0.50
$\bar{X}_3$ (3.1)	—	—	—

12.6 Analysis of Variance: Two-Way Design

So far we have focused on three groups that differ with respect to some variable. But in many situations, we are called on to test the effects of more than one variable. Suppose that we have the same three groups of college majors that we had before, but that we now divide each of these groups into two subgroups, one male and the other female. We can see clearly that these are both nominal attributes, but now we have increased our possible conclusions: we can test the effect of the academic area; or the effect of gender; or the interaction of the two, combined. This general arrangement, in which we study the effects and interaction of two variables, is known as *two-way analysis of variance*, or *factorial design*. The principle is the same as that of the one-way design, except that here we can test more than one effect. Therefore, we summarize our data according to groups, such as those presented in Table 12.4.

Table 12.4. Cumulative Grade-Point Average for Males and Females, by Major

	History	Chemistry	Sociology	Total
Male	3.1 3.2 3.4	2.7 2.4 2.3	2.8 2.8 2.9	$\Sigma X^2 =$ 73.84 $N =$ 9 $\bar{X} =$ 2.8
Female	3.5 3.3	3.4 2.2	3.3 3.7	$\Sigma X^2 =$ 64.12 $N =$ 6 $\bar{X} =$ 3.2
Total	$\Sigma X^2 = 54.55$ $N = 5$ $\bar{X} = 3.3$	$\Sigma X^2 = 34.74$ $N = 5$ $\bar{X} = 2.6$	$\Sigma X^2 = 48.67$ $N = 5$ $\bar{X} = 3.1$	$\Sigma X^2 = 137.96$ $N = 15$ $\bar{X} = 3.0$

Since the table containing data for a two-way design is presented in rows and columns, we call the effects of the two variables the *row effect* and *column effect*, respectively. This means that we are performing the one-way analysis of variance on the row and the column variables individually, and then again for their interaction. The result is three F ratios, obtained by dividing the variance estimate ($\hat{s}_w^2$) for the row, for the column, and for the row and column interaction.

As before, if one or more of the F ratios is significant, we can make comparisons to determine precisely which groups or conditions contribute to the significant differences. We do this by comparing pairs of means with a test (for instance, the one developed by Tukey).

There are other extensions that allow us to analyze the data from factorial designs involving three or more separate variables. Further modifications have been developed for use in analyzing matched groups, such as those involved in experiments that obtain a number of measures from the same subjects rather than from independent groups. We can also set up complex experiments to sample only certain variables. This leads to the use of techniques such as *random block* or *latin squares designs*. There is even a *nested* (or *hierarchical*) design that allows us to perform a two-way analysis, even though we don't have all the data. Obviously, we're not going to explore these procedures here, but you can read the Scheffé (1959) text for answers to all such questions. If we use any advanced analysis-of-variance technique, we will get an F ratio to compare against a critical F value so that we can assess whether differences among population parameters really exist.

Where We Stand

We began this chapter by observing that, as social scientists, we must have a technique for making comparisons among three or more groups, while holding the Type I error at a constant level. So we learned how to perform an analysis of variance. We even learned how to make multiple comparisons (*HSD* test) after a significant F ratio in order to pinpoint where the significant mean differences lie.

In any multigroup test of means, and even in single sample tests, we may unknowingly violate certain assumptions about variances in the population. We also may refrain from using these tests when we have reason to believe that our data are not intervally scaled. A case in point would be indices of party commitment (strong Republican, weak Republican, etc.) or various power (or influence) indices. We must rather hesitantly assume that the various points on these scales indicate equal intervals.

There are statistical tests that sidestep these problems, referred to variously as *distribution-free statistics* (as in the first case) or *nonparametric statistics* (as in the second case). Obviously, there are procedures we can follow—either to make inferences from nominal or ordinal data or to use when we can't make any assumption about the shape of the underlying populations from which our samples were drawn. Before we investigate these techniques, let's first look at the important concepts of correlation and regression, which lie at the core of much social science research.

Terms and Symbols to Remember

Interaction
Analysis of Variance
One-Way Analysis of Variance
Two-Way Analysis of Variance (Factorial Design)
Within-Group Sum of Squares (SS_w)
Between-Group Sum of Squares (SS_b)

Total Sum of Squares (SS_t)
F Ratio
F Distribution
Honestly Significant Difference Test (HSD)
Row Effect
Column Effect
Distribution-Free Statistics
Nonparametric Statistics

Exercises

1. Given three independent samples:

   A: 5, 6, 5, 5, 4
   B: 9, 8, 4, 6, 7
   C: 8, 8, 7, 10, 10

   compute:
   (a) The within-group sum of squares.
   (b) The between-group sum of squares.
   (c) The total sum of squares.

2. Are there significantly different means among the three samples of Exercise 1 at the 0.05 level of significance? If so, find the pair(s) of samples having significantly different means.

3. Conduct an analysis of variance on the data in Chapter 11, Exercise 12. Demonstrate that, for two groups, $F = t^2$.

4. Suppose that a board of trustees of a university hospital selects at random three general surgeons for study. The time, in minutes, required for each surgeon to perform a routine surgical procedure is tabulated:

Surgeon A	Surgeon B	Surgeon C
51	50	47
48	47	30
35	55	56
40	32	58
43	40	42

   (a) Apply an analysis of variance to see whether the surgeons differ significantly in their performance.
   (b) If the results show differences, apply the HSD test to find out where the differences are.

5. Test the significance of the difference between means of each of the groups in Exercise 4 by using t tests. How does this compare with the result generated with the analysis of variance and the HSD test?

6. Four cigarette companies claim they manufacture a filter cigarette that produces less tar. A random sample of the cigarettes of each manufacturer reveals the following tar levels (in milligrams). Test the null hypothesis that all three cigarettes are drawn from a common population of means. Use the 0.05 level of significance.

Brand A	Brand B	Brand C	Brand D
4.0	5.0	4.7	5.0
4.2	5.1	4.4	5.2
4.2	4.8	4.2	5.3
4.8	4.8	4.8	5.0

7. Following are data representing the scores of five groups of students on a multiple-choice examination in physics.
   (a) Set up the appropriate hypotheses.
   (b) Determine whether the means of the five groups differ significantly (assume $\alpha = 0.05$). If so, apply the HSD test to find out where the differences are.

Group 1	Group 2	Group 3	Group 4	Group 5
90	70	43	91	68
85	80	50	90	70
80	85	47	84	59
92	75	75	75	57
90	70	80	72	50

8. An experiment is conducted to see whether there is any relationship between the degree of lighting available to a person who is studying and the ability to solve problems correctly. Below is a table showing the number of problems solved correctly under different lighting conditions. Can we reject the null hypothesis at the 5 percent significance level?

		LIGHTING CONDITION		
	Too Bright	Too Yellow	Just Right	Too Little Light
Problems Solved Correctly	8	6	2	10
	6		0	8
	4		0	6
			2	

9. A study is conducted to see whether the location of newly married couples affects marital adjustment. Location is determined by how close the newly married couple lives to their parents. Marital adjustment is

measured by a scale that is treated as an interval measure. The results are given below. Do they indicate, at the 0.05 level of significance, that location makes a difference?

	Near Both Husband's and Wife's Parents	Near Wife's Parents but Far from Husband's	Near Husband's Parents but Far from Wife's	Far from Both Sets of Parents
Marital *Adjustment*	1 9 5	1 5	1 5 5 1	7 3 5

10.   Compute the *F* ratio, given the following entries:

Variation	Sum of Squares	Degrees of Freedom	Variance Estimate
Between-Group		2	8
Within-Group	250		
Total		52	

Some circumstantial evidence is very strong,
as when you find a trout in the milk.
Henry David Thoreau

Correlation

13.1 Introduction: Why Study the Relationship Between Variables?

Up to this point, we have calculated various statistics that let us describe the distribution of the values of a single variable. In the next two chapters, we will focus on *bivariate analysis*—the amount of relationship between two variables.

Although the phrase "relationship between two variables" may be unfamiliar to you, you have already encountered these relationships in your own coursework. If you are an education major, you care about the association of students' performances on various tests and such variables as intelligence and the types of curricula the students have followed. If you are a political scientist, no doubt you are interested in such relationships as the voting turnouts and literacy rates in various countries. And if you're a sociologist, you probably are concerned with the association of education and a person's socioeconomic status.

As soon as we raise questions about the relationships between variables, we enter the thought-provoking area of *correlation*. If there is a relationship between two variables, such as College Entrance Examination (CEEB) scores and college grades, we say that they are correlated. Conversely, if there is no relationship between CEEB scores and college grades, we say that there is no correlation. The statistical techniques that we will develop here and in the next chapter illustrate two major functions of correlation. First, we will develop techniques to gauge the strength (or amount) of relationship between two variables, so that a single value will tell us how two variables are related. Second, we will develop a technique that allows us to predict scores on the dependent variable from knowledge of the independent variable. For example, if CEEB scores and college grades are related, we will be able to predict the latter from knowledge of the former. The first of these functions *(correlation)* is the topic of our current discussion; the second *(regression)* will be discussed in Chapter 14.

13.2 The Concept of Correlation

Correlation, as the word itself suggests, is the relationship (or association) between two or more variables. For example, if individuals with more years of education tend to have higher incomes than individuals with fewer years of education, then we say that there is a *positive correlation* between the two variables of education and income. However, if those with more education tend to have lower incomes than those with fewer years of education, we say that a *negative correlation* exists. And when income is the same regardless of education, *no* (or *zero*) *correlation* is said to exist.

These three classes of correlation are illustrated graphically in Figure 13.1. The graphs in this figure are called *scatter plots* because they show how the points scatter over the range of possible values on two variables. Each point in a scatter plot represents two values: an individual's score on one variable (X) and the same individual's score on a second variable (Y). Usually, the independent X variable is represented along the *horizontal axis,* or *X-axis,* and the dependent Y variable is represented along the *vertical axis,* or *Y-axis.* *

If we examine scatter plot (a), we see that both variables are related proportionately; that is, individuals with low scores on one variable tend to have low scores on the other, and individuals with high scores on one also tend to have high scores on the other. This *proportional* relationship, of course, represents a positive correlation. In scatter plot (b), individuals who score low on one variable tend to score high on the other variable, and vice versa. This *inverse* relationship denotes a negative correlation between X and Y.

In contrast, scatter plot (c) shows two variables that have virtually no relationship to each other. For example, high values of X are accompanied by high values of Y, while other high values of X are paired with low values of Y. The same is true for low values of X. In short, individuals do not score with any degree of consistency in relation to X and Y. This lack of relationship between two variables typifies zero correlation.

---

*Strictly speaking, correlation is merely a measure of association between two variables and does not necessarily imply a dependent and independent variable. However, in many research situations, the researcher is using correlation in conjunction with regression analysis, which does designate the variables as independent and dependent.

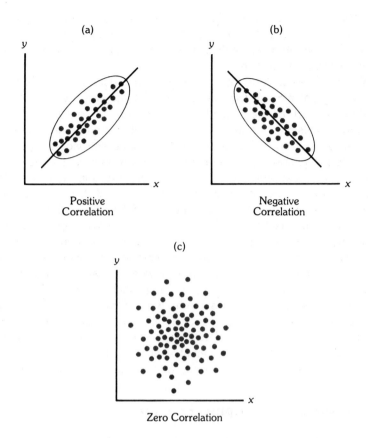

Figure 13.1.   Scatter Plots Showing Three Types of Correlation Between Two Variables (X and Y)

We've already seen that the data shown in the scatter plots exhibit different directions of correlation: positive in (a), and negative in (b). If we look at these plots even more closely, we'll find another distinguishing feature.

In scatter plot (a), we drew a straight line that falls near as many points as possible. The fact that these points do not lie far from the line indicates a very strong positive relationship between the two variables. If the relationship were a perfect positive one, all the points would fall on the line. Similarly, in a perfect negative correlation all the points would fall on a straight line, except that low values for the X variable would be paired with high values for Y, and vice versa. If we now look at the inverse relationship in scatter plot (b), we see that the points surrounding the straight line are not nearly as close as they are in scatter plot (a). Therefore these two variables are not as strongly correlated as those variables in scatter plot (a) are. By the time the correlation drops to zero, the pattern of points is almost circular and there is virtually no relationship that can be described by a straight line. In other words, a straight line drawn through the mean of the Y values parallel to the horizontal axis, or a straight line drawn through the mean of the X values parallel to the vertical axis, would do no better than a tilted line.

Therefore, we conclude that scatter plots may differ in terms of their degree of correlation, as well. Specifically, correlations vary between a perfect relationship (in which all the points fall on a straight line) to a zero relationship (in which the points are scattered so that no X value shows a greater tendency to be paired with one Y value more than any other, and vice versa).

Unfortunately, however, descriptions based on observing scatter plots are not very precise for expressing the extent of the interrelation between variables. If we want to express quantitatively the extent to which two variables are related so that we can find the precise degree of linear association, we must calculate a *correlation coefficient*. This coefficient will have a specific numerical value for any given set of paired data. It will range between 1.00, a perfect positive correlation, and -1.00, a perfect negative correlation; values near 0 will imply little or no linear relationship between variables. Since there are many types of correlation coefficients, deciding which one to use with a set of data depends upon factors such as (1) the type of data involved, and (2) the characteristic of the distribution (linear or nonlinear). In this chapter, we present two correlation coefficients—the *Pearson product-moment correlation coefficient, Pearson r, or r,* used with interval-level data; and the *Spearman rank-order correlation coefficient, Spearman r_s, or r_s,* used with ordinal-level data.*

13.3 Formulas to Compute Pearson *r*

There are many formulas for calculating Pearson *r*. We will study three types. The *machine formula* is the easiest to use, but it masks the theoretical basis of *r*. The *z score* and *mean deviation formulas* are sufficiently theoretical in design, but awkward when it comes time to calculate. We present them here primarily because they shed light on the character of *r*, as well as on the derivation of the machine score formula itself.

13.3.1 *z* Scores and Pearson *r*

Now we'll discuss the Pearson product-moment correlation coefficient in terms of *z* scores. The value of *r* is equal to the mean of the cross products, $z_x z_y$. In statistics, actually, the terms "mean" and "moment" are synonymous. This explains why Karl Pearson (1857–1936), who is remembered more for his re-

---

*We will use the three expressions of these two coefficients—that is, Pearson product-moment correlation coefficient, Pearson *r*, and *r*; and Spearman rank-order correlation coefficient, Spearman r_s, and r_s—interchangeably in this text.

markable contributions to statistics than for his equally brilliant contributions to biology, named his measure a product-moment correlation coefficient. It is written as:

$$r = \frac{\Sigma z_x z_y}{N}$$ **(13.1)**

where:

$$z_x = \frac{X - \bar{X}}{s_x},$$

$$z_y = \frac{Y - \bar{Y}}{s_y}, \text{ and}$$

$$N = \text{the total number of pairs.}$$

To compute *r*, we first determine the *z* score for each *X* and for its paired *Y*. Next, we multiply each of the paired *z* scores to obtain the cross product $(z_x z_y)$ for each pair. We then sum the cross products and divide by the total number of pairs.

Now let's see if the *r* values we get with the above formula have the same size and direction as those suggested by our earlier statements about correlation. For example, we said that a high positive correlation will occur if high values on both variables tend to be paired and if low values on both variables tended to be paired. If these conditions exist, can we expect the correlation coefficient to be positive? To answer this question, let's see how the values of *r* change for different arrangements of *z* scores.

To begin with, z_x and z_y are positive when they correspond to scores larger than the mean. However, if the score is smaller than the mean, the *z* score will be negative. In a high positive correlation, high values for the *X* variable are paired with high values on the *Y* variable so that the cross products for each pair are positive in value. Similarly, since the multiplication of a negative value by another negative value produces a positive value, when low values for the *X* variable are paired with low values for the *Y* variable, positive cross products result. Thus, the sum of the cross products divided by the total number of pairs yields a high positive correlation. Since a low positive correlation has a number of pairs of scores that are opposite in sign, their corresponding cross products are negative in value. In other words, those pairs with one score above the mean and the other score below it produce negative values for cross products, because of the multiplication of a negative by a positive *z* score. The summation of negative- and positive-value cross products yield a positive correlation that's less than when the correlation is high. We could use similar reasoning to demonstrate that a high negative correlation leads to an *r* with a negative value larger than that of a low negative correlation. Whenever the sum of negative and positive cross products cancel each other out, there is no relationship between the variables, and the Pearson *r* will, of course, equal zero.

We said that the Pearson r ranges from 1.00 (for a perfect positive correlation) to a −1.00 (for a perfect negative correlation). Why is this so? Well, a perfect correlation implies that all the points lie on a straight line, which in turn implies that in a perfect positive correlation, each z score on the X variable is identical to its corresponding z score on the Y variable; thus, $z_x = z_y = z^2$. Therefore, we can rewrite Formula (13.1) as $\Sigma z^2/N$, which is also the variance for a set of z scores. Since the standard deviation is the square root of the variance, and since a z of 1.00 is one standard deviation from the mean of a distribution, the standard deviation of a set of z scores is also 1.00. The value of r for a perfect positive correlation is obviously $\Sigma z^2/N = 1.00$. Similarly, a perfect negative correlation takes the value −1.00. So the value of the Pearson r ranges from 1.00 (for a perfect positive correlation) through 0 (for no correlation) to −1.00 (for a perfect negative correlation).

In Table 13.1, we illustrate the actual computation of Pearson r, using z scores. Note that to compute r, we don't need similar scale values. For example, we can correlate hat size with IQ, if we feel so inclined. The independence of r from specific scale values lets us directly compare relationships between an unlimited variety of interesting variables. In this respect, r is very much like z.

Table 13.1. Computational Procedures for Pearson r, Using z Scores

Individual	X	z_x	Y	z_y	$z_x z_y$
A	2	−2.0	4	−1.75	3.50
B	4	−1.0	8	−0.88	0.88
C	5	−0.5	10	−0.44	0.22
D	7	0.5	12	0.00	0.00
E	8	1.0	16	0.88	0.88
F	9	1.5	20	1.75	2.62
G	6	0.0	10	−0.44	0.00
H	5	−0.5	10	−0.44	0.22
I	6	0.0	12	0.00	0.00
J	8	1.0	18	1.32	1.32
					$\Sigma z_x z_y = 9.64$

$$r = \frac{\Sigma z_x z_y}{N} = \frac{9.64}{10} = 0.96$$

Even though it's easy for us to memorize the z score method for computing a Pearson product-moment correlation coefficient, the formula demands that we calculate separate z scores for each value. Imagine how much time and effort it would take us to calculate r if N exceeds ten cases (as it often does in actual practice)! That's why we will now turn our attention to the somewhat less difficult mean deviation method for computing r.

13.3.2 The Mean Deviation Method and Pearson r

We often use the mean deviation method for calculating a Pearson r when the number of cases is reasonably small. But with larger samples, it's just as time-consuming and awkward to use this method as it is to use standard scores. We offer it here only because it clarifies the characteristics of Pearson r and helps bridge the gap between the derivation of the machine formula (to be presented shortly) and the z score formula (presented in the last section). The formula for the Pearson r, using the mean deviation method, is:

$$r = \frac{\frac{\Sigma(X - \bar{X})(Y - \bar{Y})}{N}}{\sqrt{\frac{\Sigma(X - \bar{X})^2}{N}} \sqrt{\frac{\Sigma(Y - \bar{Y})^2}{N}}}$$

Simplified, this can be expressed as:

$$r = \frac{\Sigma(X - \bar{X})(Y - \bar{Y})}{\sqrt{\Sigma(X - \bar{X})^2} \sqrt{\Sigma(Y - \bar{Y})^2}} \qquad \textbf{(13.2)}$$

In Table 13.2, we illustrate the mean deviation method, using the figures in Table 13.1. You already know most of the computational procedures except for the *covariance*, or the sum of the cross products $[\Sigma(X - \bar{X})(Y - \bar{Y})]$ divided by N. We obtain the covariance by multiplying the deviation of each score from the mean of the X variable by the corresponding deviation on the Y variable, then summing all the cross products, and then dividing by N. In fact, our earlier discussions on the relationship between variations in $\Sigma(z_x z_y)$ and r also hold for $\frac{\Sigma(X - \bar{X})(Y - \bar{Y})}{N}$ and r.

Table 13.2. Computational Procedures for Pearson r, Using Mean Deviation Method

Individual	X	$(X - \bar{X})^2$	Y	$(Y - \bar{Y})^2$	$(X - \bar{X})(Y - \bar{Y})$
A	2	16	4	64	32
B	4	4	8	16	8
C	5	1	10	4	2
D	7	1	12	0	0
E	8	4	16	16	8
F	9	9	20	64	24
G	6	0	10	4	0
H	5	1	10	4	2
I	6	0	12	0	0
J	8	4	18	36	12
		$\Sigma(X - \bar{X})^2 = 40$		$\Sigma(Y - \bar{Y})^2 = 208$	$\Sigma(X - \bar{X})(Y - \bar{Y}) = 88$

$\bar{X} = 6$
$\bar{Y} = 12$

$$r = \frac{\Sigma(X - \bar{X})(Y - \bar{Y})}{\sqrt{\Sigma(X - \bar{X})^2} \sqrt{\Sigma(Y - \bar{Y})^2}} = \frac{88}{\sqrt{40} \sqrt{208}} = \frac{88}{91.21} = 0.96$$

13.3.3 The Machine Method and Pearson r

You may have noticed that the denominator in Formula (13.2) consists of the standard deviation of X multiplied by the standard deviation of Y. Recall that the machine formulas for calculating the respective standard deviations for X and Y are:

$$s_x = \sqrt{\frac{\Sigma X^2}{N} - \bar{X}^2} \quad \text{and} \quad s_y = \sqrt{\frac{\Sigma Y^2}{N} - \bar{Y}^2}$$

By analogy, the machine formula for the covariance of X and Y is:

$$\frac{\Sigma(X - \bar{X})(Y - \bar{Y})}{N} = \frac{\Sigma XY}{N} - \bar{X}\bar{Y}$$

Now the correlation coefficient is written as:

$$r = \frac{\dfrac{\Sigma XY}{N} - \bar{X}\bar{Y}}{\sqrt{\dfrac{\Sigma X^2}{N} - \bar{X}^2}\ \sqrt{\dfrac{\Sigma Y^2}{N} - \bar{Y}^2}}$$

This naturally leads to the simplified expression:

$$r = \frac{\Sigma XY - N\bar{X}\bar{Y}}{\sqrt{\Sigma X^2 - N\bar{X}^2}\ \sqrt{\Sigma Y^2 - N\bar{Y}^2}} \tag{13.3}$$

The procedure for calculating r by the machine method is summarized in Table 13.3. Note that the value of the correlation coefficient is exactly the same as before. All components in the formula are already familiar to you, except for the expression ΣXY. We get this value by multiplying each X by the Y value paired with it, and then summing all individual cross products.

13.3.4 Caution: On Interpreting Pearson r

There are several other things we need to note about Pearson r and its interpretation. For instance, we might at first be tempted to conclude that a very low correlation indicates little or no relationship between the two variables under study. If the Pearson r were capable of measuring every type of bivariate relationship, this would be true. But it only reflects the straight-line (or *linear*) relationship for a set of data. That is, it measures the degree to which a straight line relating X and Y can summarize the trend in a scatter plot. Many times, however, the relationship is nonlinear (or *curvilinear*) and takes on any one of

Table 13.3.   Computational Procedures for Pearson r, Using Machine Method

Individual	X	X²	Y	Y²	XY
A	2	4	4	16	8
B	4	16	8	64	32
C	5	25	10	100	50
D	7	49	12	144	84
E	8	64	16	256	128
F	9	81	20	400	180
G	6	36	10	100	60
H	5	25	10	100	50
I	6	36	12	144	72
J	8	64	18	324	144
		$\Sigma X^2 = 400$		$\Sigma Y^2 = 1{,}648$	$\Sigma XY = 808$

$\bar{X} = 6$
$\bar{Y} = 12$

$$r = \frac{\Sigma XY - N\bar{X}\bar{Y}}{\sqrt{\Sigma X^2 - N\bar{X}^2}\ \sqrt{\Sigma Y^2 - N\bar{Y}^2}} = \frac{808 - 10(6)(12)}{\sqrt{400 - 10(6)^2}\ \sqrt{1{,}648 - 10(12)^2}}$$

$$= \frac{88}{\sqrt{40}\sqrt{208}} = \frac{88}{91.21} = 0.96$$

countless curve patterns. In these instances, the Pearson r is not the appropriate measure for gauging the relationship between variables. Thus if we fail to find evidence of a relationship, it may be either because the variables are, in fact, unrelated, or because of the degree of curvilinearity in the set of data. Why the first possibility might occur goes without saying. But what about the second possibility? Let's use Figure 13.2 to illustrate. We draw two scatter plots that show different degrees of curvilinearity. Now let's see how each departure

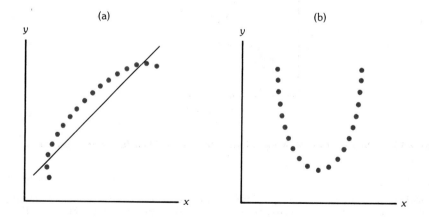

Figure 13.2.   Scatter Plots of Curvilinear Relationships

from linearity affects not only the size of the correlation coefficient but also our ability to measure the relationship with Pearson r

A glance at scatter plot (a) tells us immediately that even though these data are best described by a curved line, a straight line does come close to many points. In this case, where the curve pattern comes close to linearity, Pearson *r* will be somewhat useful for summarizing a curvilinear relationship. Scatter plot (b), in contrast, represents a curvilinear relationship so different from linearity that the computed *r* would be close to (if not actually) zero. So when computing *r*, we must always remember that the Pearson *r* applies only to patterns that are linear or reasonably close to being linear. But we must also remember that a low linear relationship does not necessarily imply a low curvilinear relationship, just as an *r* of zero does not prevent a high curvilinear relationship.*

Another situation that's likely to cause a low correlation occurs whenever we *truncate* the range of values of one or both of the variables under study. Consider the relationship between the number of years of school completed by the father (*X*) and the number of years of school completed by his child (*Y*). Suppose that we restrict the range of the *X* variable from thirteen years to seventeen years of education (that is, the college years). Let's look at Figure 13.3 to see the effect of this restriction on the correlation coefficient. Note that the inset in Figure 13.3 represents a very low correlation, whereas the overall relationship is rather high. One of the reasons for the low correlation is that a college education is more common today than it was twenty or so years ago.

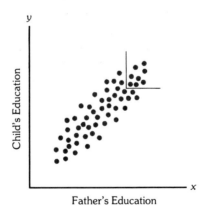

Figure 13.3. Scatter Plot Showing a Low Correlation When Range Is Restricted

---

*Statistical methods beyond the scope of this text are needed to find the degree of curvilinearity between variables. See, for example, Hays (1973: 675–698) and Loether and McTavish (1976: 246–251) for a good discussion on curvilinear techniques as they apply to correlational analysis.

Similarly, the scores for the CEEBs might produce a correlation of 0.75 with cumulative college averages at the end of the freshman year. But if we continue to compute r at the end of the sophomore, junior, and senior years, we might find a great reduction in the coefficient. The reason for the smaller r might be that students with lower CEEBs are more apt to drop out after an unsuccessful freshman year, sophomore year, and so on; naturally, the range of the scores decreases as these lower scores are removed from the computation of r. A declining r is bound to occur whenever we consider only a portion of a range of values for two variables. Although this is not an uncommon phenomenon in social science research, we should be aware of its cause.

To avoid misinterpreting the Pearson product-moment correlation coefficient (which we would do if we tried to apply it to data that are obviously not linear in their relationship with each other), we recommend that you construct a scatter plot of the data before you do the calculation. This will enable you to "eyeball" your data, which will give you an indication of whether a linear technique like Pearson r is appropriate or not.

13.3.5 Testing the Significance of Pearson r

So far, we have treated r as a descriptive statistic. This is a useful approach, to be sure, and often we may wish to know the degree of association in a set of data without needing to make inferences about the larger population. But many times we wish to draw conclusions about a population from the sample statistics; and at such times, we need to remember that a correlation coefficient based on sample data is only an estimate of the corresponding population parameter. Thus, any random sample drawn from a population in which the true linear correlation is zero may very possibly yield a high positive or negative correlation, simply by chance. So before we make any decision about the population parameter, we should consider that there may be no linear correlation at all—or, in other words, that the population correlation (ρ) is zero.

In order to test the significance of a measure of correlation, we adopt the null hypothesis that no linear correlation exists. Usually, our research hypothesis is nondirectional. Thus:

$$H_0: \rho = 0$$
$$H_1: \rho \neq 0$$

As we did in earlier chapters, we test the null hypothesis by selecting a significance level, such as 0.05 or 0.01, before the experiment. At any given level, then, the value of r required for significance turns out to be the same for all samples with the same degrees of freedom. The df is given by $N - 2$. Table G in the Appendix lists the critical values for the 0.05 and 0.01 alpha levels.

Whenever the calculated r is equal to or greater in absolute size than the tabled value, we can reject the null hypothesis of no difference and accept its alterna tive; that is, $\rho \neq 0$.

13.4 The Spearman Rank-Order Correlation Coefficient

Now that we have presented a correlational technique to use with interval-level data, we turn to the problem of finding the degrees of association for ordinal-level data—that is, data that have been rank-ordered with respect to the presence of a given characteristic.

Suppose that we want to find out whether people marry within their same socioeconomic class. One way to do this is to measure the strength of relationship between the socioeconomic status of the husband's family and the wife's family. Suppose that ten couples are ranked (from 1 to 10) on their scores on some socioeconomic status scale (Table 13.4). If all the couples in our sample have the same socioeconomic rank, the difference between their respective ranks will be zero, and there will be a perfect positive association between the rankings for the two variables (husband and wife). In this case, the correlation coefficient should register 1.00. A correlation of -1.00 will again indicate a perfect relationship between the ranks, but here the relationship will be negative. For example, if a rank of 1 on the husband's socioeconomic status is associated with a rank of 10 for the wife's socioeconomic status, a rank of 2 with a 9, . . . , a rank of 10 with a 1, the correlation coefficient should register a -1.00.

Table 13.4. Ranks for Ten Couples on Family's Socioeconomic Status

Couple	Socioeconomic Status Husband's Family (X)	Socioeconomic Status Wife's Family (Y)
A	1	3
B	2	2
C	3	5
D	4	4
E	5	1
F	6	6
G	7	8
H	8	7
I	9	10
J	10	9

Psychologist Charles Spearman (1863–1945) discussed a measure of the strength of relationship between ranks in his paper, "The Proof and Measurement of Association between Two Things." It is ironic that even though it was Galton who developed the measure when he dealt with rank-ordering peas, and Pearson who developed the present-day formula for the measure, it is Spearman whose name is borne by the rank-order correlation coefficient. In any case, Spearman r_s is merely a variant of the product-moment correlation coefficient.* It is based on the difference in rank of each pair, and is written as:

$$r_s = 1 - \frac{6\Sigma D^2}{N(N^2 - 1)} \qquad\qquad \textbf{(13.4)}$$

where:

D = the difference between pairs of ranks, and

N = the total number of pairs.

Before we compute the rank-order correlation coefficient, we need two additional columns representing the differences and squares of differences in pairs of ranks. These columns are given in Table 13.5. For our present example, then, the computed value of r_s is 0.83.

Table 13.5. Computational Procedures for Calculating r_s from Ranked Data

Couple	Socioeconomic Status Husband's Family (X)	Socioeconomic Status Wife's Family (Y)	$(X - Y)$ D	D^2
A	1	3	−2	4
B	2	2	0	0
C	3	5	−2	4
D	4	4	0	0
E	5	1	4	16
F	6	6	0	0
G	7	8	−1	1
H	8	7	1	1
I	9	10	−1	1
J	10	9	1	1
				$\Sigma D^2 = 28$

$$r_s = 1 - \frac{6\Sigma D^2}{N(N^2 - 1)}$$

$$= 1 - \frac{6(28)}{10(99)}$$

$$= 1 - \frac{168}{990}$$

$$= 1 - 0.17$$

$$= 0.83$$

*The rank-order correlation coefficient is also identified as Spearman r_ρ. The subscript ρ, however, is usually reserved for the population value of the coefficient, and r_s is usually reserved for the value of the coefficient computed from a sample.

13.4.1 The Case of Tied Ranks

Occasionally, we cannot distinguish between two scores. When this happens, we declare that the scores are tied at some rank, and we usually assign the mean value of the tied ranks to each score. Now let's come back to our example. Let's rearrange the scores so that we have some ties in the husband's socioeconomic scores. Table 13.6 shows these rankings. Note that since husband A and husband B have the two highest socioeconomic scores, they are ranked first and second, respectively. However, since both husband C and husband D received scores of 3, they are tied for the third and fourth positions. Likewise, husbands E, F, and G achieved scores of 4, placing them in a three-way tie for fifth, sixth, and seventh positions. Therefore, the position of a score of 3 that has been ranked third and fourth would be assigned the mean rank of $(3 + 4)/2$, or 3.5. Similarly, a score of 4 would be assigned the mean rank of 6. Since the rest of the scores are not tied, they are assigned the remaining ranks.

Table 13.6.   Adjusting Tied Ranks

	A	B	SOCIOECONOMIC STATUS—HUSBAND'S FAMILY					H	I	J
			C	D	E	F	G			
Score	1	2	3	3	4	4	4	5	6	7
Rank	1	2	3.5	3.5	6	6	6	8	9	10

13.4.2 Testing the Significance of Spearman r_s

Now we're ready to deal with whether or not a relationship exists between the two socioeconomic-status variables in the larger population from which we randomly selected our hypothetical data. To test the significance of our computed r_s of 0.83, we simply refer to the table of critical values of r_s (Table H), where we find the significance values of the rank-order correlation coefficient for the various levels of significance. In the present case, r_s must equal or exceed the tabled value of 0.648 and 0.794 to be significant at 0.05 and 0.01, respectively, assuming a two-tailed test. Therefore, we reject the null hypothesis that $r_p = 0$ and accept, instead, the research hypothesis ($r_p \neq 0$) that people tend to marry within their family's socioeconomic class in the population from which our sample was drawn.

Where We Stand

This chapter covered one of the most interesting and important topics in statistics—namely, correlation. Many, many studies using correlational analysis exist in the professional literature of various social science disciplines. Correlation, which is designed to find out the strength of relationship between two variables, is not only interesting but also basic to making meaningful statements about social phenomena.

However, computing correlation coefficients constitutes only one part of correlational analysis. Often we want to know how our knowledge of one variable may help us predict what the corresponding values of the other variable should be. To answer this question, let's turn to the complementary technique of simple linear regression.

Terms and Symbols to Remember

Bivariate Analysis
Correlation
Positive Correlation
Negative Correlation
Zero (No) Correlation
Scatter Plot
Horizontal Axis (X-Axis)
Vertical Axis (Y-Axis)
Correlation Coefficient
Pearson r (Product-Moment Correlation Coefficient)
z Score Formula
Mean Deviation Formula
Pearson ρ
Cross Products (Covariance)
Linearity
Curvilinearity
Nonlinearity
Truncated Range
Spearman r_s (Rank-Order Correlation Coefficient)
Spearman r_ρ

Exercises

1. Discuss the magnitude of the correlation you would expect to find between the following pairs.
    (a) IQ and income
    (b) Poverty and crime
    (c) Education and income
    (d) Domestic unrest and international conflict
    (e) Education and prejudice
2. Rank these product-moment correlation coefficients from strongest to weakest association.
    (a) 0.68   (b) −0.30   (c) 0   (d) −0.95   (e) −0.02
    (f) 1.00   (g) −0.74   (h) 0.36   (i) −0.75   (j)   0.75

3. Explain the difference between the following pairs.
   (a) $r - 0.50$ and $r - 0.85$
   (b) $r = 0.78$ and $r = -0.78$
   (c) $r = 0.09$ and $r = 0.85$

4. Suppose that you consult the *Statistical Abstract of the United States* to find the median per capita personal income, the median educational attainment of the adult population, and the voting turnout for the 1960 presidential election for ten southern states.

State	Income	Vote Turnout	Education
Alabama	1,590	35.1	9.1
Arkansas	1,510	50.7	8.9
Florida	2,100	55.0	10.9
Georgia	1,900	40.0	9.0
Kentucky	1,850	50.3	8.7
Louisiana	1,900	48.3	8.8
Maryland	2,650	59.0	10.4
Mississippi	1,380	30.0	8.9
North Carolina	1,900	54.0	8.9
South Carolina	1,600	40.6	8.7

   (a) What is the correlation between income and vote turnout?
   (b) What is the correlation between income and education?
   (c) What is the correlation between vote turnout and education?

5. After first making a scatter plot and guessing the value of r, compute the value of r for the following data on combined Scholastic Aptitude Test (SAT) scores and grade-point average.

SAT	1,400	1,350	1,200	1,100	1,000	950	1,050	1,250	1,250	1,450
GPA	3.8	2.8	3.0	2.6	2.4	2.0	2.5	3.1	2.8	2.9

6. For the table below:
   (a) Make a scatter plot of these data.
   (b) Guess the size of the relationship.
   (c) Compute the value of r.
   (d) Convert to ranks and compute r_s.

**Unemployment Rates and Auto Thefts
for Ten Rural Counties**

Percentage Unemployed	Number of Auto Thefts
6.5	25
6.8	28
7.2	35
8.1	40
9.2	52
10.5	48
11.2	40
7.8	30
8.0	45
10.0	50

7. Is the value of the above correlation coefficient unaffected if the variables "unemployment" and "auto thefts" are interchanged?

8. For a sample of 100 respondents, the z score product of two variables is 49. What is the correlation coefficient for these data?

9. If asked to compute a Pearson r for each of the following distributions:

X	Y		X	Y
100	25		25	10
150	50		5	20

should you get $r = 1.00$ and $r = -1.00$, respectively? Explain your answer.

10. How does a departure from linearity affect the magnitude of the Pearson r?

11. Give three examples in which truncating the range would result in converting a rather strong correlation into a weak one.

12. Test at a level of significance of 0.01 whether the following values of r are significant.
    (a)   0.05 and $N = 20$
    (b) -0.80 and $N = 10$
    (c)   0.86 and $N = 5$
    (d)   0.18 and $N = 72$
    (e) -0.51 and $N = 27$

13. How strong a Pearson r is needed for a sample of size 20 so that the variable can be considered correlated at the 0.05 level of significance?

14. How strong a Pearson r is needed for a sample of size 10 in order to justify the claim that the variables are related? Assume the 1 percent level of significance.

15. Assume the following data for five pairs of observations on two ordinally scaled variables: a(1, ?); b(2, ?); c(3, ?); d(4, ?); e(5, ?). Note that the ranks are given for the first variable only. Using this information, determine the ranks for the second variable so that the Spearman rank-order correlation coefficient equals:
    (a)   1.00
    (b)   0.40
    (c)   0.50
    (d) -0.55
    (e) -1.00

16. Rank the following two sets of interval data and find r_s.

Subject	A	B	C	D	E	F	G	H	I	J
Score (X)	98	62	55	55	86	76	52	93	46	85
Score (Y)	70	75	65	75	90	85	80	82	76	53

17. The data listed below reflect an economist's test to see whether there is any relationship between income and the number of times per month a person dines out. For a pretest, he selects a random sample of fifteen people and analyzes his results, using r_s. (Assume that a low rank equals a low score.)

Subject	A	B	C	D	E	F	G	H	I	J	K	L	M	N	O
Income rank	1	2	3	4	5.5	5.5	8.5	8.5	8.5	8.5	12	12	12	14	15
Dining-out rank	1	4	12	3	6	10.5	2	7	10.5	13	5	8	14	15	9

    (a) Compute the correlation coefficient.

    (b) Test the hypothesis of no correlation between income ranking and dining-out ranking. Use $\alpha = 0.01$.

18. Ten universities are selected at random and ranked according to the size of their graduate schools and according to their research accomplishments, as follows:

University	Size	Research
A	1	3
B	7	4
C	3	1
D	4	5
E	2	7
F	9	6
G	5	2
H	6	10
I	7	9
J	10	8

    (a) Compute the correlation coefficient.

    (b) Test the hypothesis of no correlation between size of graduate school and research accomplishment. Use $\alpha = 0.05$.

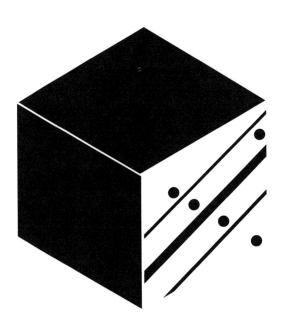

A dog starv'd at his master's gate
predicts the ruin of the state.
William Blake

Regression

14.1 Introduction: Why Study Regression?

As we learned in the last chapter, correlation measures the degree or strength of association between two variables. While this technique can give us useful information, often we wish to find out whether knowing about one variable can help us predict the values of another variable. Consider such questions as whether family income can help predict years of schooling completed, or whether the percentage of blue-collar workers in various precincts can help predict the voting turnout rates in those precincts. If two variables are measured by interval-level data, we have access to the technique of *regression,* which enables us to make predictions from our independent variable (the one used to predict the values of another variable) to our dependent variable (the one whose values we wish to predict).

Regression dates back to the late-nineteenth-century work of Sir Francis Galton (1886) (the same Galton whom we credited earlier with naming the ogive). Like his cousin Charles Darwin, Galton was interested in heredity and sought to predict from the parents' heights the stature of the offspring. After compiling hundreds of measurements, he plotted these data on the familiar X-Y coordinate axis, similar to those we already considered in Chapter 13. When Galton plotted the parent's stature against the stature of the offspring, a marked linear relationship emerged. He called his discovery "regression toward mediocrity." In other words, both the offspring of tall and short parents tended to revert to the average height of the species. Later, statisticians modified Galton's technique to apply to all lines that predicted values of one dependent variable from the knowledge of the independent variable(s).

Today, linear regression analyses fill the professional literature of the social sciences and related fields. Our study of linear regression techniques will help us by further clarifying the concept of correlation. There are also curvilinear regression techniques. In these cases, the relationships between two variables are described by curves or polynomials of the higher orders, whose equations begin with X^2 or X^3. If you're interested in their derivation and usage, you can consult Hays (1973: 675–702). However, for our purposes we will discuss only the linear regression technique whose equation is of the first order—that is, it takes a straight-line form and does not contain expressions with powers like X^2 or X^3 in them. With this in mind, we turn to some concepts that will undoubtedly bring you back to your high school days.

14.2 The Straight-Line Equation

Let's begin our discussion with an example of two variables that are usually related: Scholastic Aptitude Test (SAT) scores and Grade-Point Averages (GPA). In Table 14.1, we have listed the scores on both variables for ten students. Figure 14.1 shows these data graphically. The values for the independent variable (SAT) are plotted along the X-axis, or *abscissa,* and the values for the dependent variable (GPA) are plotted along the *ordinate,* or Y-axis, as they were in the scatter plot for the bivariate relationship. And also as before, each point in the scatter plot represents two values—an individual's SAT score and GPA score.

Table 14.1.   Scholastic Aptitude Test Scores (SATs) and Grade-Point Averages (GPAs) of Ten Students

Student	SATs	GPAs
A	800	2.6
B	900	2.4
C	950	3.3
D	1,000	2.4
E	1,050	3.2
F	1,100	3.1
G	1,150	3.0
H	1,200	2.9
I	1,300	3.6
J	1,400	3.5

As the data in Figure 14.1 suggest, we can see a pattern that can be described by a straight line. But what straight line best describes this particular

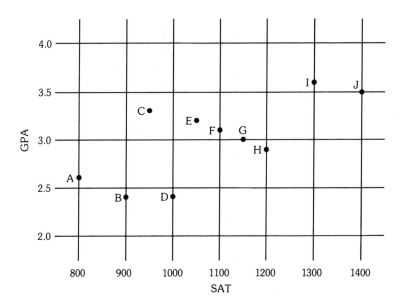

Figure 14.1. Scatter Plot of SAT and GPA Scores for Ten Students

set of scores? In elementary algebra, you learned how to compute the *point-slope formula* for a straight line as:

$$Y = mX + b$$

For our purposes, however, we follow statistical convention and write:

$$Y = a + bX$$

where *a* replaces *b* and *b* replaces *m* in the point-slope formula. The term *a*, called the *Y-intercept*, is the point where the straight line crosses the Y-axis. In other words, *a* is the value *Y* would take when $X = 0$. The other term, *b*, the *slope* of the line, is the ratio between the Y values and the X values; that is, the rate of change in *Y* (symbolized by ΔY, read delta Y) over the rate of change in X (ΔX, read delta X). Thus:

$$b = \frac{\Delta Y}{\Delta X}$$

Since any two points define a straight line, we may determine the rate of change in *X* by computing the value of:

$$\frac{Y_2 - Y_1}{X_2 - X_1}$$

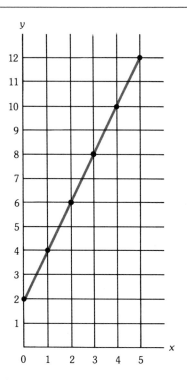

Figure 14.2.   Plotting the Straight-Line Equation Y = 2 + 2X

At this point, a simple illustration might help. Assume that all the points in a scatter plot lie on a straight line described by the equation $Y = 2 + 2X$ (see Figure 14.2). Under this condition, it would be possible to use this straight-line equation, or to read directly off the graph (called *freehand regression*) to predict the exact values of Y from X. In actual practice, however, we rarely if ever obtain perfect correlations. We are much more likely to encounter a Figure 14.1-type situation, in which a straight line describes the pattern but not all points lie on this line. Since, needless to say, our correlation is not perfect in such a situation, we must find the straight line that best fits our data and make predictions from that line. But how do we find the line that best describes the bivariate relationship between two variables when $r \neq \pm 1.00$? This is the topic of the next section.

14.3 The Regression Equation

Since most bivariate relationships are not perfectly linear, we must find the straight line (referred to as the *regression line*) from which we can predict Y from X. Statisticians have decided that this regression line meets what is known as the *least squares* criterion. Suppose that you know nothing except the values of the Y variable, and that you are required to predict a subject's score on the Y variable. What's your best guess? If you suggest the mean score of Y, you're right. In fact, when $r = 0$, our best guess of any given score on a specified variable is the mean of the distribution of that variable. You can see this point easily if you inspect Formula (14.4). When the relationship is not zero, the regression line meeting the least squares criterion will help us do better than the mean of the distribution of that variable as a best guess or predictor. Figure 14.3 illustrates this graphically.

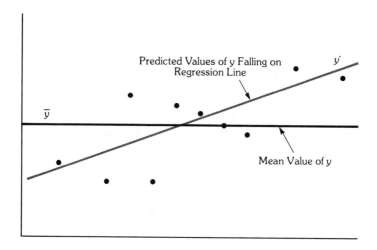

Figure 14.3.   The Regression Line versus the Mean of Y as the Best Predictor of Y

The least squares criterion requires the distance between the actual values for the Y variable and the predicted values for Y (symbolized as Y', read Y prime) to be at a minimum. In other words, the sum of the squared distances between the actual values for the Y variable and the regression line is smaller than the sum of the squared distances between those points and a line drawn to show the mean of Y. This criterion parallels what we discussed earlier; the mean is the point in a distribution that makes the sum of squared deviations from it minimal. This is why some statisticians like to think of the regression line as a sort of *floating mean.*

Using the point-slope formula, we must establish both a and b for the regression equation, subject to the least squares criterion. Using a calculus technique known as partial differentiation (which is beyond the scope of this text), statisticians have demonstrated that the least squares estimator of b reduces to the ratio of the covariance between X and Y to the variance X, which can be written symbolically as:

$$b = \frac{\Sigma(X - \bar{X})(Y - \bar{Y})}{\Sigma(X - \bar{X})^2}$$

This formula is equivalent to the Pearson r, multiplied by the standard deviation of the values for the dependent Y variable, divided by the standard deviation of the values for the independent X variable; or, in symbols:

$$b_y = r \frac{s_y}{s_x} \qquad \qquad \textbf{(14.1)}$$

The constant a or Y-intercept of the regression line is then given by the formula:

$$a = \bar{Y} - b_y\bar{X} \qquad \qquad \textbf{(14.2)}$$

Now we have all the information we need to define the least squares regression equation for a straight line. Symbolically, it is written as:

$$Y' = a + b_yX \qquad \qquad \textbf{(14.3)}$$

If we wish to combine Formulas (14.1) and (14.2) and relate the result to Formula (14.3), we arrive at the alternative formula:

$$Y' = \bar{Y} + r \frac{s_y}{s_x}(X - \bar{X}) \qquad \qquad \textbf{(14.4)}\text{*}$$

To reverse the entire process and predict scores on the X variable from values of the Y variable, we merely alter Formula (14.4), as follows:

$$X' = \bar{X} + r \frac{s_x}{x_y}(Y - \bar{Y})$$

Here, we are performing what is referred to as *regression of X on Y*, instead of the standard Y on X.

---

*If $Y' = a + b_yX$, and if $a = \bar{Y} - b_y\bar{X}$, and if $b_y = r(s_y/s_x)$, then, by simple substitution:

$$Y' = \bar{Y} - r\left(\frac{s_y}{s_x}\right)\bar{X} + r\left(\frac{s_y}{s_x}\right)X$$

or:

$$Y' = \bar{Y} + r\frac{s_y}{s_x}(X - \bar{X})$$

While both of these regression formulas may seem a bit complicated, they are merely variants of the familiar point-slope formula, with the a and b terms meeting the least squares criterion. Using the data from Figure 14.2, we compute the regression equation for predicting scores on the dependent Y variable from values of the independent X variable (see Table 14.2). The coefficient of 0.002 tells us that for every one-unit increase in X (SAT scores), there is a corresponding increase of 0.002 in Y (GPA scores). The fact that this increase is so slight comes as no surprise, since the independent variable is measured on a scale that goes up to 1,600 (double 800s) and since the college index ranges from 0.00 to 4.00. We say that b is sensitive to the scale of measurement. There are statistical procedures that remove the effect of (or standardize) different measurement units. However, in many research situations we may want to know the relationship that exists in the original units of measurement. The b coefficient also may be negative, indicating an inverse or negative relationship between the variables—that is, an additional unit of X is expected to cause a decrease in Y.

Table 14.2. Computation of the Regression Equation

	X	Y	X²	Y²	XY
A	800	2.6	640,000	6.76	2,080
B	900	2.4	810,000	5.76	2,160
C	950	3.3	902,500	10.89	3,135
D	1,000	2.4	1,000,000	5.76	2,400
E	1,050	3.2	1,102,500	10.24	3,360
F	1,100	3.1	1,210,000	9.61	3,410
G	1,150	3.0	1,322,500	9.00	3,450
H	1,200	2.9	1,440,000	8.41	3,480
I	1,300	3.6	1,690,000	12.96	4,680
J	1,400	3.5	1,960,000	12.25	4,900
	$\Sigma X = 10{,}850$	$\Sigma Y = 30$	$\Sigma X^2 = 12{,}077{,}500$	$\Sigma Y^2 = 91.64$	$\Sigma XY = 33{,}055$

$$\bar{X} = \frac{10{,}850}{10} = 1{,}085$$

$$\bar{Y} = \frac{30}{10} = 3$$

$$r = \frac{\Sigma XY - N\bar{X}\bar{Y}}{\sqrt{\Sigma X^2 - N\bar{X}^2}\ \sqrt{\Sigma Y^2 - N\bar{Y}^2}}$$

$$= \frac{33{,}055 - (10)(1{,}085)(3)}{\sqrt{12{,}077{,}500 - 10(1{,}085)^2}\ \sqrt{91.64 - 10(3)^2}} = \frac{505}{(552.49)(1.28)}$$

$$= 0.71$$

$$b_y = r\,\frac{s_y}{s_x} = 0.71\left(\frac{0.40}{174.71}\right) = 0.002 \qquad a = \bar{Y} - b_y\bar{X}$$

$$= 3 - 0.002(1{,}085) = 0.83$$

$$Y' = a + b_y X$$

$$Y' = 0.83 + 0.002X$$

To draw the regression line itself, all we need to do is take two values of X, predict Y from each of these values, and then join these two points on the scatter plot. (If you were actually required to draw a regression line on a scatter plot, it is advisable to pick two rather extreme X values in your data set. It makes the graphics easier.) This is precisely what we did in Figure 14.4 to construct the regression line for the data in Table 14.1.

Using the regression equation (or line), we can predict a person's score on the Y variable by knowing the X variable. Let's make that prediction for Student D, who had a combined SAT score of 1,000. If we substitute Student D's score into the regression equation, we get an index of 2.83. But Student D's actual index was 2.4. Although the regression equation gives us a better guess than that of the mean value of 3.00, nevertheless we have made an error. Is there any way to gauge the prediction errors that are bound to occur when we use regression equations? As you might expect, the answer is yes, and it constitutes our next topic.

14.4 The Standard Error of Estimate

As we've suggested, whenever our observed values do not exactly correspond to our predicted values, there will be error. The smaller the correlation, the greater the scatter about the regression line and the larger the prediction errors,

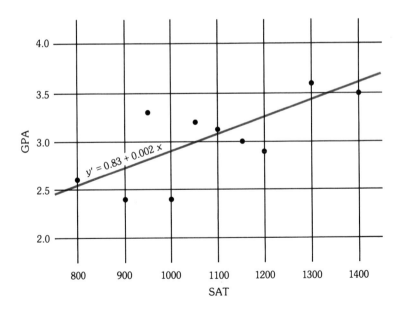

Figure 14.4. The Regression Line

called *residuals*. Conversely, the larger the magnitude of the Pearson r, the tighter the fit about the regression line and the smaller the residuals. Therefore, we can find the goodness-of-fit of the regression line to the data by determining the residual variance. But in order to do that, we must assume that the mean of the residuals measured by $\Sigma(Y - Y')/N$ equals zero, and that the residuals are not related to the values in any systematic way (that is, there is no relationship between the independent variable and the error terms). Given these two basic assumptions, now we may define the *standard error of estimate* — the square root of the residual variance — as:

$$s_{est_y} = \sqrt{\frac{\Sigma(Y - Y')^2}{N}} \tag{14.5}$$

Fortunately, there is a computational method for calculating this standard error. It is:

$$s_{est_y} = s_y \sqrt{1 - r^2} \tag{14.6}$$

As we can see from Formula (14.5), conceptually the standard error of estimate is a measure of dispersion about the regression line. It is like the standard deviation about the mean of a normally distributed distribution. Therefore, we can state that approximately 68 percent of the points in a scatter plot of two normally distributed variables will fall between $\pm 1 s_{est_y}$ from the regression line, approximately 95 percent between $\pm 2 s_{est_y}$, and approximately 99 percent between $\pm 3 s_{est_y}$.

With the aid of the data in Table 14.2, we compute a:

$$s_{est_y} = 0.40 \sqrt{1 - (0.71)^2} = 0.28$$

In Figure 14.5, we have drawn two lines parallel to the regression line for the data in Table 14.2. These lines are both one standard error of estimate (± 0.28) from the regression line. As we can see, five out of ten cases fall between $1 s_{est_y}$. As the N gets larger, however, we should expect our actual percentages falling within the standard error of estimate bands to approximate the theoretical percentages of 68, 95, and 99.

14.5 The Confidence Interval

Assuming that our data are distributed normally, we can set up a *confidence interval* for each prediction, much as we did in Section 10.6. To do this, we need the standard error of estimate that we just calculated, which allows us to

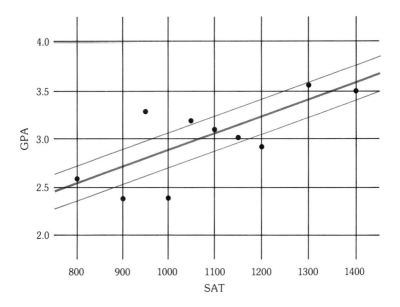

Figure 14.5.   Parallel Lines $1s_{est_y}$ Above and Below Regression Lines

define the upper and lower limits of each interval for each predicted Y value. This procedure is expressed in Formula (14.7):

$$Y' \pm z(s_{est_y}) \qquad\qquad (14.7)$$

To illustrate, let's return to Student D, who has a predicted GPA index of 2.83, based on a SAT score of 1,000. What are the limits of the interval in which we can be 95 percent sure that the predicted Y is contained? Using Formula (14.7), we find that the upper confidence limit is:

$$Y' + z(s_{est_y})$$
$$2.83 + 1.96(0.28)$$
$$3.38$$

Similarly, we find that the lower confidence limit is:

$$Y' - z(s_{est_y})$$
$$2.83 - 1.96(0.28)$$
$$2.28$$

Now that we've established the 95 percent confidence interval as 2.28 and 3.38, we can conclude that 95 percent of the time the predicted Y probably falls within the interval we have created. This time, we leave the computations involved in describing the 90 and 99 percent confidence intervals for you to do, as an exercise.

The preceding confidence interval estimation related to a specific predicted Y value. Some researchers are interested in establishing a confidence range based upon a theoretical average value of Y for given X. This type of interval estimation assumes the existence of a theoretical regression line; our least squares line is a sample estimate of this. There are procedures that consider distribution assumptions required for such interval estimation, but these go beyond the scope of this text. Check Hays (1973: 647–654) if you want to know more about them.

14.6 Regression: A Variance-Apportioning Technique

Regression analysis, like analysis of variance, is a variance-apportioning technique. Here, too, we are concerned with variation in the dependent variable—specifically, with a comparison of the scatter about the mean of Y and the scatter about the regression line. If the regression analysis is meaningful, the scatter of the points about the regression line should total less of a distance than the scatter about a line drawn to depict the mean (see Figure 14.3). In other words, the regression line has reduced the variation. What this tells us is that our knowledge of the independent variable has explained a portion of the variation in the dependent variable. Figure 14.6 graphically depicts these components of variation.

To apportion variance, it is useful to think in terms of the concept of sum of squares. Let's recall that the sum of squares is the sum of the squared deviations about the mean (that is, $\Sigma(Y - \bar{Y})^2$). Since in regression analysis we are essentially comparing the predictions based upon the mean value of Y, as opposed to knowing an independent (or X) variable that would enable us to compute a regression equation, we may think of the squared deviations about the mean as error terms. Specifically, the squared error terms (called *total sum of squares*) divided by the number of cases in the data set (N) constitutes the total variance (that is, $\Sigma(Y - \bar{Y})^2/N$). This variance can be divided into two components: *explained variance* and *unexplained variance*.

Explained variance is defined by the *regression sum of squares* ($\Sigma(Y' - Y)^2$) which is a measure of the sum of the squared distances between each predicted Y value and the mean of Y. When we divide this sum by N, we obtain the explained variance. If we have a meaningful regression, this measure should help to capture and reduce some of the error defined by the total variation. Unless our regression line is perfect (that is, all points fall on it), we should expect some scatter about the line. This is called the unexplained variance,

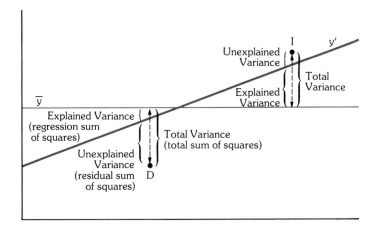

Figure 14.6. Components of Variance About the Regression Line

defined by the *residual sum of squares* $(\Sigma(Y - Y')^2)$—that is, the sum of the total squared deviations from the points in the scatter plot to the regression line. This sum divided by N gives the unexplained variance; in other words, the variation still unaccounted for by the regression line. Thus, the following mathematical expression should be true:

$$\underset{\text{Total Variance}}{\Sigma(Y - \bar{Y})^2/N} = \underset{\text{Explained Variance}}{\Sigma(Y' - \bar{Y})^2/N} + \underset{\text{Unexplained Variance}}{\Sigma(Y - Y')^2/N}$$

Since variance is defined by the respective sums of squares divided by N, we may find it useful, at this point, to work with just the numerators or variation measures—that is, sums of squares. Thus, the following mathematical expression also holds true:

$$\underset{\substack{\text{Total}\\\text{Sum of Squares}}}{\Sigma(Y - \bar{Y})^2} = \underset{\substack{\text{Regression}\\\text{Sum of Squares}}}{\Sigma(Y' - \bar{Y})^2} + \underset{\substack{\text{Residual}\\\text{Sum of Squares}}}{\Sigma(Y - Y')^2}$$

14.6.1 The Coefficient of Determination

An important question still remains: How do we know how much variation we have explained? To answer this, we must look at the *coefficient of determina-*

tion, which is defined as the proportion of explained variance to the total variance. In symbols:

$$r^2 = \frac{\Sigma(Y' - \bar{Y})^2/N}{\Sigma(Y - \bar{Y})^2/N}$$

Simplified, this can be expressed as:

$$r^2 = \frac{\Sigma(Y' - \bar{Y})^2}{\Sigma(Y - \bar{Y})^2} \qquad \textbf{(14.8)}$$

We can see that the coefficient of determination indicates the proportion of total variance that is explained in terms of the size of the Pearson *r*. When $r = 0$, the coefficient of determination (r^2) also equals zero. When $r = 0.50$, the coefficient of determination equals 0.25. In other words, 25 percent of the variance in the dependent variable is explained by the independent variable. Of course, when $r = \pm 1.00$, we have explained all the variance in our dependent variable. Obviously, then, the coefficient of determination gives us a means to measure explained variance.

The square root of the coefficient of determination also gives us another definition of *r,* namely:

$$r = \sqrt{\pm \frac{\Sigma(Y' - \bar{Y})^2}{\Sigma(Y - \bar{Y})^2}}$$

14.6.2 The Coefficient of Alienation

The complement of the coefficient of determination is the *coefficient of alienation,* which is defined as:

$$k^2 = 1 - r^2 \qquad \textbf{(14.9)}$$

Since r^2 tells us how much variation we have explained, $1 - r^2$ tells us how much variance still remains to be explained (perhaps by the introduction of a second independent variable—see Chapter 15). This is why many researchers first calculate the Pearson *r* to determine the degree or strength of relationship. Then, depending upon the research question(s), if this relationship seems interesting enough, they calculate the regression equation. For our current example, the r^2 is 0.50. This means that 50 percent of the variance is explained and that $1 - r^2$, or 50 percent, remains unexplained by the independent variable (SAT scores).

When we are dealing with sample data, we often want to know if our regression analysis can be construed as a reasonable estimate of the corresponding population parameter. This obviously means we have to test the significance of r^2. The concepts we discuss in this chapter, as well as the discussion on degrees of freedom we had earlier, enable us to perform the F test so that we can find out the significance of our coefficient of determination.

14.7 The *F* Test

Since regression analysis is a variance-apportioning technique, we can call once more on the F distribution to test the null hypothesis that both variables are unrelated in the underlying population. In effect, this says that there is no significant linear relationship in the population between the independent and dependent variables.

To perform the F test, we need to obtain the ratio between the explained and unexplained variances. The former is given by the coefficient of determination, and the latter by the coefficient of alienation. All that remains is to determine the degrees of freedom.

You'll recall that we compute the degrees of freedom by subtracting the number of independent pieces of information that restrict the results from the sample size. In order to determine the unexplained variance in the bivariate case, we had to estimate two parameters, a and b, which enabled us to draw our least squares regression line. Therefore, the degrees of freedom associated with the unexplained variation for bivariate regression analysis is $N - 2$. For explained variance, the degrees of freedom is always 1, since we have computed the sum of squares. We already know the sum of squares for both total and the unexplained variance, so all we have to do is calculate the regression sum of squares. Then we use the following formula to test the significance of r^2:

$$F_{1,N-2} = \frac{r^2(N-2)}{(1-r^2)} \tag{14.10}$$

Suppose that we use the 0.05 alpha level and compute the F ratio for our current SAT-GPA problem:

$$F_{1,8} = \frac{(0.50)(10-2)}{(1-0.50)} = 8$$

To reject the null hypothesis at the 0.05 level of significance with 1 and 8 degrees of freedom, our computed F value must equal or exceed 5.32. Since our F of 8 exceeds the established critical value, we can reject the null hypothesis and conclude that the populations from which our dependent and independent variables are drawn differ significantly, and that the relationship yielded by our regression analysis could not have happened by chance alone. (If we had selected the 0.01 level, our conclusion would have been the opposite. Once more, we see the value of selecting an alpha level before proceeding with a test of significance.)

14.8 Caution: On the Misuse of Correlation and Regression

Since the social sciences use both correlation and regression so much, we need to understand the limitations of the techniques and to interpret the results carefully. We repeat that the least squares method of regression, like the Pearson product-moment correlation coefficient, is a linear model. A straight line may represent many relationships, but not all of them. Take age and voting turnout, for example. Survey data reveal that middle-aged voters go to the polls on election day at higher rates than do any other age bracket. This type of bivariate relationship is best described by a curve rather than a straight line. Curvilinear regression techniques exist; if you are interested, see Hays (1973: 675–702). Fortunately, the linear model describes many of the empirical situations we are likely to encounter as social scientists. But the real world is not necessarily linear.

We want to emphasize that the degree of association cannot prove that one variable is the cause and that the other variable is the effect. Indeed, for centuries humans have speculated and debated about the meaning of cause and effect, and about whether experimental methods can ever yield knowledge of such a relationship. If one variable preceded the other in time, we may be in a better position to infer cause; however, a strong correlation isn't enough to prove this assertion. Although the strength of the relationship between variables is necessary, it is hardly a sufficient condition for claiming causality.

The fact that correlation is not the same as causality is illustrated by the relationship that's significant statistically but may have no meaningful relationship in nature, aptly named the *nonsense correlation* by Yule and Kendall (1950: 315–316). These British statisticians pointed to the strong statistical associations between the number of radios and the incidence of mental defectiveness in the United Kingdom during the period from 1924 to 1937. Other more humorous examples of the nonsense correlation include the strong positive association between sunspots and the increase of alcoholism, as well as the growth of grass in Ireland and the incidence of world-wide domestic turmoil.

A more serious concern for the social scientist is the *spurious correlation*. In this case, an observed relationship between two variables is really the product of a common association with a third variable. Consider the curious relationship between dollar loss from riots and whether or not policemen happen to be at the scene of the riot. It just so happens that the association is positive—strong, in fact—thus indicating to some people that policemen are actually increasing the dollar loss, rather than reducing it. But if we could somehow control for the threat involved by the riot, then we could explain away the seemingly embarrassing correlation. So if we assumed that the more explosive the situation underlying the riot, the larger the police turnout, we would observe an overall association simply because policemen and riots are related to the initial threat of a serious riot. This competing model better relates each of the original variables to the independent variable (initial threat), rather than to each other. There are techniques that allow us to deal with the more-than-two-variable case, and *multivariate analysis* is the topic of the next chapter.

We need to discuss just one more pitfall that can happen when we apply regression analysis—making predictions of values of the dependent variable, based on values of the independent variable that are outside the range of the observed data. Such predictions are commonly called *extrapolations*. Suppose, for example, that we compute the regression equation for families whose annual income ranges from $5,000 to $15,000. We'd be foolish to make a prediction of luxury expenditures for a family with an annual income of, say, $100,000 for the independent variable. Such an assumption might prove entirely invalid. Consider a regression with a negative slope, representing the number of mistakes made typing (Y) with the number of weeks of practice (X). An extrapolation for a large enough number of weeks of practice could produce a negative value for the number of typographical errors, which is an impossible result. Of course, there are instances in which a situation requires an estimate, and it is either impractical or impossible to obtain additional data. In this case, we do have to use extrapolations, but we should note the limitations and risks involved.

Where We Stand

In this chapter, we have tried to integrate the concept of correlation with a working knowledge of bivariate linear regression analysis. While simple correlation and regression techniques using two variables may yield useful information and be appropriate for some research questions, social reality is a bit more complex. To get a handle on this complexity, sometimes we have to investigate more than two variables at one time. For example, if we can include a second independent variable in our analysis, we may be able to reduce the proportion of unexplained variance in our dependent variable. In the process, we will better satisfy our concerns. Including this third variable may also allow us to find out whether the observed relationship is valid or spurious.

Multiple regression and correlation offer us methods for including two or more independent variables in the analysis, thereby reducing the unexplained variance in the dependent variable, if possible. Partial correlation, on the other hand, helps us tackle the confusing problem of spuriousness. Multivariate analysis—the simultaneous analysis of three or more variables—is the subject of the next chapter.

Terms and Symbols to Remember

Regression
Abscissa (X-Axis)
Ordinate (Y-Axis)
Point-Slope Formula
Y-Intercept (a)
Slope (b)
Regression Line
Least Squares Criterion
Regression Equation
Freehand Regression
Standard Error of Estimate (s_{est_y})
Floating Mean
Residuals
Confidence Interval
Total Variance (Total Sum of Squares)
Explained Variance (Regression Sum of Squares)
Unexplained Variance (Residual Sum of Squares)
Coefficient of Determination (r^2)
Coefficient of Alienation (k^2)
F Test
Nonsense Correlation
Spurious Correlation
Multivariate Analysis
Extrapolations

Exercises

1. Following are data on the mean number of papers published per year (X) and the number of office hours per week devoted to students (Y) for five college professors.
    (a) Compute the regression equation.
    (b) Draw the scatter plot and graph the regression line.
    (c) Find the amount of explained variance.
    (d) Find the proportion of unexplained variance.
    (e) Compute an F test (use $\alpha = 0.05$).

Professor	X	Y
A	1	3
B	2	4
C	3	2
D	4	2
E	5	4

2. Following are data on the number of friends within the academic community (X) and the number of friends outside the academic community (Y) for five graduate students at a major university. Assume interval-level measurement and compute:
   (a) The regression equation, and draw the regression line.
   (b) Y when X = 2.
   (c) Y when X = 3.
   (d) The coefficients of determination and alienation.
   (e) An F test (use $\alpha = 0.01$).

Student	X	Y
A	1	2
B	2	1
C	3	3
D	4	5
E	5	4

3. Consider the following regression equation, which represents the relationship between IQ (X) and income (Y):

$$Y' = 10{,}250 + 0.08X \qquad \text{where } r = 0.11$$

   (a) Explain the meaning of the equation as a whole.
   (b) Explain each lettered symbol and number.
   (c) Explain the value of the correlation coefficient.

4. Using the data of Chapter 13, Exercise 4:
   (a) Find the equation of the regression line Y (vote turnout) on X (income).
   (b) Find the equation of the regression line y (vote turnout) of X (education).
   (c) Which variable is a better predictor of vote turnout? Explain.

5. Suppose that the correlation between socioeconomic status and television consumption is rather high; specifically, $r = -0.78$. Analyze the meaning of this correlation.

6. Social scientists have tried to determine the impact of a host of social and political variables on the economy of a nation. Given the data below, show which of the independent variables is more closely related to unemployment (base your answer on the amount of variance in Y explained by X).

Nation	Unemployment	Rate of Urbanization	Illiteracy
Australia	4.2	0.7	1.0
Belgium	4.0	0.9	3.0
Brazil	16.2	0.3	39.0
Canada	5.9	1.4	3.0
England	7.5	1.2	0.1
Finland	3.9	0.9	0.2
Greece	10.8	0.8	17.0
New Zealand	4.0	0.9	1.0
Norway	4.0	1.5	0.1
Tunisia	12.0	0.9	61.0

7. Suppose that you consult an almanac to find the annual education expenditure per pupil (X) and the percentage of Selective Service draftees failing mental tests (Y) for 1970 for ten states.

State	X	Y
Arkansas	98	17.5
California	112	7.5
Idaho	138	6.6
Illinois	83	12.0
Kansas	120	6.0
Maine	85	10.0
Michigan	146	• 6.7
Montana	179	3.3
New Jersey	61	13.0
North Dakota	118	5.6

(a) Compute the regression equation.
(b) Compute the coefficients of determination and alienation.
(c) Might not the relationship between X and Y be spurious? Explain.

8. Can we say that an $r^2 = 0.50$ is twice as strong as an $r^2 = 0.25$? Explain.

9. Suppose that 100 families are selected at random from a list of families earning yearly incomes of $15,000 through $40,000. Assume that the regression line for predicting the yearly family expenditure on education by means of yearly income is $Y' = 400 + 0.25X$.
(a) Compute Y when X is $20,000.
(b) Compute Y when X is $30,000.

(c) Should you hesitate to use the regression equation for an income of $8,100? Explain.

10. A study concerned with the relationship between two variables (X and Y) produces the following results:

$$\bar{X} = 100, \ s_x = 10$$
$$\bar{Y} = 50, \ s_y = 4$$
$$r = 0.60, \ N = 50$$

(a) What is the value of Y when X = 120?
(b) What is the standard error of estimate of Y?
(c) What is the value of X when Y = 6.5?
(d) What is the standard error of estimate of X?
(e) Is r^2 significant at the 0.05 alpha level?

11. Following are hypothetical data on the number of male and female police employees for ten cities in New York State. Assume interval-level measurement, decide which is the Y and which is the X variable, and:
(a) Compute the equation of the regression line of Y on X.
(b) Compute the coefficients of determination and alienation, and interpret.
(c) Compute the standard error of estimate, and interpret.
(d) Predict the value of Y when X = 5, and compute the 95 percent confidence interval.
(e) Predict the value of Y when X = 7, and compute the 99 percent confidence interval.

City	Male	Female
Ashland	4	4
Canton	9	1
Dansville	10	2
Highland Falls	10	3
Larchmont	28	7
Minoa	3	0
Potsdam	15	6
Sloatsburg	1	1
Tuxedo	6	2
Woodbury	8	3

12. Using the data of Chapter 13, Exercise 5:
(a) Find the equation of the regression line Y (GPA) on X (SAT).
(b) Predict the value of Y when X = 975.
(c) Compute the standard error of estimate for Y.
(d) Predict Y when X = 1,150 and construct the 95 percent confidence interval.
(e) Test the significance of r^2 (use $\alpha = 0.01$).

13. Suppose that the slope of a regression equation of income on years of education is 2,000. Suppose further that the Pearson r is 0.40.

(a) On the average, how much will a person's income increase from completing college as opposed to from completing high school?

(b) What percentage of the variance in income is explained by years of education?

14. Demonstrate that $Y' = a$ when r is zero.

15. Give several illustrations of the following:

(a) Nonsense correlation.

(b) Spurious correlation.

(c) Extrapolations.

CHAPTER

15

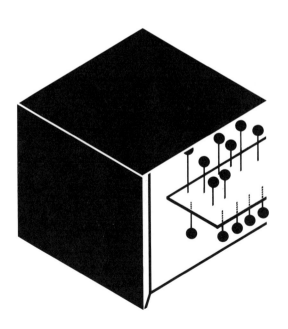

And finds, with keen, discriminating sight, black's not
black; nor white so very white.
George Canning

Multivariate Analysis

15.1 Introduction: Why Study Multivariate Analysis?

It's a complex world, statistically speaking, and although bivariate analysis is a necessary starting point for social research, it is limited by itself. We often have good reason to believe that another variable is influencing the original bivariate relationship. For the sake of illustration, let's assume that school board expenditures are found to have a high positive correlation with student performance. Let's assume further that family income is thought to affect student performance, as well. What might further analysis show? For one thing, it could reveal that school boards within more affluent communities allocate more money per pupil, which in turn affects student performance. Or we could find that the original relationship is spurious. Whatever the result, first we must be able to determine the direct impact that the two independent variables have on the dependent variable, and to remove the effect of the second independent variable. As we shall see in this chapter, partial correlation gives us the means for interpreting the impact that an additional independent variable has on the simple bivariate case.

    Let's take another example. Suppose that we are convinced that two or more variables have a considerable and independent impact upon the dependent variable. Let's make school board expenditures the dependent variable. Now, suppose that we wish to demonstrate the relative impact of family income and community participation in educational affairs upon school expenditures. We need to use multiple correlation in order to find the amount of linear statistical association between a dependent variable and two independent variables. In contrast, multiple regression will tell us how much better we do in explaining the variance in the dependent variable, using this additional independent variable.

Multivariate analysis techniques permit us to explore the multifaceted, complex reality that makes up our daily lives. As social scientists, we have very little control over phenomena; fortunately, however, the statistical procedures described in this chapter help us come to grips with this problem. Therefore, let's begin our discussion with the concept of control.

15.2 The Concept of Control

In numerous experimental settings, one group of subjects receives a treatment (such as a drug) and the control group receives a placebo. In this way, the scientist can tell whether the treatment affects the experimental group in the hypothesized manner. (Please recall our discussion of experimental and control groups in Section 11.4.2.) While this type of research design may be appropriate for the natural sciences, in general its applicability to most social science research situations is limited. Social phenomena are not easy to manipulate; most often, the social scientist does not have a controlled research situation in which to conduct experiments. Therefore, he or she can attempt to make sense of observations only after the events have occurred. Because the social researcher cannot conduct laboratory-type experiments, he or she must impose *physical controls* by using a series of cross-tabulations that break up the original bivariate relationship, according to the intervals or categories of the control variable. This allows for subgroup comparisons, taking into account any effect the control variable might have on the original relationship.

An illustration will help clarify the point. Let's come back to the school expenditure-student performance relationship. Suppose that a state education department wishes to determine whether school expenditures per pupil affect student performance, measured by high school seniors' reading level. To make this determination, the department surveys the state's 200 school districts and gathers data on the per capita student expenditure for high school students and the median high school seniors' reading level for each district (see Table 15.1).

Table 15.1. School Expenditures Per Pupil and Reading Levels for (N = 200) 200 Districts

School Expenditures	READING LEVEL Below 9.0	9.0–12.0	Above 12.0
Below $1000	30	10	10
$1000 to $2000	30	50	20
Above $2000	0	20	30

These data suggest that school districts spending relatively large amounts tend to have a higher median reading level than those spending lesser amounts. However, it can be argued that students from homes with reasonably high incomes are impressed with cultural values that stress a higher level of academic achievement. In addition, it can be argued that school districts with higher median incomes spend more on their schools. So let's further categorize these data to control for income. As we can see in Table 15.2, we break these data down into three separate cross-tabulations; each of these considers a particular category of the control variable (family income).

Table 15.2.   School Expenditure Per Pupil and Reading Levels for 200 Districts, Controlling for Median Family Income

(a) Median Family Income Above $20,000

School Expenditures	READING LEVEL		
	Below 9.0	9.0–12.0	Above 12.0
Below $1000	0	0	0
$1000 to $2000	0	10	10
Above $2000	0	10	20

(b) Median Family Income $10,000–$20,000

School Expenditures	READING LEVEL		
	Below 9.0	9.0–12.0	Above 12.0
Below $1000	20	10	0
$1000 to $2000	20	30	10
Above $2000	0	0	10

(c) Median Family Income Below $10,000

School Expenditures	READING LEVEL		
	Below 9.0	9.0–12.0	Above 12.0
Below $1000	10	0	10
$1000 to $2000	10	10	0
Above $2000	0	10	0

The use of the control variable bears out our suspicions concerning the effect of family income on the relationship between school expenditures and student performance. Matrix (a) indicates that districts with a higher median income do, in fact, tend to spend more on their schools, and this expenditure is reflected in the reading levels. Even in those school districts where expenditures are in the high and medium bracket but where the median family income is lower than in those districts represented by matrix (a), the median reading levels tend to be lower than those in higher-income districts. Clearly, family income has an effect on student performance.*

As we've just demonstrated, analyzing subtables can give us important insights into the type of effects the control variable has on the original relationship. While physical controls through the use of *contingency tables*† are useful and quite widespread in empirical research, they have two major drawbacks. First, the number of subtables that must be constructed to take the control variable into account greatly reduces the number of cases. Consider our current example, where we subdivide 200 school districts into three much smaller tables, each of which has some blank cells. Therefore, we must avoid constructing subtables that become statistically uninteresting—that is, that have too few cases. A second major drawback of contingency table analysis is the loss of information. In this respect, consider the loss of precision involved in placing a school district that spent $1,600 per student into the $1,000–$2,000 category, the school district with a median high school reading score of 10.4 into the 9.0–12.0 category, and the school district with a median income of $15,000 into the $10,000–$20,000 category. Of course, we could have increased the number of categories—but that would have made the table unwieldy and reduced our already diminishing number of cases into more cells, each of which would have a fewer number of cases.

In general, when we have interval-level measurements, we want to use all the information and the precision that this scale affords us. Another consideration, especially when we don't have a great many original cases, is how to assess the impact of the control variable without reducing the number of cases by dividing them into various subsamples. The technique of *partial correlation* allows us to control for other variables without having to deal with the drawbacks of contingency table analysis.

15.3 Partial Correlation: An Introduction

Partial correlation builds upon and utilizes the Pearson product-moment correlation coefficient. Technically, partial correlation can be defined as the product-moment correlation between the regression residuals of two variables,

---

*In passing, we note that the basic unit of analysis in this example is the school district. Therefore, we should take care to avoid the ecological fallacy—that is, we should not attribute the aggregate characteristics of the district (either reading scores or family income) to individuals or individual families who live in these respective school districts.

†A contingency table arrays the various categories of two variables into cells containing the joint frequencies of each category of the two variables. For instance, in Table 15.1 there is a cell defined by school expenditure level of $1,000 to $2,000 and a reading level of 9.0 – 12.0. This cell has an entry of 50, which means that 50 of the 200 school districts share these characteristics. In general, contingency tables are denoted by their number of rows and columns. Table 15.1 is called a 3 × 3 table, since it has three rows and three columns.

each regressed on an additional common variable. Naturally, this means that one variable is controlled statistically. A partial correlation coefficient between X and Y based on the regression residuals of a single additional variable, Z, is symbolized $r_{xy.z}$ (read r sub XY dot Z, or the partial correlation between X and Y controlling for Z). This is known as a *first-order partial correlation* (that is, one variable has been controlled), in contrast to the simple Pearson r, which is by definition a zero-order correlation (that is, no variables are controlled).

Although higher-order partial correlation coefficients, which control for two or more variables, exist, their use generally is limited to the three- or four-control variable case. The goal of scientific research is to use as few explanatory variables as possible to understand phenomena, and it's hard to get more than four conceptually distinct variables for a given research question. As a result, not only is it awkward to use a large number of control variables, but it also may lead to interpretation difficulties. Therefore, partial correlation functions best as an aid to testing well-thought-out research hypotheses, not as a method for isolating relevant variables. Such isolation must occur before we use the multivariate technique of partial correlation.

15.3.1 Four Causal Models

Once we have defined the relevant variables, then we can use partial correlation to test four common causal structures that may be quite plausible. These causal patterns are diagrammed in Figure 15.1. We will limit our discussion to the three-variable case, since it is the easiest to understand and since it provides the basis for more complex models.

The *alternative cause* pattern suggests that any association between X and Z is a result of their mutual association with Y (see Figure 15.1(a)). The plausibility of this arrangement is borne out when the partial correlation $r_{xy.z}$ approximates zero, given the moderately strong zero-order correlations between X and Y and Y and Z. (We use the term "cause" advisedly, here. Partial correlation does not test statistical causality; it merely helps us determine whether the various causal patterns we have in mind as a result of our research are plausible. *Path analysis* is a method that permits the social scientist to use explicit causal assumptions in analyzing data. For a discussion of the technique and its ramifications, see Blalock (1964 and 1969) and Land (1969).)

Partial correlation is used most often to test whether the relationship between two variables is *spurious* (see Figure 15.1 (b)). If it is, then all three sets of zero-order correlations should be reasonably strong, and the first-order partial $r_{xy.z}$ should approximate zero. The same is true of the *intervening variable* pattern described by Figure 15.1 (c). The only difference is that variable X acts through the control variable Z. Unfortunately, partial correlation cannot distinguish the spurious pattern from the intervening one, or from any other pattern, for that matter, because it hardly ever reduces an original bivariate relationship to zero. That is why we use qualifiers such as "approximate(s) zero." The focus

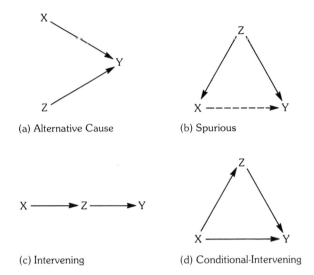

Figure 15.1. Types of Causal Patterns

of the research, the nature of the data, and the conceptualization of the variables all define the criteria by which to interpret the results.

The last causal structure we will consider is the *conditional-intervening* configuration. As Figure 15.1 (d) suggests, X influences Y both directly and through the control variable Z. Here, in contrast to the above two arrangements, controlling for Z should reduce the X-Y relationship, but not enough to yield a zero. That is, $r_{xy.z}$ should be less than r_{xy} but should not approximate zero.

15.3.2 The Partial Correlation Coefficient

At this point we need to understand the logic behind the computational formulas for partial correlation. The generalized formula for a *first-order partial correlation coefficient* is:

$$r_{12.3} = \frac{r_{12} - (r_{13})(r_{32})}{\sqrt{1 - r_{13}^2}\ \sqrt{1 - r_{32}^2}} \tag{15.1}*$$

---

*When you wish to reverse roles of the principal independent and control variables of X_2 and X_3, use Formula (15.1), as follows:

$$r_{13.2} = \frac{r_{13} - (r_{12})(r_{32})}{\sqrt{1 - r_{12}^2}\ \sqrt{1 - r_{32}^2}}$$

where:

$$X_1 = \text{the dependent variable,}$$
$$X_2 = \text{the independent variable, and}$$
$$X_3 = \text{the control variable.}$$

If we study the formula, we can readily see that the numerator removes any common variation the control variable has on the original variables X_1 and X_2, and that the denominator comprises the variation that's left unexplained by the control variable X_3.

The characteristics of the partial correlation coefficient resemble those of the zero-order product-moment correlation. First (and most important to remember), the partial correlation coefficient is a measure of linear association between variables. Thus, a zero coefficient does not necessarily mean that there is no relationship, only that there is no linear relationship. Second, the partial correlation can range from -1.00 (perfect inverse linear association controlling for X_3) through 0 (no linear association) to $+1.00$ (perfect positive linear association controlling for X_3). Third, the squared partial correlation coefficient ($r_{12.3}^2$) is similar to the coefficient of determination (r^2) and may be interpreted similarly. It reflects the proportion of variance explained by the independent variable after the controlled variable(s) has been taken into account. That is, it is the percentage of variance left unexplained by the controlled variable(s), which is accounted for by the independent variable.

We can extend the same logic that we used for first-order partials, now, to control for more than one variable. The only difference is that higher-order partials build upon lower-order partials instead of on the product-moment (zero-order) correlation. To compute a second-order partial, we use first-order partials in the formula:

$$r_{12.34} = \frac{r_{12.3} - (r_{14.3})(r_{24.3})}{\sqrt{1 - r_{14.3}^2} \; \sqrt{1 - r_{24.3}^2}} \tag{15.2}$$

where X_4 represents the second control variable and the other symbols are the same as before. Similarly, to compute a third-order partial, we use second-order partials in the formula:

$$r_{12.345} = \frac{r_{12.34} - (r_{15.34})(r_{25.34})}{\sqrt{1 - r_{15.34}^2} \; \sqrt{1 - r_{25.34}^2}} \tag{15.3}$$

where X_5 represents the third control variable and the other symbols are the same as before.

15.3.3 Partial Correlation: An Example

Now let's calculate a first-order partial correlation coefficient for the school expenditure-student performance problem to illustrate the computational procedure and the process of interpreting the results. As before, let's call the dependent variable (median high school seniors' reading level) X_1, the independent variable (school expenditures) X_2, and the control variable (median family income) X_3. Suppose that after using any one of the three formulas for computing a Pearson r (Chapter 13), we find that $r_{12} = 0.75$, $r_{13} = 0.69$, and $r_{32} = 0.67$. Given the three simple correlation coefficients, we are ready to compute the first-order partial between reading scores and school expenditures, holding median family income constant:

$$r_{12.3} = \frac{r_{12} - (r_{13})(r_{32})}{\sqrt{1 - r_{13}^2}\ \sqrt{1 - r_{32}^2}}$$

$$r_{12.3} = \frac{0.75 - (0.69)(0.67)}{\sqrt{1 - 0.69^2}\ \sqrt{1 - 0.67^2}} = 0.54$$

Let's reexamine the various correlation coefficients and see which causal model they best approximate. A glance at the simple correlations tells us immediately that the strongest relationship is between school expenditures and reading scores. Further inspection reveals that while the correlation between family income and reading scores is reasonably strong, it is not quite as strong as the other. Although the partial reduces the original bivariate relationship, it tells us that the relationship between school expenditures and reading scores is still strong.

All these results point to the conditional-intervening pattern. Family income appears to be a prior variable influencing school expenditures and acting through it to influence reading scores. It also has a secondary influence on school expenditures. It seems plausible to argue that the more affluent school districts spend more on schools, which in turn affects student performance. The correlation also indicates that median family income seems to have an independent effect, and thus that more affluent homes may provide a cultural atmosphere that encourages higher levels of school performance. Squaring the partial tells us that we are doing reasonably well in explaining slightly more than 29 percent of the variance by our independent variable, even after controlling for median family income. (We also may wish to run a partial between reading levels and median family income, holding the effects of school expenditures constant. We left this for you to do as an exercise. Your answer should come out to be approximately 0.39.)

15.3.4 Partial Correlation versus Physical Controls

In the last section, we mentioned that a major drawback of contingency-table analysis is that we must divide the sample into subcategories in order to assess the impact of the control variable. The reverse is true of partial correlation. For example, suppose that higher school expenditures had a dramatic effect on lower-income districts but didn't have very much effect on school districts characterized by higher median incomes. Partial correlation would not reveal this important difference; however, cross-tabulation would. That's why we advise you to know the nature of your data—so that you can use the appropriate technique. As a rule of thumb, though, you may want to use tabular analysis whenever you suspect that the control variable will operate differently for various subcategories of the original bivariate relationship.

The process of controlling for variables, whether by physical controls or by partial correlation, gives us important tools for assessing the nature of a relationship between two variables, independent of extraneous factors. But often we wish to consider the simultaneous effect of two independent variables on the dependent variable. Then we must turn to the related multivariate techniques of regression and correlation.

15.4 Multiple Regression: An Introduction

We have already said that one independent variable may explain a rather considerable amount of variance and yet leave an equal or greater amount unexplained. Because there are no single-factor theories of social phenomena, we have to include at least another independent variable in our regression equation. Our goal is to find the best-fitting plane in three-dimensional space that most aptly describes the data points defined by three variables. Figure 15.2

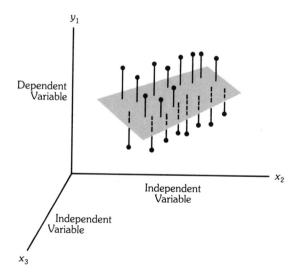

Figure 15.2.   Schematic Representation of Regression Plane in the Three-Variable Case

graphically shows such a plane. Let's start our discussion with the derivation of the *multiple-regression equation;* it describes the regression plane for predicting the dependent variable from knowledge of two independent variables.

15.4.1 The Multiple Regression Equation

We can gauge the simultaneous effects of two independent variables on the dependent variable by constructing a multiple regression equation. There are, of course, formulas that use three or more independent variables, but we would need access to a computer in order to make use of them. Therefore, we won't present complex equations; instead, we'll concentrate on the more manageable three-variable case (one dependent and two independent). This equation appears as Formula (15.4).

$$Y_1 = a_{1.23} + b_{12.3}X_2 + b_{13.2}X_3 \tag{15.4}$$

where Y_1 represents the dependent variable and X_2 and X_3 the independent variables. Here, in contrast to bivariate regression, the b coefficients each represent the slopes of the regression line for each independent variable, controlling for the other. Thus, $b_{12.3}$ reflects the change in Y_1 produced by a unit change in X_2, holding X_3 constant. Of course, the same logic applies to $b_{13.2}$. The term $a_{1.23}$ designates the intercept point on the Y-axis for both independent variables.

    Just as in the two-variable case, in the three variable case the constants of the multiple linear regression equation are subject to the least squares criterion. The respective computational formulas are:

$$b_{12.3} = \frac{b_{12} - (b_{13})(b_{32})}{1 - (b_{23})(b_{32})} \tag{15.5}$$

$$b_{13.2} = \frac{b_{13} - (b_{12})(b_{23})}{1 - (b_{32})(b_{23})} \tag{15.6}$$

$$a_{1.23} = \bar{Y}_1 - b_{12.3}\bar{X}_2 - b_{13.2}\bar{X}_3 \tag{15.7}$$

    Note that the logic behind Formulas (15.5) and (15.6) is somewhat like that behind the partial correlation: we subtract the effect of the variable held constant from variables whose direct impact we are trying to discover. Also note that since the regression coefficient is asymmetrical, the subscripts are not interchangeable; for example, $b_{23} \neq b_{32}$.

15.4.2 Multicollinearity

Formulas (14.3) and (15.4) have a lot in common. They both use the least squares criterion to minimize the average square error terms that result from the discrepancy between predicted and actual values. They also are both computed from data that are assumed to be intervally scaled. But when we use Formula (15.4), we must be careful to make sure that the independent variables are not highly correlated with each other. If they are, what we have is the condition called *multicollinearity*.

The major problem with multicollinearity is that it's hard for us to distinguish the effects of each independent variable on the dependent variable. Since the explanatory variables themselves are highly correlated, we cannot tell which of the variables explains what portion of the variance in the dependent variable. Another result is that the *b* coefficients in the regression equation become unreliable. We'll have more to say about multicollinearity later in this chapter.

15.4.3 Multiple Regression: An Example

To illustrate the computation of the multiple regression equation, we'll try to estimate the simultaneous effect of socioeconomic status (as measured by median family income) and community interest (as measured by voting turnout rates for school board elections) on school expenditures (as measured by per pupil expenditures).

Designating school expenditures (dependent variable) as Y_1, socioeconomic status as X_2, and community interest as X_3, we first establish the *b* coefficients (recall Formula (14.1)) for the following simple bivariate relationships:

$$b_{12} = 0.0662$$
$$b_{13} = 84.88$$
$$b_{32} = 0.0002$$
$$b_{23} = 499.73$$

We then plug these coefficients into Formulas (15.5) and (15.6) to obtain the partial regression coefficients:

$$b_{12.3} = 0.055$$
$$b_{13.2} = 57.55$$

To compute the constant *a*, we substitute $\bar{Y}_1 = 1,475$, $\bar{X}_2 = 15,000$, and $\bar{X}_3 = 10.5$ into Formula (15.7) to compute:

$$a_{1.23} = 40.22$$

With the obtained values of $b_{12.3}$, $b_{13.2}$, and $a_{1.23}$, the multiple regression equation for predicting school expenditures on the basis of socioeconomic status and community interest is:

$$Y_1 = 40.22 + 0.055X_2 + 57.55X_3$$

This equation tells us that we can expect school expenditures to rise a little more than $0.05 for every $1.00 increase in median family income. Likewise, a 1 percent increase in voting turnout theoretically should yield $57.55 in school expenditures. Assuming that a school district has a median family income of $15,000 and that the voting turnout in the last school board election was 10 percent, we would predict a school expenditure of:

$$
\begin{aligned}
Y_1 &= 40.22 + 0.055(15,000) + 57.55(10) \\
&= 40.22 + 825 + 575.50 \\
&= \$1,440.72
\end{aligned}
$$

15.4.4 The Beta Coefficient

The b terms in Formula (15.4) are unstandardized regression coefficients, in their present form. They reflect the net effect of each independent variable on the dependent variable, and in this sense they are easy to interpret. But variables often are measured on different scales. For instance, in our last example the two independent variables are measured by monetary and percentage units. How can we compare their relative impact on school expenditures (the dependent variable)? We must remove the effect of the measurement units; in other words, we must standardize the b coefficients so that we can compare them. We standardize the partial b coefficient by multiplying each coefficient by the ratio of the standard deviation of the independent variable in question with that of the dependent variable. The resulting coefficients are called *beta weights*, symbolized as β. The beta coefficients for independent variables X_2 and X_3 are:

$$\beta_{12.3} = b_{12.3}(s_2/s_1) \tag{15.8}$$
$$\beta_{13.2} = b_{13.2}(s_3/s_1) \tag{15.9}$$

To illustrate the use of beta coefficients, let's suppose that the standard deviation is 665.99 for school expenditure (s_1), 6,724.97 for median family income (s_2), and 4.39 for voting turnout (s_3). Substituting these values into Formulas (15.8) and (15.9) yields the following betas:

$$\beta_{12.3} = 0.055(6,724.97/665.99) = 0.5554$$

and:

$$\beta_{13.2} = 57.55(4.39/665.99) = 0.3794$$

The $\beta_{12.3}$ of 0.5554 tells us that for every increase in one standard deviation in median family income, school expenditures should increase by a 0.5554 standard deviation. Similarly, the $\beta_{13.2}$ of 0.3794 tells us that for one standard deviation in voting turnout, we should expect an increase of a 0.3794 standard deviation in school expenditures. (Of course, beta weights can be negative numbers, as well.) We can gauge the relative impact of the variables from the magnitude of their beta weights. In our present example, median family income has relatively more impact than voting turnout rate—as an inspection of their beta weights makes clear (0.5554 as compared to 0.3794).

A number of warnings are in order when we are comparing beta weights. As we've said before, the multiple regression model assumes that all relevant variables have been included explicitly in the analysis. If there are some possibly relevant unmeasured variables that affect the dependent variable and that correlate with an independent variable included in the analysis, the beta weight for the included independent variable may be inflated. And because the presence of multicollinearity makes it hard to assess the direct impact of each variable, the beta weights are unreliable.

We can really see the problem of multicollinearity and the unreliability of the beta weights when several highly correlated variables tap the same explanatory factor. The presence of several such variables may dilute the individual impact of each of these variables on the dependent variable; and because the single variable that's tapping the different explanatory factor is undiluted, it may have a larger beta weight, even if it doesn't have the strongest impact. Many researchers, suspecting this type of occurrence, drop one or two of the highly correlated variables—or, if possible, combine them into a single index. Knowing your research area and your data well is probably the best way to overcome the effects of multicollinearity on multiple regression analysis.

15.5 Multiple Correlation: An Introduction

While multiple regression is useful for purposes such as assessing the impact of each independent variable on the dependent variable, often we wish to determine the combined effects of all the independent variables acting simultaneously on the dependent variable. In short, we want a summary measure that tells us how well the multiple regression equation describes the relationship. The multiple correlation coefficient gives us this information.

15.5.1 The Multiple Correlation Coefficient

If we wanted to find out the combined influence of two independent variables on a dependent variable, we could add the two zero-order correlations (if we were certain that the two independent variables were not correlated). In other words, if $r_{23} = 0$, then the multiple correlation (R) would equal $\sqrt{r_{12}^2 + r_{13}^2}$. However, the two independent variables are almost always correlated to some extent; that is, $r_{23} \neq 0$. In these situations, we must remove the amount of intercorrelation among the variables so that we can get a measure of each independent variable's unique impact on the dependent variable. We can do this easily enough by computing the *multiple correlation coefficient*.

For the three-variable case, we may compute the multiple R directly from the formula:*

$$R_{1.23} = \sqrt{r_{12}^2 + r_{13.2}^2(1 - r_{12}^2)} \qquad (15.10)$$

where:

$$X_1 = \text{the dependent variable,}$$
$$X_2 = \text{the independent variable, and}$$
$$X_3 = \text{the independent variable.}$$

Note that in the multiple correlation coefficient, the symbol for the dependent variable is placed to the left of the period, whereas all the independent variables are placed to the right. Also note that the direction of the correlations doesn't affect our calculation of the multiple correlation coefficient. If we considered the direction, positive and negative correlations would cancel each other out, thus hiding the scope of the association. Given the absence of negative terms, the range of the multiple R is between 0 (no linear association between the dependent and independent variables) and 1.00 (a perfect linear association between the dependent variable and the designated independent variable).

The formulas for multiple Rs involving more than two independent variables resemble the equation for the three-variable case. To compute a multiple correlation coefficient for three independent variables, we use the R^2 for the three-variable case in the formula:

$$R_{1.234} = \sqrt{R_{1.23}^2 + r_{14.23}^2(1 - R_{1.23}^2)} \qquad (15.11)$$

---

*Of course, it's entirely arbitrary which variable we designate as the first independent variable, as long as we deal consistently with subsequent terms in the formula.

where X_4 represents the third independent variable and the other symbols are the same as before. Similarly, to compute a multiple R for four independent variables, we use the R^2 for the three-variable case in the formula:

$$R_{1.2345} = \sqrt{R^2_{1.234} + r^2_{15.234}(1 - R^2_{1.234})} \qquad \textbf{(15.12)}$$

where X_5 represents the fourth independent variable and the other symbols are the same as before.

These formulas indicate that including additional explanatory variables makes calculation a bit awkward. Not only do these additional variables create computational problems (which a computer can overcome, admittedly), but also there's little to gain by continually adding independent variables, beyond three or four reasonably distinct ones. Once we have controlled for, say, four or five independent variables, a sixth or seventh variable will rarely contribute uniquely to explaining the variance. And we must always watch out for multicollinearity if the correlations between the independent variables are too high. In fact, one of the symptoms of multicollinearity is the situation in which the squared multiple R is barely larger than one of the squared simple correlations (that is, if $R^2_{1.23} = r^2_{12}$ or r^2_{13}). Again, we emphasize that two or three conceptually distinct independent variables are more useful than a great many independent variables in helping us understand the dependent variable.

15.5.2 The Coefficient of Multiple Determination

Here, as we did in the case of simple bivariate regression, we compute a measure of explained variation called the *coefficient of multiple determination* (R^2). In contrast to the simple r^2, the coefficient of multiple determination measures the proportion of variance explained by more than one independent variable. For the three-variable case, the formula is:

$$R^2_{1.23} = r^2_{12} + r^2_{13.2}(1 - r^2_{12}) \qquad \textbf{(15.13)}^*$$

---

*The formula for both R and R^2 can also be written completely in terms of zero-order correlations. For R and the three-variable case, we obtain the alternative formula:

$$R_{1.23} = \sqrt{\frac{r^2_{12} + r^2_{13} - 2r_{12}r_{13}r_{23}}{1 - r^2_{23}}} \qquad \textbf{(15.14)}$$

Removing the radical sign in the above formula gives us the alternative equation for:

$$R^2_{1.23} = \frac{r^2_{12} + r^2_{13} - 2r_{12}r_{13}r_{23}}{1 - r^2_{23}} \qquad \textbf{(15.15)}$$

As we see, the coefficient of multiple determination indicates the proportion of total variation that's explained in terms of the magnitude of the multiple correlation coefficient. For example, when $R = 0.60$, the coefficient of multiple determination is 0.36. In other words, 36 percent of the total variation is accounted for by the two independent variables. (Note that the computation of R^2 is the same as that of the multiple correlation coefficient, except that the radical sign is removed. Thus, computation of one gives the other.)

The important point to remember is that R is a measure of the strength of association between one dependent variable and a linear combination of two (or more) independent variables. R is also a correlation between the dependent variable's actual values and the predicted values obtained by the multiple regression equation. The coefficient of multiple determination (R^2), on the other hand, tells us how much variance our prediction equation has explained. Because both these summary measures give the researcher useful and readily interpretable results, they are used widely in social analysis.

15.5.3 Multiple Correlation: An Example

Let's return to the school expenditure example, since it offers us a way to illustrate the calculation and interpretation of both the multiple R and the related R^2 coefficient. Suppose that we want to determine the linear association between school expenditures (X_1) and the two independent variables, median family income (X_2) and voting turnout rates for school board elections (X_3). Table 15.3 shows the zero-order correlations in matrix form.

Table 15.3   Correlation Matrix for School Expenditures (X_1), Median Family Income (X_2), and Voting Turnout Rates for School Board Elections (X_3)

	X_1	X_2	X_3
X_1	—	0.67	0.56
X_2	—	—	0.33
X_3	—	—	—

We could use the computational Formula (15.14) to compute the multiple correlation coefficient directly from the correlation matrix in Table 15.3. However, for the sake of conceptual clarity, we'll use Formula (15.10). This formula, as you will recall, requires us to compute the partial $r_{12.3}$. Thus, using the zero-order coefficients listed in the table, we find that:

$$r_{13.2} = \frac{r_{13} - (r_{12})(r_{32})}{\sqrt{1 - r_{12}^2}\ \sqrt{1 - r_{32}^2}} = 0.48$$

With this information, we may now proceed to the computation of:

$$R_{1.23} = \sqrt{r_{12}^2 + r_{13.2}^2(1 - r_{12}^2)} = 0.76$$

Thus $R^2 = 0.58$. (When doing hand calculations, researchers often find it easier to compute R^2 first. Do you agree?)

One interpretation of the above finding is that each independent variable is tapping a somewhat different explanatory factor. How can we tell? By the fact that there is an increase in the multiple $R_{1.23}$ of 0.76, as compared with the simple zero-order r_{12} of 0.67. Although the independent variables are correlated to some extent, voting turnout contributes so uniquely to the analysis that its inclusion in the multiple regression equation is justified.

The coefficient of multiple determination tells us that approximately 58 percent of the variation in our dependent variable, school expenditures, is jointly explained by median family income and voting turnout. This is an improvement over the variation that's explained by median family income alone ($r_{12}^2 = 0.45$). Of course, the unexplained variation, or *the coefficient of multiple alienation* (K^2), is $1 - R_{1.23}^2$, which in this case equals approximately 42 percent. Can you think of another independent variable that would be justified in the analysis and would help to reduce the amount of unexplained variance? Remember to consider the effect of intercorrelations among the independent variables that would reduce their unique contribution to the proportion of variance explained; and of course take note of the possible presence of multicollinearity.

15.5.4 A Correction Factor

When we use sample data instead of the entire population, the previous formula may tend to overestimate the multiple correlation coefficient. This bias can occur either when the sample size is relatively small or the number of independent variables is rather large. In these cases, a *correction factor* can compensate for the resulting overestimation of the computed multiple correlation coefficient. The formula for the corrected value of R, which we label R' (read R prime), is:

$$R' = \sqrt{1 - \frac{N - 1}{N - k}(1 - R^2)} \qquad \textbf{(15.16)}$$

where:

$N =$ the sample size,
$k =$ the number of independent variables, and
$R^2 =$ the coefficient computed by either Formula (15.10) or Formula (15.14).

R and R' will reasonably approximate each other, except where the above-mentioned causes for bias are present. More often than not, the correction factor doesn't have to be used.

15.6 Significance Tests

As we know, when we get our data from a sample rather than from the entire population, we deal with estimates rather than with the parameters themselves. Therefore, in these situations, we undoubtedly will want to test for the significance of our correlation coefficients.

The F test, a ratio between explained and unexplained variance, allows us to test the significance of the multiple and partial correlation coefficients. As we learned in Chapter 14, the F test takes both the sample size and the strength of relationship into account. In order to use the F test, we first must determine the degrees of freedom associated with the figures in Table D. When we're testing for the significance of the multiple R, the degrees of freedom associated with the explained variance is always the number of independent variables k. The degrees of freedom associated with the unexplained variance is the sample size minus the number of parameters that must be estimated for the regression equation, which is equivalent to the number of independent variables plus the dependent variable, or $k + 1$. Therefore, the degrees of freedom for the unexplained variance component of the F test is $N - k - 1$. Thus, we write:

$$F_{k,\ N\ -\ k\ -\ 1} = \frac{R^2(N - k - 1)}{(1 - R^2)k} \tag{15.17}$$

Returning to our problem, we find that $R^2 = 0.58$, $N = 200$, and there are two independent variables (median family income and voter turnout rate). Thus our computed F with 2 and 197 degrees of freedom is:

$$F_{2,197} = \frac{(0.58)(200 - 2 - 1)}{(1 - 0.58)2} = 136.02$$

Turning to Table D in the Appendix, we find that the F associated with our degrees of freedom are 3.04 and 4.71, respectively, for the 0.05 and 0.01 significance levels. Since our computed F of 136.02 exceeds both of these tabled values, we may conclude that our multiple R is significant at either level of significance.

The computational formula for the partial r also takes into account the sample size and the proportion of variation explained. The relevant degrees of freedom are 1 and $N - k - 1$, where k equals the number of independent and control variables. The generalized formula for a test of significance for the partial correlation coefficient thus is:

$$F_{1,\ N\ -\ k\ -\ 1} = \frac{r^2_{12.3\ \dots\ N}(N - k - 1)}{1 - r^2_{12.3\ \dots\ N}}$$ **(15.18)**

Using the results of the analysis in Section 15.3.3, which involved one dependent variable (reading scores) and two independent variables (school expenditures and median family income), the F test is:

$$F_{1,197} = \frac{r^2_{12.3}(N - 3)}{1 - r^2_{12.3}} = 81.09$$

Our computed value of 81.09 far exceeds both the 0.05 tabled value of 3.89 and the 0.01 tabled value of 6.76 (at df of 1,197), telling us that our partial correlation coefficient far exceeds the critical values in Table D. This, of course, means that most likely our observed relationship in the sample data did not happen by chance.

Strictly speaking, however, we didn't really have to compute the F value for this problem, since presumably we are dealing with all of the state's school districts; in other words, we have the entire population. Statisticians have debated whether significance tests are warranted in situations like this. Some say that significance tests are inappropriate where all the cases of defined universe or population have been included in the analysis (as in our example). Advocates of significance tests argue that the defined population can be seen as a subuniverse of an even larger population. For example, in our problem, the state's school districts could be seen as a subset of all school districts. This argument implicitly assumes that all school districts are reasonably homogeneous with respect to certain characteristics. As the foregoing discussion shows, a great deal depends upon how we define our universe. If you're interested in reading a discussion of this and other matters relating to significance tests, consult Morrison and Henkel (1970).

15.7 Caution: On the Misuse of Multivariate Analysis

From time to time in the course of this chapter, we've called your attention to possible distortions of multivariate analysis. Now we want to summarize a few of the basic cautionary notes about the use of multivariate regression/correlation, and to add a few others, as well. First, because we've been dealing with the more widely used linear models, we cannot discover nonlinear relationships. While some social scientists use curvilinear techniques, many of them refrain from fitting polynomials to data because it is so hard to give meaning to higher-order equations. One way around this problem is to use logarithmic procedures to transform the data, so that they approximate those appropriate to a linear model. (See Gurr (1972: 108–111), Cohen and Cohen (1975: 244–252 and 261–262), and Tufte (1974: 108–132) for a preliminary discussion of these techniques.) Another alternative is to use cross-tabulation procedures where feasible, such as those discussed in Section 15.2—perhaps in conjunction with statistics (to be discussed in the next chapter) that don't make the strict assumptions of the linear correlational model.

An implicit assumption of multiple correlation and regression models, as well as of their bivariate counterparts, is that the data are measured on the interval scale. But this assumption is often violated, especially in attitudinal research involving election studies, for example. The statistical purist may insist upon interval-level measurements, but the practical researcher will maintain that the "technique is robust." Robustness, in this context, means that the violation of the interval scale assumption does not affect the procedure. In fact, the use of ordinal-type measurements in these procedures tends to understate the statistical relationships among variables. It's quite true that interval measurements yield more accurate estimates; however, in social research, ordinal-type measurements are often the best we can do. Many researchers prefer to use the correlational techniques with this type of data, keeping the warning in mind.

As we know, multicollinearity results when the independent variables are highly correlated. What constitutes high correlations is somewhat relative, since we expect *some* intercorrelations among variables in the social sciences. Nevertheless, when the intercorrelations among the explanatory factors exceed 0.80, we should be aware of possible confusing effects on our partial regression coefficients and beta weights. The best guard against multicollinearity is good theoretical conceptualization. Independent variables should be reasonably distinct if they are to represent different explanatory factors. Variables whose indicators turn out to be more or less linear transformations of each other do not help us theoretically or mathematically. The best defense is to think through the research question and to select the indicators designed to measure the explanatory variables.

One more warning regarding significance tests: sometimes we tend to equate statistical significance with substantive significance. As we realize, statistically significant correlation coefficients are a function of sample size and the computed strength of association. If we experience difficulties in finding quantitative indicators that truly tap the conceptual variable we are trying to measure, we can end up rejecting the research hypothesis. Thus, because we have failed to meet a statistical criterion at the predetermined alpha level, we also can end up concluding that there is no theoretical significance. But the failure is not necessarily in the statistical stage of the analysis; it may lie in the prior theoretical and operational stage. And the reverse is always true; often correlations that meet the statistical criteria for significance do so because of a very large sample size. Therefore, we may accept moderately supported results because they meet a certain significance level, and yet reject a theoretically rich hypothesis because it fails to meet a certain criterion. When interpreting the F ratio for R, R^2, or the partial r, remember not only sample size but also the indicators used to measure your theoretical variables.

Where We Stand

Most of the hypothesis-testing procedures we've discussed up until now involve inferences regarding population parameters, such as the mean. Therefore, we call these tests *parametric,* and their test statistics *parametric statistics.* The sampling distributions for the statistics used in these tests usually depend upon assumptions regarding the populations for which the samples are obtained. The assumption that the population is distributed normally is particularly strict. Equally strict is the necessity for interval measurement. Fortunately, a great many social science data sets either meet or approximate these assumptions sufficiently to warrant the widespread use of parametric statistics such as correlation or regression analysis.

Although slight deviations from normality may be tolerated in using the t and the F tests, gross distortions of normality and the lack of interval measurement are not welcome. Yet they found their way into some social scientists' data sets, especially psychological and sociological research that's primarily concerned with the study of small-group behavior. Statisticians recognized the need for alternatives to the parametric measures, and they developed a host of nonparametric alternatives, beginning in the early years of this century and reaching a period of unusually rapid development in the 1940s and 50s. These days, nonparametrics continue to be a major focus of statistical research. These so-called *sturdy* statistics, to which we now turn, have become part of the standard repertoire of the social science researcher.

Terms and Symbols to Remember

Multivariate Analysis
Physical Control
Contingency Table
Partial Correlation
First-Order Partial Correlation

Causal Patterns: Alternative Cause, Spurious, Intervening,
   Conditional-Intervening
Partial Correlation Coefficient
Multiple Regression
Multiple Regression Equation
Multicollinearity
Beta Coefficient (β Weights)
Multiple Correlation
Multiple Correlation Coefficient
Coefficient of Multiple Determination (R^2)
Coefficient of Multiple Alienation (K^2)
F Test
Correction Factor (R')
Parametric
Parametric Statistics

Exercises

1. Discuss the differences between bivariate and multivariate correlational analysis.

2. A political researcher wishes to find out whether there is a relationship between party competition, as measured on a scale ranging from 0 (no competition) to 100 (maximum competition), and public spending for social well-being, as measured by state expenditures for health care per capita. From the data listed below:
    (a) Construct a contingency table with state expenditures as your row variable (low: below $50; medium: $50–$100; high: above $100) and party competition as your column variable (low: below 30; medium: 30–70; high: above 70).
    (b) What inferences can you draw from the contingency table?
    (c) Some researchers argue that state expenditures for health care is a function of the economic wealth of the state, as measured by its ability to produce tax revenue (state tax burden). Control for the following categories of the state's revenue-producing capabilities: below $250, $250–$500, above $500.
    (d) Does the control variable affect the original relationship between state expenditures and party competition?
    (e) Compute the partial correlation between state expenditures and party competition, controlling for tax burden. Based upon the inspection of the contingency table, are your suspicions confirmed?

State	Health Care	Party Competition	Tax Burden
A	40	75	200
B	45	30	240
C	60	55	280
D	75	20	350
E	110	75	525
F	125	80	510
G	40	70	325
H	80	40	510
I	45	65	480
J	130	50	400
K	42	85	230
L	75	80	420
M	120	35	550
N	125	15	502
O	48	15	245
P	40	20	280
Q	139	60	265
R	160	75	600
S	85	50	500
T	45	35	475

3. Give examples illustrating the alternative cause, spurious, intervening, and conditional-intervening causal patterns.

4. Designating the dependent variable as X_1 and the independent variables as X_2 and X_3, assume the following zero-order correlations: $r_{12} = 0.60$, $r_{13} = 0.50$, $r_{23} = 0.30$. Assuming that X_1 always occurs after X_2 and X_3 and that the time order of X_2 and Y_3 is unclear, which causal pattern(s) make the most sense? Explain your answer.

5. Using the same symbols for the variables as we used in the above exercise, assume that X_2 comes before X_3, which in turn precedes X_1, and that the zero-order correlations are as follows: $r_{12} = -0.55$, $r_{13} = -0.65$, $r_{23} = -0.35$. Select the best causal pattern(s). Explain your answer.

6. Given the following correlation matrix:

	X_1	X_2	X_3
Income (X_1)	—	0.50	0.70
Number of books read (X_2)	—	—	0.80
Education (X_3)	—	—	—

(a) What is the correlation between income and number of books read, controlling for education?

(b) What does this mean?

7. You are doing a study of suicide and hypothesize that, on a cross-national basis, suicide is a function of social disorganization. In addition to social disorganization and suicide, you measure a number of other variables for control purposes. Given the fact that a control variable must be correlated with the dependent variable and the other independent variable, which partial correlation would you compute to test your hypothesis? Suppose that the following matrix of zero-order correlations applied.

	X_1	X_2	X_3	X_4
Suicide (X_1)	—	0.50	0.30	−0.01
Disorganization (X_2)	—	—	0.40	0.03
Per capita income (X_3)	—	—	—	−0.30
Inequality (X_4)	—	—	—	—

8. In a study of 100 cases sampled at random, the correlations among three variables (X_1, X_2, and X_3) were $r_{12} = 0.40$, $r_{13} = 0.50$, and $r_{23} = -0.20$.
    (a) Compute $r_{12.3}$.
    (b) Compute $r_{23.1}$.
    (c) Test the significance of these partials.

9. In a multiple-regression equation, how many dependent variables may there be?
    (a) 0
    (b) 2
    (c) 1
    (d) More than 2

10. In a multiple-regression equation, how many independent variables may there be?
    (a) 2
    (b) 1
    (c) 0
    (d) More than 3

11. Given two simple regression equations:

$$Y'_1 = a + bX_2$$
$$Y'_1 = a + bX_3$$

Describe the circumstances under which both bs, above, will be the same as the multiple regression equation:

$$Y'_1 = a + bX_2 + bX_3$$

12. A criminologist wishes to predict the number of violent crimes. This prediction will be made from a regression equation, using police personnel and population density. For a sample of five states, the following data are obtained:

Violent Crimes (per million population)	Police Personnel (per million population)	Population Density (per square mile)
200	75	114
350	150	378
180	70	450
100	50	88
270	135	392

(a) Compute the regression equation.
(b) Explain, in your own words, the meaning of the coefficients $b_{12.3}$ and $b_{13.2}$.

13. A college admissions officer uses high school GPA (X_2) and CEEB scores (X_3) to predict college GPA (Y_1) and arrives at the regression equation: $Y'_1 = 0.10 + 0.007X_2 + 0.002X_3$. Compute the predicted college GPA for each of the following students.
(a) $X_2 = 85$ and $X_3 = 1,200$
(b) $X_2 = 70$ and $X_3 = 1,050$
(c) $X_2 = 75$ and $X_3 = 1,223$
(d) $X_2 = 83$ and $X_3 = 1,157$
(e) $X_2 = 92$ and $X_3 = 1,000$

14. Using the data of Chapter 13, Exercise 4:
(a) Compute the partial correlation between vote turnout and income, controlling for education.
(b) Compute the partial correlation between vote turnout and education, controlling for income.
(c) Compute the multiple-regression equation for these data, taking vote turnout as the dependent variable. Determine the relative effect of each independent variable on vote turnout by comparing beta coefficients. How does this compare with your previous evaluation of Exercise 4(c), Chapter 14?
(d) What percentage of the variance in vote turnout is explained by the two independent variables, taken together?

15. Given the following unstandardized and standardized regression coefficients for a multiple-regression analysis, compare the effects of independent variables X_2 and X_3. Also, determine which has the greatest independent impact on the dependent Y variable.

Independent Variable	Unstandardized b coefficients	Standardized β coefficients
X_2	$b_2 = 3.4$	$\beta_2 = 2.2$
X_3	$b_3 = 1.7$	$\beta_3 = 2.4$

16. Using the data of Chapter 14, Exercise 6, determine what proportion of the nation's variation in unemployment is explained by the additive effects of "urbanization" and "illiteracy."

17. Suppose that the correlation between grades in a first-year course in dentistry and a test of medical ability is 0.65, and with a test of mechanical dexterity 0.45. Suppose, also, that the correlation between the latter two is 0.68. Compute R between course and medical ability test scores and the combined effect of medical ability and mechanical dexterity test scores.

18. In a study of fifty cases sampled at random, the correlations among three variables $(X_1, X_2, \text{and } X_3)$, are $r_{12} = 0.60$, $r_{13} = 0.50$, and $r_{23} = 0.30$.
    (a) Compute R.
    (b) Compute R^2.
    (c) Test the significance of the multiple correlation coefficient.

19. Compute the multiple correlation coefficient between the two independent variables and the dependent variable in Exercises 4 and 5. Also, compute the multiple coefficients of determination and alienation for Exercises 4 and 5.

Nonparametric Statistics

16.1 Introduction: Why Study Nonparametric Statistics?

In this chapter, we present statistical tests that make no assumptions about the shape of the population distributions. The statistics involved are called *distribution-free*. Many of them require no assumptions about population parameters, and are therefore called *nonparametric statistics*. (The Spearman rank-order correlation coefficient was a nonparametric statistic.) And since nonparametric statistics also don't require interval-level measurement, we can test the difference between frequencies or ranks in much the same way as we perform a *t* test or analysis of variance on interval-level data. The binomial test, which we dealt with in an earlier chapter, was a nonparametric test of a dichotomous nominal variable.

Nonparametric statistics have many advantages over their parametric counterparts. One such advantage is their ease of calculation. Often, they have sampling distributions that the most fundamental laws of probability can explain easily. But their major advantage is that they require fewer restrictive assumptions. This is particularly important when samples must be small. For example, nonparametric statistics can really benefit the social researcher, who rarely deals with large samples when doing long-range research involving people.

Parametric tests make explicit assumptions about the shape of a population and the level of measurement. Some tests make further assumptions about the parameters that are not under investigation. As you recall, we assumed equal variances when we used the Student t ratio for testing the difference between population means with independent samples. However, with nonparametric statistics, we cannot use equal variances. Another disadvantage of parametric tests is that they are preoccupied with the mean. But when we cannot meet the assumption of a symmetrical distribution, the median is often a better measure of central tendency than the mean is. Some nonparametric tests do well in testing hypotheses regarding medians; the popular parametric tests, on the other hand, are limited to the mean.

Of course, there are disadvantages to using nonparametric statistics. One major drawback is that there are well over forty nonparametric measures from which to choose. A researcher often spends more time on design efficiency than on the statistical test itself. Another disadvantage is that nonparametric tests are less precise than parametric tests, which makes them generally less efficient, as well. But it has been argued that, in most cases, some data do not offer the scientist the choice of using the parametric counterparts.

In this chapter, we will introduce some of the more popular and relatively efficient nonparametric tests. Most of these will be nonparametric counterparts to procedures that we discussed earlier. Probably the best place to start is with *chi square*, the most widely used of the nonparametric techniques, particularly in the social sciences.

16.2 Chi Square: An Introduction

As you know, it's often impossible to satisfy the requirements of parametric tests. In the first place, much of the data available to the social scientist are at the nominal level of measurement. Secondly, we cannot always be sure that the characteristics we intend to study are really distributed normally in the population. In instances such as these, the social scientist might find the *chi-square test of significance* considerably valuable.

In one-variable applications, the chi-square test is called a "goodness-of-fit" measure because it allows us to determine whether a significant difference exists between the observed and expected number of cases for each category of a nominal variable. In other words, chi square lets us determine how well our observed distribution fits the theoretical distribution.

However, more often our concern is with the interrelationship between and among nominal variables. Here, chi square is of great value, too: it allows us to test for the significance of a difference between two nominal variables. When we use chi square in this way, we call it a test of *independence*.

Before we study these chi-square tests, we want to clarify that chi square is the term for both a statistical test of significance and a sampling distribution. The sampling distribution is like the t distribution in that there is a whole family of distributions. The distribution that's used to test an hypothesis then will depend on the number of degrees of freedom involved. Like the t table (Table C), the chi-square distribution (Table I in the Appendix) is set up with degrees of freedom in the first column and selected probabilities in the other columns. Unlike the t test, however, chi square is used to make comparisons between frequencies rather than between mean scores. As a result, the null hypothesis for the chi-square test states that the populations do not differ with respect to the frequency of a given characteristic. The research hypothesis, on the other hand, says that sample differences reflect actual population differences in the relative frequency of a given characteristic.

16.2.1 Chi Square: The One-Variable Case

An industrial sociologist samples 600 factory workers for a large plant to find out whether they prefer the traditional five-day work week, or a lengthened four-day week, or a schedule that enables the worker to arrange his or her own working hours. This type of problem is best answered with the chi-square one-variable test. As we've said, the one-variable application of the chi-square test is called a "goodness-of-fit" test. Under the null hypothesis of no difference, it lets us determine whether a significant difference exists between the observed and expected number of frequencies in each category of a nominal variable. In other words, chi square lets us determine the extent to which our observed distribution fits the theoretical one.

What we need, then, is (1) a null hypothesis that lets us discover the frequencies that would be expected for each category of the nominal variable, and (2) to test the null hypothesis. Such a test is given by the chi-square formula for the one-variable case:

$$\chi^2 = \sum_{i=1}^{k} \frac{(f_o - f_e)^2}{f_e} \tag{16.1}$$

where:

k = the number of categories,
f_o = the observed frequency in any category, and
f_e = the expected frequency in any category.

As the formula indicates, the chi-square test is essentially a comparison of the actual results with the theoretical expectation set forth in the null hypothesis. Thus, as the discrepancies between $f_o - f_e$ increase, so does the value of chi square—which, in turn, makes the rejection of the null hypothesis possible. A decrease in discrepancies will, of course, make it more difficult to reject the null hypothesis.

In the above example, the null hypothesis states that there is an equal preference for each type of work schedule—that is, 600/3, or 200, is the expected frequency in each category. However, the observed frequencies for the work schedule preferences were 175 (traditional work week), 240 (four-day week), and 185 (flexible work week). Thus:

$$\chi^2 = \frac{(175 - 200)^2}{200} + \frac{(240 - 200)^2}{200} + \frac{(185 - 200)^2}{200}$$
$$= 3.12 + 8.00 + 1.12$$
$$= 12.24$$

Since the chi-square distribution is like the t distribution, it also varies as a function of the degrees of freedom. But with chi square, the degrees of freedom are a function of the number of categories, not of the size of the sample. And as we said earlier, degrees of freedom is the number of parameters (or categories with nonparametric statistics) that are allowed to vary. Thus, the degrees of freedom for the one-variable case is determined by the number of categories being tested, minus one; in symbols, $k - 1$. Table I in Appendix III lists the critical values of chi square for various degrees of freedom and levels of significance for two-tailed tests.

Since there are three categories in this example, two categories are left free to vary; in other words, we have $3 - 1$ or $df = 2$. Using a two-tailed test at $\alpha = 0.01$, we find that Table I indicates that a chi-square value of 9.21 or greater is required for statistical significance. Since our calculated value of 12.24 exceeds the tabled value, we reject the null hypothesis and conclude, instead, that there might be a difference in the preferences among these factory workers concerning work schedule.

16.2.2 Chi Square: The Two-Variable Case

By far, chi square is used most often in contingency tables—which, as you will recall, array data on two variables. Used in this way, chi square is a test of independence, asserting that the two variables represent different populations and that they therefore bear no statistical relationship to each other. This assertion of independence means that the column variable has no effect on the way the row variable distributes itself in the cells of the cross classification. The same logic applies to the effect of the row variable on the distribution of the column variable. The chi-square statistic lets us find out whether the departures from a pure-chance distribution of frequencies that we observe are enough to make us reject the null hypothesis of independence and thereby accept the alternative hypothesis (that is, the relationship in the table probably couldn't have occurred by chance alone).

Table 16.1.   Responses to a Public Opinion Poll on the Legalization of Marijuana

	For	Against
Male	190	380
Female	210	400

To help us understand this application of chi square, consider a 1978 public opinion poll in which the respondents expressed an opinion on the legalization of marijuana (see Table 16.1). The researchers wish to test whether men and women responded differently to the prospect of legalizing marijuana (see Table 16.2). To make this test, we must use the notion of the joint occurrence of independent events (see Section 8.10) in order to arrive at the expected values for each cell. Probability theory tells us that the joint occurrence of independent events is equal to the product of their separate probabilities. Thus, each cell will represent the intersection of its corresponding row category and column category. For example, the expected frequency (or value) for the first row/first column constitutes the product of the probability of a joint occurrence of being a male respondent and approving the legalization of marijuana, which is 570/1,180, or 0.4831, and 400/1,180, or 0.3390, respectively. Therefore, the probability of their joint occurrence is (0.4831)(0.3390), or 0.1638. Then we must multiply this probability (or proportion) by 1,180 to get the expected value for row 1/column 1, which is 193.28. Incidently, we can arrive at approximately the same value with much less effort if we multiply the row marginal by the column marginal in which the target cell is located and then divide by the size of the sample N. Thus, the expected value for the cell defined by row 1/column 1 is: (570)(400)/1,180, or 193.22.

Table 16.2.   2 × 2 Contingency Table Showing the Number of Men and Women Expressing an Opinion on the Legalization of Marijuana

	For	Against	Row Marginal
Male	190	380	570
Female	210	400	610
Column Marginal	400	780	1180

Computing by either of these methods will produce the same results (allowing, of course, for rounding errors). Table 16.3 presents the obtained data; the expected cell frequencies are set off in brackets in each cell.

Table 16.3. Number of Men and Women Expressing an Opinion on the Legalization of Marijuana

	For	Against	Row Marginal
Male	190 [193.22]	380 [376.78]	570
Female	210 [206.78]	400 [403.22]	610
Column Marginal	400	780	1180

The formula for computing the chi-square value for the two-variable case is:

$$\chi^2 = \sum_{i=1}^{i} \sum_{j=1}^{j} \frac{(f_o - f_e)^2}{f_e} \tag{16.2}$$

The basic rationale for this formula is obviously the same as in the one-variable case; that is, we want to find out whether the difference between observed and expected value is enough to warrant the rejection of the null hypothesis. Substituting the observed and expected values of Table 16.3 into Formula (16.2), we find:

$$\chi^2 = \frac{(190 - 193.22)^2}{193.22} + \frac{(380 - 376.78)^2}{376.78} + \frac{(210 - 206.78)^2}{206.78}$$
$$+ \frac{(400 - 403.22)^2}{403.22}$$
$$= 0.05 + 0.03 + 0.05 + 0.03 = 0.16$$

Now we need one more piece of information before we can determine the critical value from the chi-square table: the degrees of freedom. Degrees of freedom in contingency tables are dependent on the number of cells in the table, which in turn depends on the number of categories into which we have classified our data. The fact that our marginal frequencies are fixed also restricts the degrees of freedom. The general formula for the degrees of freedom in $r \times c$ contingency tables is: $df = (r - 1)(c - 1)$. In the 2×2 table, this formula yields one degree of freedom. Table 16.3 will show this readily. Since the marginals are fixed, as indicated below, once one expected frequency is computed (for example, 193.22 in row 1/column 1), then all others are determined. Therefore, since 193.22 is placed in its cell, 376.78 and 206.78 must be placed in row 1/column 2 and row 2/column 2, respectively. Once these cells are determined, we know that 403.22 must occupy row 2/column 2 in order to equal the marginal totals.

In the present example, $df = (2 - 1)(2 - 1) = 1$. Using a two-tailed test at $\alpha = 0.01$, we find that the critical region consists of all values of $x^2 \geqslant 6.64$. Since our computed value of $0.16 < 6.64$, we cannot reject the null hypothesis. Therefore we must conclude that attitudes concerning the legalizing of marijuana and sex of respondents do appear to be independent.

At this point, we want to mention a warning about 2×2 contingency tables (or, more generally, about whenever $df = 1$). The theoretical distribution of chi square is a continuous distribution, whereas the observed values are always discrete. In 2×2 tables where N is relatively small and the expected value of any cell is less than 10, it's advisable to use a correction factor in order to more accurately approximate the continuity of the theoretical distribution. The *Yates* (1934) *correction factor* can be used in such cases. This factor reduces the absolute difference $|f_o - f_e|$ by 0.5 before squaring. Thus, in the one-degree-of-freedom situation, the formula for computing chi square becomes:

$$x^2 = \sum_{i=1}^{2} \sum_{j=1}^{2} \frac{(|f_o - f_e| - 0.5)^2}{f_e} \qquad \textbf{(16.3)}$$

Including this correction, which reduces the computed value of chi square, can make a significant chi square near the critical region nonsignificant. Since, in our example, the N and the expected values are quite large, we needn't bother with the Yates correction factor. (But do compute chi square for this example, using the correction factor as an exercise.)

In actual research, however, we often have to subdivide the nominal variable into more than two categories. For example, we can extend our present survey problem to include the three-attitude categories of "for," "against," and "no opinion" (see Table 16.4). The expected frequencies of the first four cells differ from those in the 2×2 case because of the inclusion of the "no opinion" category, which has changed the marginals. Of course, the procedure for obtaining the expected values is the same as the one in the 2×2 case. Thus, the substitution of the observed and expected values of Table 16.4 into Formula (16.2) yields:

$$x^2 = \frac{(190 - 200)^2}{200} + \frac{(380 - 390)^2}{390} + \frac{(180 - 160)^2}{160}$$
$$+ \frac{(210 - 200)^2}{200} + \frac{(400 - 390)^2}{390} + \frac{(140 - 160)^2}{160}$$
$$= 0.50 + 0.26 + 2.50 + 0.50 + 0.26 + 2.50$$
$$= 6.52$$

Table 16.4. 2 × 3 Contingency Table Showing the Number of Men and Women Answering a Survey on the Legalization of Marijuana

	For	Against	No Opinion	Row Marginal
Male	190 [200]	380 [390]	180 [160]	750
Female	210 [200]	400 [390]	140 [160]	750
Column Marginal	400	780	320	1500

For the 2 × 3 contingency table, *df* is $(2 - 1)(3 - 1)$, or 2. Again using a two-tailed test at $\alpha = 0.01$, we find that Table I in the Appendix indicates that a $\chi^2 \geq 9.21$ is required for significance. Since $6.52 < 9.21$, we cannot reject the null hypothesis and must conclude that the whole range of attitudes concerning the legalization of marijuana and sex of respondent do appear to be independent.

16.2.3 Caution: Some Restrictions on the Use of Chi Square

If we are to use chi square properly (that is, as we would any other test of significance), we have to make certain assumptions. One of the most fundamental is that we can use chi square only with nominal (or frequency) data. For example, it is incorrect to use chi square to test the deviation between the observed mean on some trait for a group and some expected mean. These are interval measurements; and one reason why chi square cannot be used with such data is that the computed chi-square value varies with the size of the unit of measurement. Thus, length may be measured either in inches or in feet. But if the measurement is recorded in inches rather than in feet, the chi-square value representing the difference between observed and expected lengths will be different even for the same subjects. How can we determine which chi-square value is correct?

In addition to the assumption about the level of measurement, authorities agree that no expected frequency should be smaller than five, except in situations where the degrees of freedom is more than one. In such instances, at least 80 percent of the cells must meet the requirement of an expected value of five or more. The basic reason is that when the expected frequencies in a cell are smaller than five, the chi-square distribution cannot sufficiently approximate the empirical or observed distribution of the variable under investigation; therefore, the chi-square value becomes unreliable. This problem is sometimes eliminated when a small expected frequency in one cell is combined with another more populated cell. For example, consider two cells, one containing four strong

Democratic party identifiers and the other containing six moderate Democratic party identifiers. We can combine the cells into one labeled "Democrat," with an expected frequency of ten. (Remember that we then would have to reduce the degrees of freedom correspondingly to account for the combining of cells.) Though it makes us lose both information and our handle on the intensity of party identification, *cell collapsing,* as it is called, counteracts the problem inherent in small samples.

In addition to having these limitations, the chi-square test also assumes random sampling and frequencies that are independent of one another. Consider the case where 100 people are asked to comment on a political figure. There is no reason to think that the answer of any one of the respondents depends upon the answer of another respondent. But it could be incorrect, for example, to ask each of the 100 respondents to comment on the same figure over two different time periods, and then to claim that the total frequency was 200. Clearly, since their responses given at the second interview could be expected to bear some relation to their views on the previous occasion, these responses could not be independent of each other. There are other tests for determining before-and-after responses on the same subject. If you're interested, see Siegel (1956) for a detailed discussion.

The computation of chi square has one further requirement: the sum of the observed frequencies must equal the sum of the expected frequencies, which in turn must equal the marginal totals. This restriction allows for the possibility of obtaining a computed chi square of zero. In testing for independence, a chi square of zero (where $f_o - f_e = 0$ for all cases) must be a distinct possibility.

16.3 Associational Measures Based on Chi Square

Researchers often want a summary measure that conveys the degree to which two variables are related. But associational measures must be comparable, and so chi square fails here as a measure of association. Chi-square values are not comparable, because they are sensitive to the sample size. For example, a larger sample increases the chances of getting a significant value. Given the same proportional distribution of frequencies in two contingency tables, the chi-square value that's computed from the table with the larger N will have a greater probability of being statistically significant (see Table 16.5.).

Note that the two contingency tables would have identical distributions if we converted the cell frequencies into percentages. Note, also, that subtable (a) is based on a sample size of 50, with a chi-square value of 2, whereas subtable (b) has a chi-square value of 4, based on a sample of 100. (But please verify these chi-square values for yourself.) Thus Table I in Appendix III shows that for $df = 1$, for a two-tailed test at $\alpha = 0.05$, we can reject the null hypothesis for the computed chi-square value of 4 but not for subtable (a), since $2 < 3.84$.

Table 16.5. 2 × 2 Contingency Tables Showing the Impact of Sample Size on Statistical Significance

(a) $\chi^2 = 2$

		Y		
	15	10	25
	[12.5]	[12.5]	
X	10	15	25
	[12.5]	[12.5]	
	25	25	50

(b) $\chi^2 = 4$

		Y		
	30	20	50
	[25]	[25]	
X	20	30	50
	[25]	[25]	
	50	50	100

Clearly, there must be some way to find out the strength of relationship between two variables in a 2 × 2 contingency table, without the distortion that the sample size brings about. What we want is a standardized or normed measure that can make the results obtained for these two tables comparable. The *phi coefficient,* to which we now turn, can accomplish this easily.

16.3.1 The Phi Coefficient

The *phi coefficient* standardizes the value of the computed chi-square value by dividing it by the sample size. This removes the distortion that a large sample size may have. Thus, the phi coefficient is defined as:

$$\phi = \sqrt{\frac{\chi^2}{N}} \tag{16.4}$$

Phi varies between 0 (which indicates complete independence, or no relationship) and 1.00 (which is equivalent to saying that there is a perfect association between the categories of two nominal variables).

It is interesting to note the identical phi coefficients of 0.20 for both subtables in Table 16.5. It's also interesting to note that while subtable (b) yields a significant chi-square value at the 0.05 level, the degree of association between variables is quite weak. This further demonstrates that statistical significance does not necessarily equate with statistical association. One type of measure tells us whether our results are due to chance alone; the other type conveys the strength of relationship. Therefore, it is possible to get weak associations that are statistically significant and strong associations that are nonsignificant. Significance and association measures are designed to yield different types of information. They should not be confused.

Some study of Formula (16.4) will reveal that phi has an upper limit of 1.00 but that this can only occur in 2 × 2 tables. In other words, its value can exceed 1.00 in tables larger than 2 × 2. This problem is easily solved with the introduction of *Cramér's V.*

16.3.2 Cramér's V

While the phi coefficient is limited to 2×2 tables, *Cramér's V* provides a summary associational measure for tables larger than 2×2. Its formula also uses chi square, and is written as:

$$V = \sqrt{\frac{\chi^2}{N(k - 1)}} \qquad \textbf{(16.5)}$$

where:

> $k =$ the number of rows or columns, whichever is smaller (if the number of rows equals the number of columns, as in a 3×3 table, either number can be used for k).

The basic rationale behind Cramér's V is to establish an upper limit of 1.00 and still maintain the characteristic of comparability, a necessary element of a measure of association. In the case of the 2×2 table, the upper limit of 1.00 can be reached if we divide the chi-square value by N, where there are maximum diagonal formations with the off-cells empty. In $r \neq c$ tables, the maximum upper limit can be reached if we multiply the divisor N by the number of rows or columns, whichever is smaller, minus one. As you can see from Formula (16.5), Cramér's V adds this correction factor to the phi coefficient to account for the larger tables, which may not be square.

Table 16.4 gives us a 2×3 table. Calculation of Cramér's V is appropriate for this table. Referring back to Section 16.2.2, we see that our computed chi-square value and N are 6.52 and 1,500. We substitute these values into Formula (16.5) and compute a Cramér's V of 0.07.

16.3.3 The Interpretability of Chi-Square-Based Association Measures

Chi-square-based measures have a number of drawbacks, which center mainly on their interpretability. First, their sensitivity to tabular dimensions and marginal distributions gives more weight to rows and columns with the smaller marginals. Secondly, while we can interpret the lower and upper limits of 0 and 1.00 as "no" and "perfect association," respectively, we can't decipher the meaning of the numbers in between quite as clearly. All we can say is that the greater the magnitude, the stronger the association. For example, we can't interpret these measures as being equivalent to the variance-apportioning measures based on the linear regression model. (For a further discussion of some of these issues, consult Reynolds (1977: 30–32, 45–52).)

Both phi and Cramér's V are symmetrical measures that are based on the model of statistical independence. Calling them symmetrical merely indicates that they do not distinguish between dependent and independent variables. As we know from our discussion of chi square, the model of statistical independence is designed to test whether the observed data depart from a hypothetical distribution. Consequently, chi-square-based measures compare the observed distributions in the cells to the distribution that would have occurred had there been no association. These measures attempt to convey the amount of departure from the hypothesized lack of association. For many purposes, chi-square-based measures yield useful information; but, for many other purposes, we need an alternative model.

16.4 *PRE* Measures of Association: Some Preliminaries

A group of techniques that have been labeled *proportional reduction of error (PRE)* measures overcome some of the interpretability problems inherent in the chi-square-based associational measures. The logic behind *PRE* is quite simple and clever. As the term "proportional reduction of error" indicates, these procedures are designed to measure the reduction of prediction error that results when we know an independent variable. We accomplish this measurement by using two prediction rules:

> *Rule I: Prediction is based on the categorical distribution of the dependent variable alone.*

> *Rule II: Prediction is based on the information supplied by the independent variable.*

We can express the prediction errors that result from the use of these rules as a ratio, which is essentially the summary *PRE* coefficient.

The basic rationale becomes readily apparent once we think this process through. Assume that we have no information other than the distribution of a nominal-level dependent variable into its various categories. The "best guess" for any subject's classification on the dependent variable would be the modal category, or a calculated expected value based upon random placement into the various categories. These prediction devices make up Rule I. We may symbolize the errors that result from using this rule as E_1. A bivariate contingency table gives us additional information: namely, the manner in which the dependent variable is distributed among the various categories of the independent variable. This new information should help us reduce the number of errors that result from Rule I if the two variables are not statistically independent. We use the symbol E_2 to identify the number of errors that result when using the additional information supplied by the independent variable. A *PRE* measure determines the association between two variables by discovering what proportion of errors is reduced when we use Rule II, as opposed to Rule I. We can express the proportional reduction as a ratio of their respective number of errors. In symbols:

$$PRE = \frac{E_1 - E_2}{E_1}$$

(16.6)

where:

E_1 = the number of errors that result from prediction Rule I, and

E_2 = the number of errors that result from prediction Rule II.

Associational techniques that come under the heading of "proportional reduction of error" share some basic characteristics. According to the logic of the procedure, $E_1 \geq E_2$. Where there is statistical association between the two variables, we expect E_1 to be greater than E_2. This is like saying that if the independent variable is related to the dependent variable, it should help reduce prediction error. Conversely, if the two variables are statistically independent (that is, if they bear no relationship to each other), we would expect knowledge of the independent variable to be no help in predicting the categorical distribution or rank (with ordinal data, that is) of the dependent variable. Therefore, in these situations, we should expect an equal number of errors, no matter which prediction rule (that is, $E_1 = E_2$) we use. When this occurs, the *PRE* measure, as we might expect, is zero, since there is no reduction in error using E_2. In symbols:

$$\frac{E_1 - E_2}{E_1} = \frac{0}{E_1} = 0$$

PRE measures have definable limits; they vary between 0 and 1.00. The worst case is where there is no reduction in error (that is, where there is complete statistical independence); the best case is where the two variables are completely related. When that happens, $E_2 = 0$, since we make no errors using Rule II (symbolically, $(E_1 - 0)/E_1 = 1.00$). The key advantage of *PRE* measures is the interpretability of the values between 0 and 1.00. These intermediate values directly resemble the coefficient of determination (r^2) of the linear regression model. Although we cannot say anything meaningful about reduction of variance for nominal-level data, we can speak of a proportional reduction of error. Thus, a *PRE* value of 0.50 means that our knowledge of the independent variable has helped us reduce the number of errors by 50 percent.

Even though symmetrical versions for some *PRE* measures exist, most are asymmetrical, which means that we must designate the dependent and independent variables. Therefore, *PRE* values are not interchangeable: when X is seen as the dependent variable and Y as the independent variable, the reduction of prediction errors associated with dependent variable Y, using information supplied by independent variable X, generally will not equate with the proportional reduction of errors.

One final general note, which will become apparent when we discuss the first *PRE* measure, *lambda:* in certain situations, a *PRE* measure will be zero even where the two variables are not statistically independent. This will occur in cases where knowing the independent variable will not help us to predict the dependent variable because the modal categories of the dependent variable are highly skewed. We'll say more about this in the next section, when we discuss lambda.

16.4.1 A Nominal-Level *PRE* Measure of Association: Lambda

Nominal-level variables, as we know, consist of frequency counts within specific categories. Thus, if we drew a case at random, to which category of the variable would it belong? Our best guess, of course, would be the modal category. A bivariate contingency table arranges the data on the basis of two classification schemes: one for the dependent variable and the other for the independent variable. If the two variables are related, knowing how the dependent variable distributes itself according to the categories of the independent variable should help us reduce the number of prediction errors. This is what the *PRE*-based *lambda coefficient* is designed to measure.

Assume that a state legislature is considering a formula for aiding church-related schools. In the state senate, the proposal is approved by a vote of 35 to 25. Knowing nothing else about a legislator's vote except the vote tally, we would predict that the random legislator voted approval, since the category is the modal one. Using the mode as a basis of prediction constitutes Rule I of the *PRE* scheme and results in an error rate of 25/60, or 42 percent.

Suppose that we receive an additional piece of information; namely, the religious backgrounds of the legislators. Now the question becomes whether knowing the legislator's religious background helps us predict his/her vote. The data displayed in Table 16.6 reflect how the two categories of the dependent variable are distributed among the three categories of the independent variable. (Conventionally, the dependent variable is designated as the row variable.)

Table 16.6. 2 × 3 Contingency Table Showing Votes on Aid to Church-Related Schools and Religious Backgrounds of Senators

Vote on Aid	RELIGIOUS BACKGROUNDS			
	Protestant	Catholic	Jewish	
Yes	13	(18)	4	(35)
No	(15)	2	(8)	25
	28	20	12	60

Note: Modal categories are circled.

The underlying logic behind lambda is to compare modes. Specifically, we compare the mode of the dependent variable (vote on the aid proposal) to the mode of the various categories of the independent variable, which in this case is 15 (for Protestants), 18 (for Catholics), and 8 (for Jews). In other words, we compare prediction Rule I (modal frequency of the dependent variable) with prediction Rule II (modal frequencies of the independent variable). This allows us to define the *asymmetric lambda* coefficient as:

$$\lambda_a = \frac{\Sigma \ max \ f_c - max \ F_r}{N - max \ F_r} \qquad (16.7)$$

where:

$max \ f_c$ = the modal frequency within each category of the column or independent variable, and

$max \ F_r$ = the modal frequency of the row or dependent variable.

Returning to our example and substituting the appropriate frequencies, we find that:

$$\lambda_a = \frac{(15 + 18 + 8) - 35}{60 - 35} = 0.24$$

This tells us that we have reduced our error rate by 24 percent. Now let's verify this. Without knowing the independent variable, our error rate is 25/60, or 42 percent. When we predicted on the basis of the categorical modes of the independent variable, we made 19 errors. Specifically, on the basis of the categorical mode for a Protestant legislator, we would have predicted a "no" vote and thereby have made 13 errors. The Catholic and Jewish categories would have provided us with 2 and 4 errors, respectively. The ratio 19/60 [13/28 + 2/20 + 4/12 = 19/60] is equal to 32 percent. Thus, our knowledge of the independent variable reduced our error percentage from 42 to 32, or 10 percent. This constitutes a 24 percent reduction, as indicated by the lambda coefficient ((0.10/0.42)(100) = 24 percent).

Lambda is an asymmetric measure that we use to determine the degree to which the column variable can help us predict the row variable. Of course, if the research question warrants it, we can also determine the row variable's ability to predict the column variable by reversing the designations of the two variables, thereby treating the column variable as the dependent variable and the row variable as the independent variable. But when we switch the roles of the variables, we become concerned with the overall modal frequency of the column variable and the categorical modal frequencies of the row variable. For exercise purposes, compute the lambda coefficients (λ_a and λ_b) for Table 16.7. Your values should be $\lambda_a = 0.125$ and $\lambda_b = 0.10$, which points out that values obtained from asymmetrical measures are not interchangeable, and that the variables must be designated as dependent and independent.

Table 16.7. 2 × 3 Contingency Table

	B			
A	10	15	35	60
	5	20	15	40
	15	35	50	100

Very often, however, we have two variables that we have reason to believe mutually influence each other. There is a symmetric version of lambda that's equivalent to combining the formulas for the lambda coefficients for the situations in which row and column variables are each treated as the dependent variable. The general formula for *symmetric lambda* is:

$$\lambda_s = \frac{(\Sigma \ max \ f_c + \Sigma \ max \ f_r) - (max \ F_r - max \ F_c)}{2N - (max \ F_r + max \ F_c)} \qquad (16.8)$$

where:

$max \ f_c$ = the modal frequency within each category of the column variables, treated as the independent variable,

$max \ f_r$ = the modal frequency within each category of the row variables, treated as the independent variable,

$max \ F_r$ = the modal frequency of the row marginals, and

$max \ F_c$ = the modal frequency of the column marginals.

To illustrate symmetric lambda, we present data on husbands' and wives' party preferences in Table 16.8. We will use these data to test the political socialization hypothesis that marriage partners mutually influence each other's party identification. After using Formula (16.8), we find:

$$\lambda_s = \frac{(65 + 35 + 25 + 65 + 35 + 15) - (100 + 90)}{2(200) - (100 + 90)} = 0.24$$

Note the weak relationship between the two variables.

Table 16.8. 3 × 3 Contingency Table Showing Marital Partner's Party Preferences

Husband's Party Preference	Wife's Party Preference			
	Democrat	Republican	Independent	
Democrat	65	10	25	100
Republican	15	35	10	60
Independent	10	15	15	40
	90	60	50	200

We use the symmetric version of lambda when we don't want to designate the direction of the relationship. The symmetric lambda is not used as often as its asymmetric counterpart. The ability to state the proportion of error reduction when each variable is used simultaneously as a basis for predicting the other is a bit difficult to interpret, since we cannot readily disentangle the effect of each variable. Thus, the measure is deemed to be limited in usefulness.

At this point we'd like to say a word about the term "guess," used in explaining measures such as lambda. Unfortunately, "guess," associated with prediction Rule I, suggests a drama in which the plot unfolds gradually. But as researchers, we don't play a guessing game and then receive a new piece of information in the form of an independent variable. No, we have all the information at hand before we compute the measure. The terms "guess" and "predictions" merely help us define the logic of this particular associational measure. We say that we act *as if* we make these predictions on the basis of two rules. Needless to say, our real intention is to determine the degree of association according to the *PRE* logic. This subtlety sometimes confuses the novice.

As a measure of association, lambda has one serious defect: it will tend to be insensitive to a bivariate association where the distribution of the dependent variable is highly skewed as it distributes itself among the categories of the independent variable (see Table 16.9). Consequently, the computed coefficient for the table values is zero, since each of the categorical modes lies in the same row as the overall mode of the dependent variable. Since lambda is a comparison of prediction rules, switching to prediction Rule II in situations like the one in Table 16.9 does not help us to reduce errors. Therefore, according to the logic of *PRE,* lambda is zero. In general, unless at least one of the categorical modes of the independent variable lies in a different row from the mode of the dependent variable, lambda will be zero. To overcome this situation, we must use something other than modal comparisons as a criterion. The *Goodman-Kruskal tau* provides us with such a criterion for nominal-level data.

Table 16.9.   2 × 3 Contingency Table Showing Skewness Among the Categories of the Independent Variable

	Variable B			
Variable A	(50) 30	(75) 30	(75) 40	(200) 100
	80	105	115	300

Note: Modal categories are circled.

16.4.2 A Nominal-Level *PRE* Measure of Association: Goodman-Kruskal Tau

The *Goodman-Kruskal tau* manages to overcome the problems inherent in skewed categorical distributions. How? By using a different criterion as a basis for prediction—the random assignment of subjects into categories rather than modal comparison. Specifically, the logic of the measure requires the use of all the marginal categories of the dependent variable and all the within-category cells of the independent variable, not just those cells that have the highest frequency for a given category.

As in all *PRE* measures, we make a comparison between two prediction rules: one when we know only the categorical distribution of the dependent variable, and the other when we receive additional information in the form of an independent variable.

The first prediction rule of the Goodman-Kruskal tau requires all subjects N to be assigned randomly into categories of the dependent variable, while maintaining the marginal distribution. As an illustration, assume that a randomly selected sample of 500 women polled on their attitude toward the Equal Rights Amendment (ERA) reveals that 300 expressed approval for it and 200 expressed disapproval (see Table 16.10). If each of these 500 women were assigned on a random basis and in accord with only the marginal totals, we would assign 300 to the approval category and 200 to the disapproval category.

Table 16.10.   2 × 2 Contingency Table Showing the Work Status of Women and their Attitudes Toward ERA

	Work Outside Home	Do Not Work Outside Home	
Approve	130	170	300
Disapprove	50	150	200
	180	320	500

The random assignment prediction rule (Rule I) naturally will lead to some errors. In the long run, random assignment of 300 women to the approval category would result in an error rate of 0.4, since 200 of the 500 women expressed disapproval ($200/500 = 0.4$), and we would expect some of these to be included in the approval category if the assignment were made randomly. Likewise, we can expect an error rate of 0.6 by assigning 200 women to the disapproval category on a random basis, since many of the women might actually be in the other category (as indicated by the proportion 300/500, or 0.6). We multiply these error rates by the number assigned to each category, and then add them. The result constitutes E_1 (that is, the number of errors we make using prediction Rule I). In our example, E_1 is $(300(200/500) + 200(300/500))$, or 240.

Now suppose that we have all the information in Table 16.10 and that we wish to test the hypothesis that the work status of women will influence a woman's attitude toward ERA (the dependent variable). Here, random assignment is made according to the categories of the independent variable (work status). Now we are subject to the constraints imposed by the marginal totals of the column categories. So if we randomly assign 130 women who work outside the home to the "approve ERA" category, we should expect an error rate of 50/180, or 0.28, since 50 out of 180 women in column 1 disapprove of ERA. Likewise, if we assign 50 of the women who have jobs outside the home to the "disapprove" category, our error rate will be 130/180, or 0.72. Then we multiply the frequency in each cell by the appropriate error rate. According to the way in which the dependent variable distributes itself in the first column of the independent variable, we should expect $(130(50/180) + 50(130/180))$, or 72.22 errors in the long run. The logic is extended to the second column of the independent variable. The error rates for misclassified women who do not work outside the home are, respectively, $150/320 = 0.47$ and $170/320 = 0.53$. When these rates are multiplied by the appropriate cell frequencies, they yield expected errors of 79.69 for each cell. Thus, E_2 (expected errors that result from using prediction Rule II) is $(130(50/180) + 50(130/180) + 170(150/320) + 150(170/320))$, or 231.6.

Then, as in all *PRE* measures, we compare E_2 (231.6) to E_1 (240) by computing the difference $E_1 - E_2$ and then dividing by E_1 to standardize the measure to yield a coefficient between 0 and 1.00. In the present example, this results in a Goodman-Kruskal tau of 0.035, which tells us that with the knowledge supplied by the independent variable, we have reduced our expected number of errors by 3.5 percent. Our interpretation is that the work status of women does not seem to have a very considerable impact on their attitude toward the Equal Rights Amendment. However, it is important to note that the comparable lambda coefficients for the data computed from this contingency table would be zero. Why? (Hint: see Section 16.4.1.)

We can generalize these computational procedures into formulas for the calculation of E_1 and E_2, which in turn enable us to calculate the Goodman-Kruskal tau directly. These formulas merely represent the intuitive procedures we have just examined. The formula associated with prediction Rule I is:

$$E_1 = \Sigma \left[\frac{N - F_r}{N} (F_r) \right] \tag{16.9}$$

where:

F_r = the marginals of the various categories of the dependent variable. (Remember, we have designated the row variable as the dependent variable.)

The complementary formula associated with prediction Rule II is:

$$E_2 = \Sigma \left[\frac{F_c - f_c}{F_c} (f_c) \right] \qquad (16.10)$$

where:

F_c = the column marginals (independent variable), and

f_c = the cell frequencies of the dependent variable within the various categories of the independent variable.

Now we can define the Goodman-Kruskal tau (τ) as the proportional reduction of errors. Thus:

$$\tau_r = \frac{\begin{array}{c}\text{Number of errors made on the} \quad \text{Number of errors made on} \\ \text{basis of knowing the dependent} \; - \; \text{the basis of knowing the} \\ \text{variable alone} \qquad\qquad \text{independent variable}\end{array}}{\begin{array}{c}\text{Number of errors made on the basis of} \\ \text{knowing the dependent variable alone}\end{array}}$$

or, in symbols:

$$\tau_r = \frac{E_1 - E_2}{E_1} \qquad (16.11)$$

(We leave the computation of the Goodman-Kruskal tau, using Formula (16.11), to you as an exercise. Remember, the answer is 0.035.)

Tau can also be computed from the general formula:

$$\tau_r = \frac{N\Sigma \dfrac{f_c^2}{F_c} - \Sigma F_r^2}{N^2 - \Sigma F_r^2} \qquad (16.12)$$

where the row variable is designated as the dependent variable. Using the appropriate cell frequencies and marginals, we compute the Goodman-Kruskal tau for our present example to be:

$$\tau_r = \frac{500 \left[\left(\dfrac{130^2}{180} \right) + \left(\dfrac{50^2}{180} \right) + \left(\dfrac{170^2}{320} \right) + \left(\dfrac{150^2}{320} \right) \right] - (300^2 + 200^2)}{500^2 - (300^2 + 200^2)} = 0.035$$

The Goodman-Kruskal tau is an asymmetric measure of association, which means that we must designate the roles of the variables as dependent or independent before we do our computation. But sometimes we wish to treat the column variable as the dependent variable and the row variable as the independent variable. In this case, we compute τ_c instead of τ_r. We can do this by first working with the column marginals and then the within-cell frequencies of the row variable. When using the formulas, we substitute the appropriate column marginals and cell frequencies instead of the row designations (that is, we use F_c instead of F_r and f_c instead of f_r, where required). Except for 2×2 tables, where τ_r always equals τ_c, the row and column taus generally will not be equal to each other.

The Goodman-Kruskal tau has several advantages for contingency table analysis. First, like lambda but unlike phi, it can be computed for any size table. Second, its intermediate values between the lower and upper limits of 0 and 1.00 are readily interpretable. Third, it avoids the insensitivity to highly skewed distributions that lambda is subject to. Generally speaking, a *PRE* interpretation is highly desirable for measures seeking to determine the degree of association between nonintervally-scaled variables.

16.5 An Ordinal-Level *PRE* Measure of Association: Gamma

Probably the best way to explicate the *gamma coefficient,* which Goodman and Kruskal proposed in their 1954 paper, is to work through a problem for which it is aptly suited. Suppose that a sociologist who's interested in education and in work and occupations conducts a study of 200 university professors in a particular discipline. This sociologist wants to know whether the data in Table 16.11, showing the rank of the department from which a professor receives the Ph.D., have a bearing on where it is that he or she now teaches.

Table 16.11.   4 × 4 Contingency Table Showing Rank of Ph.D. Program and Present Affiliation of Its Graduates

Rank of Program	PRESENT AFFILIATION				
	Highly Prestigious	Rated	Adequate Plus	Adequate	
Highly Prestigious	20	10	5	5	40
Rated	10	25	15	10	60
Adequate Plus	5	10	25	20	60
Adequate	2	10	18	10	40
	37	45	63	45	200

Gamma is a symmetric measure of association. This means that the contingency table will yield only one coefficient, regardless of which variable is logically or theoretically dependent. In this case, it would be logical for the column variable ("present affiliation") to be the dependent variable. As a pair-wise comparison measure, gamma makes us look at all relevant pairs and determine whether each pair is concordant or discordant. For example, take one of the twenty subjects occupying the cell defined by row 1/column 1. This subject occupies the highest rank on both variables ("highly prestigious"/"highly prestigious"). When we compare it with a subject occupying the cell defined by row 2/column 2 ("rated"/"rated"), we can see that the second subject ranks lower on both variables. Thus, there is *concordance* between the variables on the way this particular pair is ranked. Let us now consider the rank of a professor who graduated from an "adequate-plus" Ph.D. program and now teaches in a "rated" department. When we compare this pair, we find discordance, since there are opposite rankings on the row and column variables.

The total number of unique pairs that can possibly be formed for a given N is: $N(N - 1)/2$, which, in our example, translates into $[(200)(199)]/2$, or 19,900. Fortunately, we don't have to perform the Herculean task of listing all relevant pairs in order to compute the gamma coefficient. In fact, since gamma ignores tied pairs, the twenty subjects occupying row 1/column 1 are not compared with each other.

To determine the number of *concordant* and *discordant pairs,* we return to the data of Table 16.11 and locate the *focal cell* that has the highest rankings on both variables. All pairs consisting of subjects in that cell and any other relevant cell will form concordant pairs. The member from the other cell will always be ranked lower on both variables. Next, we locate the focal cell that's defined by the highest rank on one variable and the lowest ranking on the other variable. It is one of the cells that's defined by row 1/column 4 or row 4/column 1. When members of this cell are paired with subjects from any relevant cell, they will form discordant pairs; that is, one member will rank higher on one variable and lower on the other.

Once we have located the starting focal cells, we proceed step by step to arrive at all concordant and discordant pairs. Here is a convenient method of enumeration for concordant cells: First, eliminate any cell that would result in tied pairs. In other words, eliminate any cell that's located in the same column or row as the focal cell. In the shaded cells of Figure 16.1(a), we can see all the relevant cells whose members are to be paired with the members of the first focal cell. Second, multiply the focal-cell frequency by the sum of the frequencies in all the relevant cells, which in this case results in $[20(25 + 15 + 10 + 10 + 25 + 20 + 10 + 18 + 10)]$, or 2,860 pairs. Then move right to the new focal cell defined by row 1/column 2 (see Figure 16.1 (b)). Remember to look for cells whose subjects are defined by ranks consisting of less than the highest

(a)

20			
	25	15	10
	10	25	20
	10	18	10

(b)

	10		
		15	10
		25	20
		18	10

Figure 16.1. Focal Cells for the Computation of Concordant Pairs

ranking on the row variable and less than the second-highest ranking on the column variable. Continue this process by multiplying each new focal cell by the sum of the frequencies of all cells lower on both variables—which, as we can tell by inspecting the table, are situated down and to the right of each new focal cell—until all relevant cells are exhausted. Lastly, sum the concordant pairs as follows:

$$20(25 + 15 + 10 + 10 + 25 + 20 + 10 + 18 + 10)$$
$$+ 10(15 + 10 + 25 + 20 + 18 + 10)$$
$$+ 5(10 + 20 + 10)$$
$$+ 10(10 + 25 + 20 + 10 + 18 + 10)$$
$$+ 25(25 + 20 + 18 + 10)$$
$$+ 15(20 + 10)$$
$$+ 5(10 + 18 + 10)$$
$$+ 10(18 + 10)$$
$$+ 25(10)$$
$$= 7,965$$

Similarly, we obtain the number of discordant cells by selecting the two cells defined by the highest ranking on one variable and the lowest ranking on the other; for example, the cell defined by row 4/column 1. Then we multiply this cell frequency by the sum of the frequencies of all relevant cells that are situated up and to the right, as shown by Figure 16.2 (a). In this case, the term "relevant cells" means that the subjects in those cells must rank higher on one variable and lower on the other than the subjects in each of the successive focal cells. Next, we move to the right in order to establish the new focal cell defined by row 4/column 2, and we locate cells whose subjects are rated above "adequate" on the row variable and below "rated" on the column variable

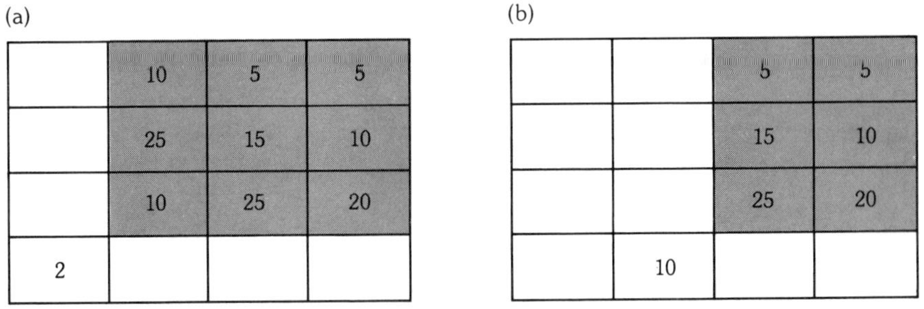

(a)

(b)

Figure 16.2. Focal Cells for the Computation of Discordant Pairs

(see Figure 16.2 (b)). We continue this process until all discordant pairs are accounted for,* and sum as follows:

$$
\begin{aligned}
&2(10 + 25 + 20 + 25 + 15 + 10 + 10 + 5 + 5) \\
&+ 10(25 + 20 + 15 + 10 + 5 + 5) \\
&+ 18(20 + 10 + 5) \\
&+ 5(25 + 15 + 10 + 10 + 5 + 5) \\
&+ 10(15 + 10 + 5 + 5) \\
&+ 25(10 + 5) \\
&+ 10(10 + 5 + 5) \\
&+ 25(5 + 5) \\
&+ 15(5) \\
&= 3{,}280
\end{aligned}
$$

The gamma coefficient is defined by Formula (16.13):

$$
\gamma = \frac{C - D}{C + D} \tag{16.13}
$$

where:

$$
C = \text{the number of concordant pairs, and}
$$
$$
D = \text{the number of discordant pairs.}
$$

---

*It is possible to arrive at the same number of concordant pairs, starting with the cell defined by row 4/column 4 as the first focal cell and then moving left and up. In addition, we can obtain the number of discordant pairs by starting with the cell defined by row 1/column 4. The logic is the same, except that when we start with the cell defined by the lowest ranks, the agreement is that subjects in all relevant cells paired in this direction will rank higher than the successive focal cells.

As this formula shows, the range of gamma is from −1.00 (perfect inverse relationship) through 0 (no association) to 1.00 (perfect positive relationship).

In the present example, the coefficient is:

$$\gamma = \frac{7,965 - 3,280}{7,965 + 3,280} = 0.42$$

The positive coefficient of 0.42 tells us that concordant pairs are dominant, and that subjects tend to be ranked in the same order on both variables. In this example, it means that the graduate program from which a university professor received the doctorate has a direct bearing on what type of department that professor is presently affiliated with.

Gamma also has a *PRE* interpretation (see Costner (1965)), which is based upon the probability of randomly selecting concordant and discordant pairs. According to *PRE* logic, as we recall, we make a comparison between two prediction rules. To arrive at prediction Rule I (yielding E_1), we play our hypothetical guessing game: supposedly, we don't know the independent variable. All we presume to know is the number of relevant pairs expressed by ($C + D$). Since we don't know which is dominant, we predict that all pairs are the same, either concordant or discordant. In the long run, our error rate based on this prediction rule will be 0.5. Thus, E_1 is derived from the following equation:

$$E_1 = \tfrac{1}{2}(C + D) \tag{16.14}$$

Once we know the independent variable, we can predict that any random pair will be of the dominant type, which in the present example is the concordant one. This prediction rule based on this knowledge (Rule II) will result in the number of errors equal to the number of pairs of the other type; in this case, the number of discordant pairs. Thus:

$$E_2 = min(C,D) \tag{16.15}$$

where:

$min(C,D)$ = the value of the concordant or discordant pairs; that is, whichever is least.

Use the familiar *PRE* equation:

$$\gamma = \frac{E_1 - E_2}{E_1} \tag{16.16}$$

we can arrive at the gamma coefficient for our problem:

$$\frac{5{,}622.5 - 3{,}280}{5{,}622.5} = 0.42$$

which equals the gamma coefficient that we computed before. Our interest in the *PRE* interpretation relates only to the absolute value of gamma, not to the direction of the relationship. Thus, 0.42 tells us that our knowledge of the independent variable has enabled us to reduce our number of errors by 42 percent. The *PRE* interpretation makes gamma an attractive measure for researchers whose data are ordinally scaled.

Nevertheless, gamma has a serious disadvantage: since it does not take tied pairs into account, it tends to overstate the strength of relationship between two variables. In fact, it is possible to achieve a perfect association of ±1.00 with only one untied pair if all the others are tied on one or both variables. When there are numerous ties in the data, it may be advisable to move down a scale of measurement and compute a lambda coefficient, if we want a *PRE* measure. A number of measures, specifically *Kendall's tau* and *Somer's d,* do take tied pairs into account, but generally they don't have a *PRE* interpretation. (*See* Weiss (1968) and Blalock (1972: 418–427) for a discussion of these measures.)

16.6 Yule's *Q*: A Special Gamma

At the turn of the century, a statistician named George U. Yule developed an associational measure for a 2 × 2 contingency table, which became known as *Yule's Q.* The *Q* measure was used widely in sociological literature, and it still is used by political scientists doing roll-call analysis of legislative bodies. In legislative studies, Yule's *Q* is used with such nominal-level data as determining an association between roll-call votes and specific categories (for example, party affiliation or region). After gamma was developed in 1954, it became apparent that Yule's *Q* could also be seen as a case of gamma, when ranked data are arranged in a 2 × 2 table. The cells in a 2 × 2 table with reference to Yule's *Q* conventionally are labeled:

$$\begin{array}{cc} a & b \\ c & d \end{array}$$

and the coefficient itself is generated by the formula:

$$Q = \frac{ad - bc}{ad + bc} \tag{16.17}$$

where, with ordered data:

$$ad \;=\; \text{the concordant pair, and}$$
$$bc \;=\; \text{the discordant pair.}$$

If we inspect the Q measure, we will see that it is equivalent to the formula for gamma. Since the development of gamma, the Q measure generally has not been used with ranked data. The more general gamma coefficient (of which the Q measure is a special case) is reported instead. However, the Q measure still is used when nominal-level data are involved.

16.7 Significance Tests for Ordinal-Level Variables

As we have indicated in this chapter, statisticians and statistically oriented social scientists have used nonparametric statistics extensively since the 1940s. Two important reasons for the wide use of nonparametrics are that they make more lenient assumptions and that they apply to many practical research situations. This is especially true of significance tests designed for the ordinal-level variables that social scientists, professionals, and people working in applied social science fields often encounter.

In the sections to come, we present some tests paralleling those designed for nominal- and interval-level data. Specifically, the one-variable Kolmogorov-Smirnov test is like the one-variable chi-square goodness-of-fit test for a nominal-level variable. The Mann Whitney, which is appropriate for ranked data and nominal categories, is just like the difference between means test used with interval data. Finally, the Kruskal-Wallis test, which allows us to determine significance with an ordinal and nominal variable, is just like the analysis of variance test. We use this test when we have three or more nominal categories and an intervally scaled variable.

16.7.1 Kolmogorov-Smirnov: The One-Variable Case

When rank-ordered data are involved, the *Kolmogorov-Smirnov test* for goodness-of-fit (commonly referred to as the *Smirnov test*) is a useful counterpart to the one-variable chi-square test. Another attraction of the Smirnov test is that it can be used with smaller-sized samples than chi square can. And as in chi square, the null hypothesis asserts that the variable is distributed uniformly among the various categories. But as an ordinal-level test, Kolmogorov-Smirnov uses a characteristic that's associated with this type of data—namely, the ability to cumulate frequencies.

To conduct the one-variable Smirnov test, we compare the observed cumulative frequency distribution with the hypothetical one. We make this comparison by obtaining the maximum absolute difference between the two cumulative distributions. In symbols:

$$D = max \left| Cum \, f_o - Cum \, f_e \right| \qquad (16.18)$$

where:

$Cum \, f_o$ = the observed cumulative frequency distribution, and
$Cum \, f_e$ = the expected cumulative frequency distribution.

As in all significance tests, if the obtained value is equal to or greater than the critical value (listed in Table J of the Appendix), we reject the null hypothesis.

For example, suppose that an educational researcher takes a random sample of forty professors who had articles published in the leading journals of a particular field (see Table 16.12). The question the researcher poses is whether the rating of the university with which the journal contributor is affiliated had some bearing on the article's publication in these prestigious outlets.

Table 16.12. Number of Publications and Departmental Affiliation for Forty Professors

Departmental Ranking	Number of Published Articles	$Cum \, f_o$	$Cum \, f_e$	$max \left\| Cum \, f_o - Cum \, f_e \right\|$
Highly prestigious	20	20/40	10/40	10/40
Rated	9	29/40	20/40	9/40
Adequate plus	7	36/40	30/40	6/40
Adequate	4	40/40	40/40	—

First we have to locate the maximum (or absolute) difference between the cumulative relative frequencies that represent the empirical and hypothesized distributions, to arrive at $max \left| Cum \, f_o - Cum \, f_e \right|$. A glance at the table reveals that the largest absolute difference is 10/40, or 0.25. Next we have to obtain the critical value for an alpha level of, say, 0.05, which is 0.22. (We arrived at the critical value, in this case, by using the formula $1.36/\sqrt{N}$. Critical values for sample sizes below 35 are obtained directly from the table.) Since the computed value of 0.25 exceeds the critical value of 0.22, we reject the null hypothesis. We assert instead that the proportion of articles accepted in the prestigious journals is not distributed uniformly among professors affiliated with variously ranked departments.

We ought to bear a few warnings in mind when we use the one-variable Smirnov test. The statistical purist may advise using the test only with continuously distributed data, and may say that grouping makes it more difficult to reject the null hypothesis. But unfortunately, in most realistic social science research situations the variables lend themselves to grouping. When you have good reason to suspect that the null hypothesis is not being rejected because of the grouping, it might be worthwhile to move down a level of measurement and use the chi-square test, which explicitly requires the use of grouped data.

16.7.2 Kolmogorov-Smirnov: The Two-Sample Case

We may extend the Smirnov test for goodness-of-fit to the situation involving two samples. In the last section, we compared an observed cumulative frequency distribution with an expected cumulative frequency distribution to find out whether these frequencies could have occurred by chance. Now we want to determine whether two random samples are drawn from the same population. The null hypothesis, which asserts no difference between the samples, would lead us to suspect that the cumulative frequency distribution for a given characteristic should be similar. Large differences between the samples would suggest rejection of the null hypothesis and acceptance of the alternative hypothesis that the samples come from different populations. In practice, the difference between the one-variable and the two-sample Kolmogorov-Smirnov tests is that we now compare two empirical cumulative distributions instead of an observed distribution and a random allocation.

The following problem illustrates the Smirnov test for independence. Suppose that a state university's political research institute asks us to conduct a survey of party identifiers in the state. We want to know whether Democratic and Republican party identifiers show a significant difference in terms of their educational attainment levels. So we rank-order the educational attainment levels of the party adherents into the four categories listed in Table 16.13.

In order to compute the test statistic labeled K_D (we use this label to distinguish it from the one-variable statistic), we convert the two cumulative frequency distributions into percentages. Then we compute the maximum absolute difference between the two cumulative frequencies, using the formula:

$$K_D = max \ |Cum \ f_1 - Cum \ f_2| \qquad \textbf{(16.19)}$$

where $Cum \ f_1$ and $Cum \ f_2$ indicate the respective cumulative frequency distributions.

Using the data from Table 16.13, we obtain a maximum difference of $|0.469 - 0.375|$, or 0.094. We then compare this value with the critical region in Table L (Table K is used for samples $\leqslant$ 40). Using a two-tailed test at $\alpha = 0.05$, the critical region is:

$$1.36 \ \sqrt{\frac{N_1 + N_2}{N_1 N_2}} = 1.36 \ \sqrt{\frac{320 + 400}{(320)(400)}} = 0.102$$

Table 16.13. Party Identification and Educational Attainment of Individual Within a State

Educational Attainment	Democrat	Cum f	Cum %	Republican	Cum f	Cum %	max $|Cum f_1 - Cum f_2|$
College	50	50	12.50	60	60	18.70	0.062
Some college	100	150	37.50	90	150	46.90	0.094
High school	210	360	90.00	140	290	90.60	0.006
Some high school	40	400	100.00	30	320	100.00	—

Since $0.094 < 0.102$, we cannot reject the null hypothesis. Instead, we say that there is no reason to maintain that the educational attainment levels of Democratic and Republican party identifiers in this particular state differ significantly. Of course, this is statistically equivalent to saying that the two samples are drawn from the same underlying population.

16.7.3 The Mann-Whitney U Test

The *Mann-Whitney U test,* first proposed in the 1940s, is a nonparametric test that we can use instead of the Student t test when the measurements fail to achieve interval scaling, or when we want to avoid the assumptions of its parametric counterpart. Most often, the Mann-Whitney U test is used in experimental situations where we want to determine whether two groups exposed to different treatments differ significantly—for example, if one group undergoes a training program and the other doesn't. Thus, the null hypothesis is that the ranks in question are distributed randomly between the two groups. The research hypothesis, against which we test the null hypothesis, is that the groups differ—for example, that one group shows higher ranks than the other (one-tailed research hypothesis). A two-tailed or nondirectional research hypothesis would not specify direction.

Consider two groups of college students randomly selected from students taking the Law School Admissions Test (LSAT). One group of students (Group A) took a preparation course for the examination, and the other (Group B) did not (see Table 16.14). Does the prep course make a difference?

Table 16.14. Ranks of Two Groups of College Students Participating in an Experiment to Judge the Effectiveness of an LSAT Preparatory Course

GROUP A (TOOK COURSE)		GROUP B (DID NOT TAKE COURSE)	
LSAT Score	Rank	LSAT Score	Rank
760	1	700	2
680	3	600	5
650	4	550	6
500	8	520	7
480	10	490	9
$(N_1 = 5)$		450	11
		420	12
		$(N_2 = 7)$	

To determine whether one group tends to have a disproportionate number of high or low rankings, we array the data in Table 16.14, as follows:

Rank	1	2	3	4	5	6	7	8	9	10	11	12
Group	A	B	A	A	B	B	B	A	B	A	B	B

Now we are ready to compute the number of times that a rank in a designated category, in this case Group A, precedes a subject identified with the other category (in our case, Group B). Inspection reveals that the highest-ranking A (rank 1) precedes seven Bs, the second and third As (ranks 3 and 4, respectively) each precede six Bs, and so on, until we reach the last A (rank 10), which precedes two Bs. Hence, the number of As preceding the Bs is 7 + 6 + 6 + 3 + 2, or 24. For B, we find the number of Bs preceding A to be 4 + 2 + 2 + 2 + 1, or 11. The smaller of these two values is designed U; the larger is designated U'.

The following equations should hold if we compute the U values properly:

$$U = N_1 N_2 - U' \tag{16.20}$$
$$U' = N_1 N_2 - U \tag{16.21}$$

where N_1 and N_2 are the sizes of the respective samples. (Convention generally designates the larger sample size as N_2.) And in the current problem, both U and U' check out:

$$U = (5)(7) - 24 = 11$$
$$U' = (5)(7) - 11 = 24$$

Table M (in Appendix III) contains two critical values for each pair of sample sizes. The upper value refers to U, and the bottom value refers to U'. Thus, we must reject the null hypothesis $(U = U')$ if the computed U is less than or equal to the selected critical tabled value. We can also use U as a point of comparison; in this case, if the obtained U' is equal to or greater than the critical tabled value (bottom value) for the particular sample sizes, we reject the null hypothesis.

Using a one-tailed test at $\alpha = 0.05$, we compare the obtained U value of 11 with the critical value for sample sizes $N_1 = 5$ and $N_2 = 7$, which is 6. Since 11 is greater than 6, we cannot reject the null hypothesis. If we wish to use U', we can readily see that the bottom value for this pair of sample sizes is 29. Since the critical value of 29 is greater than our computed value of 24 for U, we retain the null hypothesis.

When the sample sizes become reasonably large, it may be more convenient to use the following computational formulas rather than the intuitive counting procedure we've used thus far:

$$U = N_1 N_2 + \frac{N_1(N_1 + 1)}{2} - \Sigma R_1 \tag{16.22}$$

and:

$$U = N_1 N_2 + \frac{N_2(N_2 + 1)}{2} - \Sigma R_2 \tag{16.23}$$

where ΣR_1 and ΣR_2 are the sum of the ranks for sample size N_1 and N_2, respectively. As before, the smaller computed value becomes the U and the larger U' when compared with the table of critical values. Referring to Table 16.16, we see that $\Sigma R_1 = 26$ and $\Sigma R_2 = 52$. Substituting the appropriate values into Formulas (16.22) and (16.23) will yield a U' of 24 and a U of 11, which confirms our previous results.

As the sample sizes increase, the sampling distribution of U will approximate a normal distribution. In such situations, the theoretical sampling distribution of U will have a mean of $\mu_U = (N_1 N_2)/2$ and a standard deviation σ_U defined by $\sqrt{N_1 N_2(N_1+N_2+1)/12}$. In these situations, the obtained U value is converted into a z score, as follows:

$$z = \frac{U - (N_1 N_2)/2}{\sqrt{N_1 N_2(N_1 + N_2 + 1)/12}} \quad \text{(16.24)}$$

We can then use Table B to determine statistical significance. While the U statistic will begin to approximate the normal distribution when each sample size is greater than eight, the normal approximation for U ordinarily is not used unless one of the samples exceeds twenty.

Generally, we use the Mann-Whitney U test when there are no ties across groups, although there is a correction factor for ties when we use the normal approximation for U. However, some researchers advise that an alternative to U, such as Kolmogorov-Smirnov, be used when the number of ties across categories is large. You can find a discussion of tied ranks in connection with the Mann-Whitney U test in Siegel (1956: 123–125) and Blalock (1972: 259–263).

16.7.4 The Kruskal-Wallis Test

The *Kruskal-Wallis test* is a direct extension of the Mann-Whitney U test for more than two independent samples. Just as Mann-Whitney parallels the t test, Kruskal-Wallis parallels the F test and may be considered a one-way analysis of variance with rank-order data. Computing the Kruskal-Wallis H statistic is straightforward, and its logic is easy to interpret.

Let's take an example. A social scientist who's interested in community power obtains a list of a local community's business, civic, and minority group leaders from an impartial panel of judges. Four leaders are selected randomly from each group and subjected to in-depth interviews. One series of questions is designed to get at their sense of political effectiveness—that is, the feeling that they make a significant contribution to the community's decision-making process. The responses are coded, and each leader is assigned an efficacy score on a scale ranging from a low of 0 to a high of 100. The leaders are then ranked on the basis of their scores, as indicated in Table 16.15.

Table 16.15. Political Effectiveness Scores for Three Different Groups of Community Leaders

	BUSINESS		CIVIC		MINORITY GROUP	
	Score	Rank	Score	Rank	Score	Rank
	94	1	90	2	70	8.5
	85	3	80	5	65	10
	83	4	75	7	60	11
	78	6	70	8.5	50	12
$R_i =$		14		22.5		41.5

The basis for the comparison among the categories, as in Mann-Whitney, is the sum of the respective group rankings. If there is no significant difference among the groups, we would expect the sums of the group rankings to be reasonably similar and differ only by sampling error. If, however, the ranked sums differ by more than sampling error, which can be determined by the known distribution of the H statistic (Table N in Appendix III), we reject the null hypothesis and conclude that the samples came from different populations. We test the null hypothesis by:

$$H = \frac{12}{N(N + 1)} \frac{\Sigma R_i^2}{N_i} - 3(N + 1) \qquad \textbf{(16.25)}$$

where:

N = the number of cases among all groups,
R_i = the ranks in any given group, and
N_i = the number of cases in any given group.

When we have tied ranks, H should be divided by the correction factor:

$$C = 1 - \frac{(t^3 - t)}{(N^3 - N)} \qquad \textbf{(16.26)}$$

where t is the number of ranks tied at any given rank.

As we do with all significance tests, we compare the computed value with that in the tabled critical value. For the Kruskal-Wallis test, there are actually two tables of critical values that we can consult. If all the sample sizes are 5 or less, we consult Table N at the significance level desired. For larger samples, we may use the standard chi-square table (Table I) by substituting H for the chi-square value. We find the degrees of freedom needed to determine the critical value from Table I by subtracting the number of groups by one (symbolically, $df = k - 1$).

In the present example, the value of H is:

$$\frac{12}{12(12 + 1)} \left[\frac{14^2}{4} + \frac{22.5^2}{4} + \frac{41.5^2}{4}\right] - 39 = 7.62$$

Since we have a pair of tied scores, we compute a correction factor of:

$$1 - \frac{(2^3 - 1)}{(12^3 - 12)} = 0.9965$$

and obtain a corrected H statistic of 7.62/0.9965, or 7.65.

Using a two-tailed test at $\alpha = 0.05$ for sample sizes of 4, 4, 4, we learn from Table N that our computed value must equal or exceed 5.69. Since the computed H of 7.65 does this, we reject the null hypothesis and conclude instead that the population distributions are not identical.

The Kruskal-Wallis test is a useful alternative to the standard one-way analysis of variance when the interval-level assumptions required for the parametric test are not met. One of the attractive features of H is that it is not affected seriously by tied ranks (as the above example shows) unless an unusually large number of them exists. In fact, tied ranks make the Kruskal-Wallis test conservative in this sense: when the correction for ties is made, we are less likely to commit a Type I error.

This rather lengthy chapter doesn't begin to exhaust all the nonparametric measures that have been developed, some for very specific types of problems.* And although the Kruskal-Wallis test concludes our survey of nonparametrics, we'd like to say a few general words about nonparametric measures before we end.

16.8 Caution: Some Notes on Power and Efficiency

When we use significance tests, we automatically involve risk factors. That is, we bring in the probability of making an incorrect decision. As we know, these errors fall into two general patterns: (1) rejecting the null hypothesis when it is in fact true (Type I errors), and (2) failing to reject the null hypothesis when it is false (Type II errors). We have direct control over the risk we are taking in making a Type I error, and we almost always use the reasonably low 0.01 and 0.05 alpha levels. If we want to be more conservative, we use an alpha level of 0.001 instead. But unfortunately, as we increase the alpha level (thereby minimizing the risk of making a Type I error), we also increase the probability of

---

*If you wish to know more about nonparametrics, refer to such specialized texts as Siegel (1956), Mosteller and Rourke (1973), and Lehmann (1975).

making a β or Type II error. We generally cannot compute a beta level for each particular test of a null hypothesis, since the parameters needed for such a computation are not known, and therefore the risk associated with Type II errors is compounded further. However, statisticians have computed probabilities of making Type II errors for hypothetical values of parameters. We don't really need to concern ourselves with the procedures for calculating the probability of making Type II errors, since they belong to the field of mathematical statistics. But we do need to become acquainted with the concept of the *power of a test.*

The power of a test is defined as $1 - \beta$. In other words, it is the probability of rejecting a false null hypothesis. When we compare statistical tests, if we say that one test is more powerful than another, it's like saying that such a test is more likely to discover any significant differences that may exist. In general, parametric tests are more powerful than nonparametric tests.

We express comparisons between parametric and nonparametric significance tests in the form of the latter's *power efficiency,* which is defined as the power of a parametric test relative to the power of a nonparametric test for a given sample size. The power efficiency of nonparametric tests usually ranges between 0.63 and 1.00. For example, when we say that a nonparametric test has a power efficiency of 0.65, we mean that using a nonparametric test, it would take a sample of 100 to achieve a specific power, whereas a parametric test would need a sample of only 65 to achieve the same result.

Power comparisons have a heuristic value, but we shouldn't let them mislead us. They are made on sample data that meet the assumptions of both parametric and nonparametric tests. A power efficiency of under 1.00 merely means that if we have a choice—that is, if our data meet the assumptions of a parametric test (usually the *t* test, since we probably are dealing with small samples)—we should use the more powerful parametric test instead of a nonparametric alternative. But most often, our data don't give us this choice. Remember, nonparametrics were developed to overcome the rather strict requirements of parametric tests. Researchers invariably use them when the assumptions required for the parametric tests are lacking. Therefore, don't worry that nonparametrics are said to be "less powerful" than parametrics. In most instances, this isn't true.

Where We Stand

By now, you have taken a sometimes tedious but (we trust) worthwhile journey into what is known as applied statistics. We hope that you have a better appreciation of the professional literature (which uses the concepts and symbols discussed in this text) in your major, whether that major is education or any of the social and behavioral sciences. The techniques you have mastered this term should enable you to conduct your own research projects in the same way as professionals in your field. Now stop a moment. Can you remember how you used to feel about the term "statistics" just a short time ago (that is, before you read this book)? Your understanding of the term and its ramifications is much greater now, isn't it?

As this book comes toward its end, we say: there is more to learn—much more. And it can be fascinating. You may want to learn about applied regression analysis, path analysis (especially if you are a sociology or political science major), or perhaps time-series analysis (if your major is economics). If you're a psychologist or interested in educational tests and measurements, you might want to study factor analysis and the uses of factor scores in doing regression analysis. The growing area of nonparametrics also has more specialized techniques for those whose work involves small samples and categorical data. The "Bibliography" section at the back of this text offers you a starting point for further exploration. You also might want to consult with your instructor about the statistical literature that's appropriate for your field. And now it's time for us to part company and congratulate you on completing "stat."

Terms and Symbols to Remember

Distribution-Free Statistics
Nonparametric Statistics
Chi Square (χ^2)
Goodness-of-Fit Test
Test of Independence
Chi Square (One-Variable Case)
Chi Square (Two-Sample Case)
Yates Correction Factor
Cell Collapsing
Phi Coefficient (ϕ)
Cramér's V
Proportional Reduction of Error *(PRE)*
Lambda (λ)
Asymmetric Lambda (λ_a)
Symmetric Lambda (λ_s)
Goodman-Kruskal Tau (τ)
Gamma (γ)
Concordant Pairs
Discordant Pairs
Focal Cells
Yule's Q
Kolmogorov-Smirnov (D)
Kolmogorov-Smirnov (K_D)
Mann-Whitney U Test (U and U')
Kruskal-Wallis Test (H)
Power of a Test
Power Efficiency

Exercises

1. A study is conducted by taking a simple random sample of fifty case files in a large social agency. Each of the selected case files is classified by the sex of the caseworker who worked with the client. The data for this study appear below. Can we reject the null hypothesis (there is no difference in the sex distribution of clients seen by male and female caseworkers) at the 0.01 level of significance? Assume a two-tailed test.

	CASE WORKER	
Client	Male	Female
Male	5	15
Female	10	20

2. A survey is conducted with a simple random sample of 400 students at a large university. The table below shows the type of change most desired by students, according to majors.
   (a) On the basis of this table, can we reject the null hypothesis (there is no difference in the distribution of desired change for students with different majors)? Use a two-tailed test at $\alpha = 0.05$.
   (b) Consider the science majors as one sample and test the null hypothesis that there is no difference in the distribution of desired change for science majors. Use a two-tailed test at $\alpha = 0.01$.

	MAJOR		
Desired Change	Science	Humanities	Other
Lower tuition	40	10	100
Smaller classes	90	10	50
Greater freedom	20	30	50

3. A survey is conducted with a simple random sample of 500 registered Democrats in New York State. The survey includes questions about place of residence, education, and preferred Democratic candidate in the 1972 presidential election. The data for this survey appear below. Using a two-tailed test at $\alpha = 0.01$, test the following hypotheses:
   (a) There is no difference in the candidate preference distribution of Upstate New York and New York City Democrats.
   (b) There is no difference in the candidate preference distribution of college-educated and non-college-educated Democrats in New York State when we control for place of residence (New York City-Upstate New York).

| Preferred Candidate | NEW YORK CITY | | UPSTATE NEW YORK | |
	Not College-Educated	College-Educated	Not College-Educated	College-Educated
Muskie	50	30	25	20
Kennedy	100	25	50	15
Humphrey	25	25	25	10
Other	25	20	50	5

4. A survey is conducted with a simple random sample of 600 registered voters in a large city. Answer the following questions, using the data from this survey (given below).

    (a) Using a two-tailed test at $\alpha = 0.05$, can we reject the null hypothesis that there is no difference in the party registration distribution of Protestants, Catholic, and Jewish voters in the given city?

    (b) Using a two-tailed test at $\alpha = 0.05$, can we reject the null hypothesis that there is no difference in the party registration distribution of Catholic and Jewish voters in the given city?

    (c) Using a two-tailed test at $\alpha = 0.05$, can we reject the null hypothesis that there is no difference in the party registration distribution of Protestant and other voters (Catholic and Jewish combined) in the given city?

    (d) Given the results of your analysis in questions 4(a), 4(b), and 4(c), what conclusions would you come to about the difference among party registration distributions for the three religious groups in this city?

Party Registration	Protestant	Catholic	Jewish
Democratic	90	134	76
Republican	180	44	16
Other	30	22	8

5. A city is considering repealing its Sunday blue-law ordinance. The local newspaper takes a sample of the residents. Respondents' religious preferences are cross-tabulated with their approval or disapproval of the repeal proposal. The data appear below. Using a two-tailed test at $\alpha = 0.05$, determine whether there is a significant difference among the religious groups on this question.

Repeal of Blue Law	Protestant	Catholic	Jewish
Approve	120	55	50
Disapprove	80	65	30

6. Compute the appropriate measure of association for the data in Exercise 1.

7. Compute the appropriate measures of associations for the data in Exercises 2, 4, and 5.

8. Explain the differene between chi-square-based associational measures and lambda.

9. Suppose that for the data given below, you want to find out the association of the age variable with a person's tendency to approve the legalization of marijuana. Select an appropriate PRE measure of association and explain your answer.

Legalization of Marijuana	AGE GROUP					
	13–19	20–29	30–39	40–49	50–59	above 60
Approve	50	80	55	35	20	10
Disapprove	10	20	50	45	40	40

10. A political scientist who's interested in the question of working-class authoritarianism wants to find out the degree of association between socioeconomic status (SES) and commitment to democratic ideals (measured on a "democraticness" scale). Compute an associational measure. (Assume that both variables are ordinally scaled.)

Commitment to Democratic Ideals	SES		
	High	Medium	Low
High	60	30	10
Medium	35	35	30
Low	20	30	50

11. Compute an appropriate associational measure of intergroup differences for this legislative problem.

	Yea	Nay
Republicans	30	10
Democrats	15	45

12. A professor who has experimented for years with statistics test questions makes this hypothesis: for a particular version of his exam, the distribution of grades from A to F will be in the ratio of 1:3:4:1:1. He gives the exam to 200 students. The number of students falling into the respective grade levels of A, B, C, D, F are 25, 50, 80, 35, 10. Do the professor's

actual results fit his hypothesized distribution? Test the null hypothesis at the 0.05 level of significance. (Hint: two different goodness-of-fit tests can be used to test this hypothesis. Use both, and compare your results.)

13. A social psychologist wants to test this hypothesis: working-class people are more likely to be intolerant of deviant groups. So he samples both the socioeconomic elite and the mass public in a city, and ranks them according to a tolerance scale ranging from highly tolerant to intolerant. The data appear below. Using a two-tailed test at $\alpha = 0.01$, test the null hypothesis that there is no difference between these two samples.

	Socioeconomic Elite	Mass Public
Highly tolerant	65	30
Somewhat tolerant	80	100
Minimally tolerant	30	120
Intolerant	25	50

14. An evaluations specialist wants to find out the impact of a federal job-training program for the unemployed. He gives a test, designed to measure basic job skills, to a randomly selected group of individuals who went through the training program (experimental group) and to another sample of unemployed people who didn't go through the training program (control group). The evaluator hypothesizes that the group with the training will do better. Based on the data below, perform a one-tailed Mann-Whitney test at the 0.05 significance level to test the evaluator's hypothesis.

EXPERIMENTAL		CONTROL	
Score	Rank	Score	Rank
96	1	90	3
95	2	85	5
89	4	75	9
83	6	70	10
80	7	65	13
78	8	60	15
68	11	58	16
67	12	55	17
63	14	50	18

15. An educational researcher employed by the state samples a number of school districts in the state. She gathers data on the median reading scores for high school seniors and rank-orders the districts. The data appear below. Using a two-tailed test at the $\alpha = 0.05$, determine whether

you would reject the null hypothesis (which is, there is no significant difference among the samples classified as urban, rural, and suburban).

	URBAN DISTRICTS		RURAL DISTRICTS		SUBURBAN DISTRICTS	
	Score	Rank	Score	Rank	Score	Rank
	9.8	21	9.5	23	11.3	10
	11.0	12	9.3	24	12.1	2
	11.5	7	10.9	13	12.9	1
Median	10.8	14	11.4	8.5	10.7	15
Reading Scores	10.5	17	10.6	16	10.1	20
	11.6	5.5	11.8	4	11.6	5.5
	9.7	22	11.1	11	11.4	8.5
	10.4	18	10.3	19	12.0	3

Bibliography

Alder, A. L., and Roessler, E. B. *Introduction to Probability and Statistics.* 5th ed. San Francisco: W. H. Freeman, 1972.

Alker, H. R. *Mathematics and Politics.* New York: Macmillan, 1965.

Baggaley, A. R. *Intermediate Correlational Methods.* New York: John Wiley, 1964.

Bartlett, M. S. "Some Experiments of Statistical Research in Agriculture and Applied Biology." *Journal of the Royal Statistical Society* 4(1937): 137–170.

Blalock, H. M. *Causal Inferences in Nonexperimental Research.* Chapel Hill: University of North Carolina Press, 1964.

———. *Social Statistics.* 2nd ed. New York: McGraw-Hill, 1972.

———. *Theory Construction: From Verbal to Mathematical Formulations.* Englewood Cliffs, N.J.: Prentice-Hall, 1969.

Boneau, C. A. "A Note on Measurement Scales and Statistical Tests." *American Psychologist* 16(1961): 160–161.

Box, G. E. P. "Non-Normality and Tests of Variances." *Biometrika* 40(1953): 318–335.

Bradley, J. V. *Distribution-Free Statistical Tests.* Englewood Cliffs, N.J.: Prentice-Hall, 1968.

Brugger, R. M. "A Note on Unbiased Estimation of the Standard Deviation." *American Statistician* 23 (1969): 32.

Carroll, J. B. "The Nature of the Data, or How to Choose a Correlation Coefficient." *Psychometrica* 26(1961): 347–372.

Cochran, W. G. "The Comparison of Percentages in Matched Samples." *Biometrika* 37(1950): 256–266.

———. *Sampling Techniques.* New York: John Wiley, 1953.

———. "Some Methods for Strengthening the Common Tests." *Biometrics* 10 (1954): 417–451.

Cohen, J., and Cohen, P. *Applied Multiple Regression/Correlation Analysis for the Behavioral Sciences.* Hillsdale, N.J.: Laurence Erlbaum, 1975.

Coleman, J. S. *Introduction to Mathematical Sociology.* New York: Free Press, 1964.

Coombs, C. H. *A Theory of Data.* New York: John Wiley, 1964.

——— ; Dawes, R. M.; and Tversky, A. *Mathematical Psychology: An Elementary Introduction.* Englewood Cliffs, N.J.: Prentice-Hall, 1970.

Costner, H. L. "Criteria for Measures of Association." *American Sociological Review* 30(1965): 341–353.

Cowden, D. J. "A Procedure for Computing Regression Coefficients." *Journal of the American Statistical Association* 53 (1958): 144–150.

Cramér, H. *Mathematical Methods of Statistics.* Princeton, N.J.: Princeton University Press, 1946.

Davis, J. A. *Elementary Survey Analysis.* Englewood Cliffs, N.J.: Prentice-Hall, 1971.

Deming, W. E. *Some Theory of Sampling.* New York: John Wiley, 1950.

Draper, N. R., and Smith, H. *Applied Regression Analysis.* New York: John Wiley, 1966.

Edwards, A. L. *Experimental Design in Psychological Research.* New York: Holt, Rinehart and Winston, 1960.

——— . *Statistical Analysis.* New York: Holt, Rinehart and Winston, 1954.

——— . *Statistical Methods for the Behavioral Sciences.* New York: Holt, Rinehart and Winston, 1954.

Ezekiel, M., and Fox, K. A. *Methods of Correlation and Regression Analysis.* New York: John Wiley, 1960.

Fisher, R. A. *The Design of Experiments.* 8th ed. Edinburgh: Oliver and Boyd, 1966.

——— . *Statistical Methods for Research Workers.* 13th ed. Edinburgh: Oliver and Boyd, 1958.

——— , and Yates, F. *Statistical Tables for Biological Agricultural and Medical Research.* 6th ed. London: Longman, 1974.

Freund, J. E. *Mathematical Statistics.* Englewood Cliffs, N.J.: Prentice-Hall, 1962.

——— . *Modern Elementary Statistics.* 2nd ed. Englewood Cliffs, N.J.: Prentice-Hall, 1962.

Galton, F. "Regression Toward Mediocrity in Hereditary Stature." *Journal of the Anthropological Institute of Great Britian and Ireland* 15(1886): 246–263.

Gold, D. "Statistical Tests and Substantive Significance." *American Sociologist* 4(1969): 42–46.

Goodman, L. A. "A General Model for the Analysis of Surveys." *American Journal of Sociology* 77(1972): 1035–1086.

——— . "How to Ransack Social Mobility Tables and Other Kinds of Cross-Classification Tables." *American Journal of Sociology* 75(1969): 1–39.

——— . "A Modified Multiple Regression Approach to the Analysis of Dichotomous Variables." *American Sociological Review* 37(1972): 28–46.

——— . "The Multivariate Analysis of Qualitative Data: Interactions Among Multiple Classifications." *Journal of the American Statistical Association* 65(1970): 225–256.

——— , and Kruskal, W. H. "Measures of Association for Cross Classification." *Journal of the American Statistical Association* 49(1954): 732–764.

——— . "Measures of Association for Cross Classification, II: Further Discussion and References." *Journal of the American Statistical Association* 54(1959): 123–163.

——— . "Measures of Association for Cross Classification, III: Approximate Sampling Theory." *Journal of the American Statistical Association* 58(1963): 310–364.

Green, B. F., and Tukey, J. W. "Complex Analyses of Variance: General Problems." *Psychometrika* 25(1960): 127–152.

Grizzle, J. E. "Continuity Correction in the Chi-Square Test for 2 × 2 Tables." *American Statistician* 21(1967): 28–32.

Gurr, T. R. *Politimetrics: An Introduction to Quantitative Macropolitics.* Englewood Cliffs, N.J.: Prentice-Hall, 1972.

Hays, W. L. *Statistics for the Social Sciences.* 2nd ed. New York: Holt, Rinehart and Winston, 1973.

Hodges, J. L., and Lehmann, E. L. "Testing the Approximate Validity of Statistical Hypotheses." *Journal of the Royal Statistical Society* 16(1954): 261–268.

Hoel, P. G. *Introduction to Mathematical Statistics.* 3rd ed. New York: John Wiley, 1962.

Hogg, R. V., and Craig, A. T. *Introduction to Mathematical Statistics.* 2nd ed. New York: Macmillan, 1965.

Huff, D. *How to Lie with Statistics.* New York: W. W. Norton, 1954.

Johnston, J. E. *Econometric Methods.* 2nd ed. New York: McGraw-Hill, 1972.

Kemeny, J. G.; Snell, J. L.; and Thompson, G. L. *Introduction to Finite Mathematics.* Englewood Cliffs, N.J.: Prentice-Hall, 1956.

Kendall, M. G. *Rank Correlation Methods.* New York: Hafner, 1955.

Kirkpatrick, S. A. *Quantitative Analysis of Political Data.* Columbus, Ohio: Charles E. Merrill, 1974.

Kish, L. *Survey Sampling.* New York: John Wiley, 1965.

Kruskal, W. H. "Ordinal Measures of Association." *Journal of the American Statistical Association* 53(1958): 814–861.

———, and Wallis, W. A. "Use of Ranks in One-Criterion Variance Analysis." *Journal of the American Statistical Association* 47(1952): 583–621.

Kyburg, H. E., and Smokler, H. E. *Studies in Subjective Probability.* New York: John Wiley, 1964.

Labovitz, S. "Criteria for Selecting A Significance Level: A Note on the Sacredness of .05." *American Sociologist* 3(1968): 220–222.

Land, K. C. "Principles of Path Analysis." In *Sociological Methodology 1969,* edited by E. Borgatta, ch. 1. San Francisco: Jossey-Bass, 1969.

Laplace, M. (Simon, P.) *A Philosophical Essay on Probabilities.* New York: Dover, 1951.

Lehmann, E. L. *Nonparametrics: Statistical Methods Based on Ranks.* San Francisco: Holden-Day/McGraw-Hill, 1975.

———. *Testing Statistical Hypotheses.* New York: John Wiley, 1959.

Lieberman, B. *Contemporary Problems in Statistics.* New York: Oxford University Press, 1971.

Loether, H. J., and McTavish, D. G. *Descriptive and Inferential Statistics: An Introduction.* Boston: Allyn and Bacon, 1976.

Maxwell, A. E. *Analysing Qualitative Data.* London: Methuen, 1961.

McNemar, Q. *Psychological Statistics.* 3rd ed. New York: John Wiley, 1962.

Miller, D. C. *Handbook of Research Design and Social Measurement.* 3rd ed. New York: David McKay, 1977.

Morris, R. N. "Multiple Correlation and Ordinally Scaled Data." *Social Forces* 48(1970): 299–311.

Morrison, D. F. *Multivariate Statistical Methods.* New York: McGraw-Hill, 1967.

———, and Henkel, R. E., eds. *The Significance Test Controversy.* Chicago: Aldine, 1970.

Moses, E. L. "Non-Parametric Statistics of Psychological Research." *Psychological Bulletin* 49(1952). 122–143.

Mosteller, F., and Rourke, R. E. K. *Sturdy Statistics.* Reading, Mass.: Addison-Wesley, 1973.

————, and Thomas, G. B. *Probability with Statistical Applications.* Reading, Mass.: Addison-Wesley, 1961.

Mueller, J. H.; Schuessler, K. F.; and Costner, H. L. *Statistical Reasoning in Sociology.* 3rd ed. Boston: Houghton Mifflin, 1977.

Nagel, E. *Principles of the Theory of Probability.* Chicago: University of Chicago Press, 1939.

Natrella, M. G. "The Relations Between Confidence Intervals and Tests of Significance." *American Statistician* 14(1960): 20–22.

Olds, E. G. "The 5 Percent Significance Levels of Sum of Squares of Rank Differences and Correlation." *The Annals of Mathematical Statistics* 20(1949): 117–118.

Owen, D. B. *Handbook of Statistical Tables.* Reading, Mass.: Addison-Wesley, 1962.

Parzen, E. *Modern Probability Theory and its Applications.* New York: John Wiley, 1960.

Plackett, R. L. "The Continuity Correction on 2 × 2 Tables." *Biometrika* 5(1964): 327–337.

Quenouille, M. H. *Rapid Statistical Calculations.* New York: Hafner, 1959.

Rai, K. B., and Blydenburgh, J. C. *Political Science Statistics.* Boston: Holbrook, 1973.

Reichmann, W. J. *Use and Abuse of Statistics.* Baltimore: Penquin, 1961.

Reynolds, H. T. *Analysis of Nominal Data.* Beverly Hills, Calif.: Sage, 1977.

Robinson, W. S. "Ecological Correlations and Behavior of Individuals." *American Sociological Review* 15(1950): 351–357.

Rosenberg, M. *The Logic of Survey Analysis.* New York: Basic Books, 1968.

Sandler, J. "A Test of the Significance of the Difference Between the Means of Correlated Measures, Based on Simplification of Student's t." *British Journal of Psychology* 46(1955): 225–226.

Scheffé, H. *The Analysis of Variance.* New York: John Wiley, 1959.

————. "A Method for Judging All Contrasts in the Analysis of Variance." *Biometrika* 40(1953): 87–104.

Schmid, C. F. *Handbook of Graphic Presentations.* New York: Ronald Press, 1954.

Selvin, H. "A Critique of Tests of Significance in Survey Research." *American Sociological Review,* 22(1957): 519–527.

Senders, V. L. *Measurement and Statistics.* New York: Oxford University Press, 1958.

Siegal, S. *Nonparametric Statistics.* New York: McGraw-Hill, 1956.

Simon, H. A. "Spurious Correlation: A Causal Interpretation." *Journal of the American Statistical Association* 49(1954): 467–479.

Slonim, M. J. *Sampling.* New York: Simon and Schuster, 1960.

Snedecor, G. W., and Cochran, W. G. *Statistical Methods.* 6th ed. Ames: Iowa State University Press, 1967.

Somers, R. H. "A New Asymmetric Measure of Association for Ordinal Variables." *American Sociological Review* 27 (1962): 799–811.

Spears, M. E. *Charting Statistics.* New York: McGraw-Hill, 1952.

Stephan, F. F., and McCarthy, P. J. *Sampling Opinions.* New York: John Wiley, 1958.

Stevens, S. S. "Measurement, Statistics and the Schemapiric View." *Science* 161(1968): 849–856.

————. "On the Theory of Scales of Measurement." *Science* 103(1946): 677–680.

————, ed. *Handbook of Experimental Psychology.* New York: John Wiley, 1951.

Stillson, D. W. *Probability and Statistics in Psychological Research and Theory.* San Francisco: Holden-Day, 1966.

Student. (Gosset, W. S.) "On the Error of Counting with a Haemacytometer." *Biometrika* 5(1907): 351–360.

————. "The Probable Error of a Mean." *Biometrika* 6(1908): 1–25.

Tanur, J. M., *et al. Statistics: A Guide to the Unknown.* San Francisco: Holden-Day, 1972.

Thorndike, R. L., and Hagger, E. *Measurement and Evaluation in Psychology and Education.* New York: John Wiley, 1969.

Torgerson, W. *Theory and Methods of Scaling.* New York: John Wiley, 1958.

Tufte, E. R. *Data Analysis for Politics and Policy.* Englewood Cliffs, N. J.: Prentice-Hall, 1974.

————, ed. *The Quantitative Analysis of Social Problems.* Reading, Mass.: Addison-Wesley, 1970.

Tukey, J. W. "Comparing Individual Means in the Analysis of Variance." *Biometrics* 5(1949): 99–114.

————. "One Degree of Freedom for Non-Additivity." *Biometrics* 5(1949): 232–242.

————. "The Problem of Multiple Comparisons." Mimeographed. Princeton University, 1953.

Walker, H. *Studies in the History of Statistical Methods.* Baltimore: Williams and Wilkins, 1929.

Walker, H. M., and Lev, J. *Statistical Inference.* New York: Henry Holt, 1953.

Wallis, W. A., and Roberts, H. V. *The Nature of Statistics.* New York: Free Press, 1965.

Walsh, J. E. *Handbook of Nonparametric Statistics.* Princeton, N.J.: Van Nostrand, 1962.

————. *Handbook of Nonparameteric Statistics, II.* Princeton, N.J.: Van Nostrand, 1965.

Weiss, R. S. *Statistics in Social Research.* New York: John Wiley, 1968.

Westergaard, H. L. *Contributions to the History of Statistics.* New York: A. M. Kelly, 1932.

Wilcoxen, F., and Wilcox, R. A. *Some Rapid Approximate Statistical Procedures.* Pearl River, N.Y.: Lederle Laboratories, 1964.

———— ; Katte, S.; and Wilcox, R. A. *Critical Values and Probability Levels for the Wilcoxen Rank Sums Test and the Wilcoxen Signed Rank Test.* New York: American Cyanamid, 1963.

Wilks, S. S. *Mathematical Statistics.* New York: John Wiley, 1962.

Wilson, T. P. "A Critique of Ordinal Variables." *Social Forces* 49(1971): 432–444.

Winch, R. F., and Campbell, D. T. "Proof? No. Evidence? Yes. The Significance of Tests of Significance." *American Sociologist* 4(1969): 140–143.

Wonnacott, R. J., and Wonnacott, T. H. *Econometrics.* New York: John Wiley, 1970.

Wonnacott, T. H., and Wonnacott, R. J. *Introductory Statistics.* New York: John Wiley, 1969.

Woolf, H., ed. *Quantification: A History of Measurement in the Natural and Social Sciences.* Indianapolis: Bobbs-Merrill, 1961.

Yates, F. "Contingency Tables Involving Small Numbers and the Chi Square Test." *Journal of the Royal Statistical Society* 1(1934): 217–235.

Yule, G. U., and Kendall, M. G. *An Introduction to the Theory of Statistics.* 14th ed. London: Charles Griffin, 1950.

Zeisel, H. *Say It With Figures.* 5th ed. New York: Harper and Row, 1968.

Appendixes

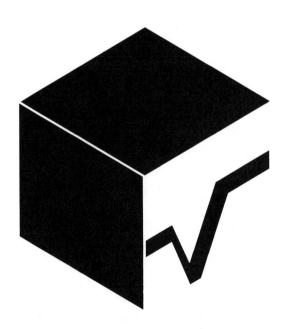

Appendix I
Glossary of Symbols

This book uses mathematical, Greek, and English symbols throughout. The numbers in parentheses indicate the page where the symbol was first defined.

Mathematical Symbols

$\approx$	nearly equal to (3)
$\neq$	not equal to (17)
$<$	less than (11)
$>$	greater than (11)
$\leq$	less than or equal to (106)
$\geq$	greater than or equal to (143)
$\sqrt{}$	square root (78)
X^a	X raised to the ath power (21)
$|X|$	absolute value of X (77)
Σ	(capital Greek sigma) sum of (17)
π	(small greek pi) mathematical constant, equal approximately to 3.14159265 (3)
Δ	(capital Greek delta) indicates a finite incremental change (209)

Greek Symbols

α (small Greek alpha) size of the critical region (125); probability of a Type I error (127)

β (small Greek beta) probability of a Type II error (127); beta weights (240)

γ (small Greek gamma) gamma coefficient (280)

λ_a (small Greek lambda) asymmetric lambda coefficient (271)

λ_s symmetric lambda coefficient (272)

μ (small Greek mu) population mean (Figure 7.2) (92)

ρ (small Greek rho) population Pearson product-moment correlation coefficient (197); as subscript, population Spearman rank-order correlation coefficient (199)

σ^2 (small Greek sigma) population variance (152)

σ population standard deviation (Figure 7.2) (92)

$\sigma_{\bar{X}}$ standard error of the sample means (136)

$\sigma_{\bar{X}_1 - \bar{X}_2}$ standard error of the difference between means (152)

τ_r (small Greek tau) Goodman-Kruskal tau, with row variable designated as the dependent variable (276)

τ_c Goodman-Kruskal tau, with column variable designated as the dependent variable (277)

ϕ (small Greek phi) phi coefficient (266)

χ^2 (small Greek khi) chi square (259)

English Symbols

A Sandler's test (162)

a Y-intercept in a bivariate regression equation (209)

ad concordant pair (Yule's Q) (283)

$a_{1.23}$ Y-intercept for multiple regression equation (two independent variables) (238)

b_y slope in a bivariate regression equation (212)

bc discordant pair (Yule's Q) (283)

$b_{12.3}$ slope in a multiple regression equation, holding X_3 constant (238)

$b_{13.2}$ slope in a multiple regression equation, holding X_2 constant (238)

C number of concordant pairs (Gamma) (280)

c number of columns in a contingency table (262)

$cum\ f_e$ expected cumulative frequency distribution (Kolmogorov-Smirnov) (284)

$cum\ f$ cumulative frequency (Table 3.4) (32)

$cum\ f_o$ observed cumulative frequency distribution (Kolmogorov-Smirnov) (284)

$cum\ f_{ll}$ cumulative frequency at the lower true limit of the interval containing X (53)

$cum\ \%$ cumulative percent (Table 3.4) (32)

D index of dispersion (82); number of discordant pairs (280); Kolmogorov-Smirnov test for goodness of fit (284)

D_1 first decile; D_2 equals the second decile, and so on (56)

D^2 squared difference (Sandler's A test) (162); squared difference between pairs of ranks (Spearman r_s) (199)

$\bar{D}$ mean difference across a pair having matched score (means test for matched samples) (160)

df degrees of freedom (141)

E_1 errors resulting from prediction based on the categorical distribution of the dependent variable alone (268)

E_2 errors resulting from prediction based on the information supplied by the independent variable (268)

F ratio of variances (158)

F_c column marginals (independent variable) (Goodman-Kruskal) (276)

F_r marginals of the various categories of the dependent variable (Goodman-Kruskal) (275)

f frequency (Table 3.2) (28)

f_c cell frequencies of the dependent variable (Goodman-Kruskal) (276)

f_e expected frequency in a given category (chi square) (259)

f_o observed frequency in a given category (chi square) (259)

H	Kruskal-Wallis test (290)
H_0	null hypothesis (122)
H_1	alternative (or research) hypothesis (122)
HSD	honestly significant difference test (178)
i	width of the class interval (29)
k	constant (18); number of categories or groups (82); number of independent and control variables (247); number of rows or columns (267)
k^2	coefficient of alienation (219)
K^2	coefficient of multiple alienation (245)
K_D	Kolmogorov-Smirnov test for independence (285)
MD	mean deviation (77)
MP	midpoint of the class interval (64)
$max\ F_c$	modal frequency of the column marginals (symmetric lambda) (272)
$max\ F_r$	modal frequency of the row marginals (symmetric lambda) (272)
$max\ F_r$	modal frequency of the row or dependent variable (asymmetric lambda) (271)
$max\ f_c$	modal frequency within each category of the column or independent variable (asymmetric lambda) (271)
$max\ f_c$	modal frequency within each category of column variable, treated as the independent variable (symmetric lambda) (272)
$max\ f_r$	modal frequency within each category of the row variable, treated as the independent variable (symmetric lambda) (272)
N	total number of scores or quantities in a data set (17)
N_i	number of cases in any given group (Kruskal Wallis) (290)
P_1	first percentile; P_2 equals the second percentile, and so on (56)
p	probability (104); proportion (104); probability of obtaining the favored event (121)
$p(A)$	probability of A (104)
$p(B/A)$	probability of B, given that A has occurred (109)
PR	percentile rank (51)

PRE	proportional reduction of error (268)
Q	Yule's associational measure (282)
Q_1	first quartile; Q_2 equals the second quartile (or the median), and so on (56)
q	probability of obtaining the nonfavored event (121)
$q\alpha$	tabled Studentized range value for a given alpha level (178)
$R_{1.23}$	multiple correlation (two independent variables) (242)
$R_{1.234}$	multiple correlation (three independent variables) (242)
$R_{1.2345}$	multiple correlation (four independent variables) (243)
$R_{1.23}^2$	coefficient of multiple determination (243)
R'	corrected value of R (245)
R_i	ranks in any given group (Kruskal-Wallis) (290)
R_1	rank for sample size N_1 (Mann-Whitney) (288)
R_2	rank for sample size N_2 (Mann-Whitney) (288)
r	Pearson product-moment correlation coefficient (190)
r_s	Spearman rank-order correlation coefficient (190)
r^2	coefficient of determination (219)
$r_{12.3}$	first-order partial correlation coefficient (234)
$r_{12.34}$	second-order partial correlation coefficient (235)
$r_{12.345}$	third-order partial correlation coefficient (235)
$r_{12.3}^2$	coefficient of determination for first-order partial (235)
s	sample standard deviation (78)
s^2	sample variance (79)
$\hat{s}$	unbiased estimate of population standard deviation (140)
$\hat{s}^2$	unbiased estimate of population variance (142)
s_D^2	variance of the differences (160)
$s_{\bar{x}_1 - \bar{x}_2}$	standard error of the difference of means (154)
$\hat{s}_b^2$	between-group variance estimate (175)
$\hat{s}_w^2$	within-group variance estimate (175)
s_{est_y}	standard error of estimate when predicting Y from X (215)

SS_b between-group sum of squares (173)

SS_w within-group sum of squares (172)

SS_t total sum of squares (174)

t Student t ratio (141)

U statistic (Mann-Whitney) (288)

U' statistic (Mann-Whitney) (288)

V Cramér's associational measure (267)

X quantities or scores of variables (16)

X_i specific quantities of X, indicated by the subscript i (17)

X_j specific quantities of X, indicated by the subscript j (17)

X_N last quantity or score in the data set (17)

$\bar{X}$ arithmetic mean (64)

$\bar{X}_w$ weighted mean (66)

$\bar{X}_t$ overall group mean (173)

X_{ij} individual score (172)

X_{ll} score at the lower true limit of the interval containing X (53)

$\Sigma(X - \bar{X})^2$ sum of the squared deviations (67)

Y quantities or scores of variables (16)

$\bar{Y}$ arithmetic mean (191)

Y_i specific quantities of Y, indicated by the subscript i (17)

Y_j specific quantities of Y, indicated by the subscript j (17)

Y' score predicted by regression equation (211)

$\Sigma(Y - \bar{Y})^2$ total sum of squares (217)

$\Sigma(Y' - \bar{Y})^2$ regression sum of squares (217)

$\Sigma(Y - Y')^2$ residual sum of squares (218)

z deviation of a score from the mean expressed in standard deviation units (93); z ratio (140)

Appendix II
List of Formulas

The formula number, used throughout the text, identifies a formula. The page number indicates where the formula first appears in the text.

Number	Formula	Page
(4.1)	$PR = \dfrac{Cum\ f}{N}\ (100)$	51
(4.2)	$PR = \dfrac{Cum\ f_{ll} + \left(\dfrac{X - X_{ll}}{i}\right)f}{N}(100)$	53
(4.3)	$Cum\ f = \dfrac{(PR)(N)}{100}$	54
(4.4)	$X = X_{ll} + \left(\dfrac{Cum\ f - Cum\ f_{ll}}{f}\right)i$	55
(5.1)	$\text{Median} = X_1 + \left(\dfrac{N/2 - Cum\ f_{ll}}{f}\right)i$	63
(5.2)	$\bar{X} = \dfrac{\Sigma X}{N}$	64
(5.3)	$\bar{X} = \dfrac{\Sigma fX}{N}$	64

(5.4)
$$\bar{X} = \frac{\Sigma fMP}{N}$$
64

(5.5)
$$\bar{X}_w = \frac{\Sigma N\bar{X}}{\Sigma N}$$
66

(6.1)
$$MD = \frac{\Sigma |X - \bar{X}|}{N}$$
77

(6.2)
$$s = \sqrt{\frac{\Sigma(X - \bar{X})^2}{N}}$$
78

(6.3)
$$s^2 = \frac{\Sigma(X - \bar{X})^2}{N}$$
79

(6.4)
$$s = \sqrt{\frac{\Sigma X^2}{N} - \bar{X}^2}$$
80

(6.5)
$$s = \sqrt{\frac{\Sigma fMP^2}{N} - \left(\frac{\Sigma fMP}{N}\right)^2}$$
81

(6.6)
$$D = \frac{k(N^2 - \Sigma f^2)}{N^2(k - 1)}$$
82

(7.1)
$$z = \frac{X - \bar{X}}{s}$$
93

(8.1)
$$p(A) = \frac{\text{Outcomes Favoring Event } A}{\text{Total Number of Outcomes}}$$
104

(8.2)
$$p(A \text{ or } B) = p(A) + p(B) - p(A \text{ and } B)$$
107

(8.3)
$$p(A \text{ or } B) = p(A) + p(B)$$
$$\textit{if A and B are mutually exclusive}$$
108

(8.4)
$$p(A \text{ and } B) = p(A)p(B)$$
$$\textit{if A and B are independent}$$
108

(8.5)
$$p(A \text{ and } B) = p(A)p(B/A) = p(B)p(A/B)$$
109

(9.1)
$$(p + q)^N = p^N + \frac{N}{1}p^{N-1}q + \frac{N(N-1)}{(1)(2)}p^{N-2}q^2$$
$$+ \frac{N(N-1)(N-2)}{(1)(2)(3)}p^{N-3}q^3 + \ldots + q^N$$
121

(10.1)
$$z = \frac{\bar{X} - \mu}{\sigma/\sqrt{N}}$$
140

(10.2)
$$t = \frac{\bar{X} - \mu}{s/\sqrt{N-1}}$$
142

(10.3)
$$z = \frac{\bar{X} - \mu}{\hat{s}/\sqrt{N}}$$
144

(10.4)
$$\mu = \bar{X} \pm z(\hat{s}/\sqrt{N})$$
145

(10.5)
$$\mu = \bar{X} \pm t(s/\sqrt{N - 1})$$
145

(11.1)
$$z = \frac{\bar{X}_1 - \bar{X}_2}{\sqrt{\dfrac{\sigma_1^2}{N_1} + \dfrac{\sigma_2^2}{N_2}}}$$
153

(11.2)
$$t = \frac{\bar{X}_1 - \bar{X}_2}{\sqrt{\dfrac{s_1^2}{N_1 - 1} + \dfrac{s_2^2}{N_2 - 1}}}$$
154

(11.3)
$$t = \frac{\bar{X}_1 - \bar{X}_2}{\sqrt{\left(\dfrac{N_1 s_1^2 + N_2 s_2^2}{N_1 + N_2 - 2}\right)\left(\dfrac{1}{N_1} + \dfrac{1}{N_2}\right)}}$$
155

(11.4)
$$F = \frac{\hat{s}^2}{\hat{s}^2}$$
158

(11.5)
$$t = \frac{\bar{D}}{\sqrt{\dfrac{s_D^2}{N - 1}}}$$
160

(11.6)
$$t = \frac{\bar{X}_1 - \bar{X}_2}{\sqrt{s_1^2 + s_2^2 - 2rs_1 s_2}}$$
160

(11.7)
$$A = \frac{\Sigma D^2}{(\Sigma D)^2}$$
162

(12.1)
$$SS_w = \sum_{j=1}^{k} \sum_{i=1}^{N_j} (X_{ij} - \bar{X}_j)^2$$
172

(12.2)
$$SS_w = \Sigma \left[\Sigma X_j^2 - \frac{(\Sigma X_j)^2}{N}\right]$$
173

(12.3)
$$SS_b = \sum_{j=1}^{k} N_j (\bar{X}_j - \bar{X}_t)^2$$
173

(12.4)
$$SS_b = \Sigma \left[\frac{(\Sigma X_j)^2}{N_j}\right] - \frac{(\Sigma\Sigma X_{ij})^2}{N_t}$$
173

(12.5)
$$SS_t = \sum_{j=1}^{k} \sum_{i=1}^{N_j} (X_{ij} - \bar{X}_t)^2$$
174

$$(12.6) \qquad SS_t = \Sigma X_{ij}^2 - \frac{(\Sigma X_{ij})^2}{N_t} \qquad 174$$

$$(12.7) \qquad F = \frac{\hat{s}_b^2}{\hat{s}_w^2} \qquad 175$$

$$(12.8) \qquad HSD = q\alpha \sqrt{\frac{\hat{s}_w^2}{N}} \qquad 178$$

$$(13.1) \qquad r = \frac{\Sigma z_x z_y}{N} \qquad 191$$

$$(13.2) \qquad r = \frac{\Sigma(X - \bar{X})(Y - \bar{Y})}{\sqrt{\Sigma(X - \bar{X})^2}\ \sqrt{\Sigma(Y - \bar{Y})^2}} \qquad 193$$

$$(13.3) \qquad r = \frac{\Sigma XY - N\bar{X}\bar{Y}}{\sqrt{\Sigma X^2 - N\bar{X}^2}\ \sqrt{\Sigma Y^2 - N\bar{Y}^2}} \qquad 194$$

$$(13.4) \qquad r_s = 1 - \frac{6\Sigma D^2}{N(N^2 - 1)} \qquad 199$$

$$(14.1) \qquad b_y = r\,\frac{s_y}{s_x} \qquad 212$$

$$(14.2) \qquad a = \bar{Y} - b_y\bar{X} \qquad 212$$

$$(14.3) \qquad Y' = a + b_y X \qquad 212$$

$$(14.4) \qquad Y' = \bar{Y} + r\frac{s_y}{s_x}(X - \bar{X}) \qquad 212$$

$$(14.5) \qquad s_{est_y} = \sqrt{\frac{\Sigma(Y - Y')^2}{N}} \qquad 215$$

$$(14.6) \qquad s_{est_y} = s_y\sqrt{1 - r^2} \qquad 215$$

$$(14.7) \qquad Y' \pm z(s_{est_y}) \qquad 216$$

$$(14.8) \qquad r^2 = \frac{\Sigma(Y' - \bar{Y})^2}{\Sigma(Y - \bar{Y})^2} \qquad 219$$

$$(14.9) \qquad k^2 = 1 - r^2 \qquad 219$$

$$(14.10) \qquad F_{1, N-2} = \frac{r^2(N - 2)}{(1 - r^2)} \qquad 220$$

$$(15.1) \qquad r_{12.3} = \frac{r_{12} - (r_{13})(r_{32})}{\sqrt{1 - r_{13}^2}\ \sqrt{1 - r_{32}^2}} \qquad 234$$

$$(15.2) \qquad r_{12.34} = \frac{r_{12.3} - (r_{14.3})(r_{24.3})}{\sqrt{1 - r_{14.3}^2}\ \sqrt{1 - r_{24.3}^2}} \qquad 235$$

(15.3) $$r_{12.345} = \frac{r_{12.34} - (r_{15.34})(r_{25.34})}{\sqrt{1 - r_{15.34}^2}\ \sqrt{1 - r_{25.34}^2}}$$ 235

(15.4) $$Y_1 = a_{12.3} + b_{12.3}X_2 + b_{13.2}X_3$$ 237

(15.5) $$b_{12.3} = \frac{b_{12} - (b_{13})(b_{32})}{1 - (b_{23})(b_{32})}$$ 238

(15.6) $$b_{13.2} = \frac{b_{13} - (b_{12})(b_{23})}{1 - (b_{32})(b_{23})}$$ 238

(15.7) $$a_{12.3} = \bar{Y}_1 + b_{12.3}\bar{X}_2 - b_{13.2}\bar{X}_3$$ 238

(15.8) $$\beta_{12.3} = b_{12.3}(s_2/s_1)$$ 240

(15.9) $$\beta_{13.2} = b_{13.2}(s_3/s_1)$$ 240

(15.10) $$R_{1.23} = \sqrt{r_{12}^2 + r_{13.2}^2\ (1 - r_{12}^2)}$$ 242

(15.11) $$R_{1.234} = \sqrt{R_{1.23}^2 + r_{14.23}^2(1 - R_{1.23}^2)}$$ 242

(15.12) $$R_{1.2345} = \sqrt{R_{1.234}^2 + r_{15.234}^2(1 - R_{1.234}^2)}$$ 243

(15.13) $$R_{1.23}^2 = r_{12}^2 + r_{13.2}^2(1 - r_{12}^2)$$ 243

(15.14) $$R_{1.23} = \sqrt{\frac{r_{12}^2 + r_{13}^2 - 2r_{12}r_{13}r_{23}}{1 - r_{23}^2}}$$ 243

(15.15) $$R_{1.23}^2 = \frac{r_{12}^2 + r_{13}^2 - 2r_{12}r_{13}r_{23}}{1 - r_{23}^2}$$ 243

(15.16) $$R' = \sqrt{1 - \frac{N - 1}{N - k}(1 - R^2)}$$ 245

(15.17) $$F_{k, N - k - 1} = \frac{R^2(N - k - 1)}{(1 - R^2)k}$$ 246

(15.18) $$F_{1, N - k - 1} = \frac{r_{12.3 \dots N}^2{}^{(N - k - 1)}}{1 - r_{12.3 \dots N}^2}$$ 247

(16.1) $$\chi^2 = \sum_{i=1}^{k} \frac{(f_o - f_e)^2}{f_e}$$ 259

(16.2) $$\chi^2 = \sum_{r=1}^{i} \sum_{j=1}^{j} \frac{(f_o - f_e)^2}{f_e}$$ 262

(16.3)
$$\chi^2 = \sum_{r=1}^{2} \sum_{j=1}^{2} \frac{(|f_o - f_e| - 0.5)^2}{f_e}$$
263

(16.4)
$$\phi = \sqrt{\frac{\chi^2}{N}}$$
266

(16.5)
$$V = \sqrt{\frac{\chi^2}{N(k - 1)}}$$
267

(16.6)
$$PRE = \frac{E_1 - E_2}{E_1}$$
269

(16.7)
$$\lambda_a = \frac{\sum max\, f_c - max\, F_r}{N - max\, F_r}$$
271

(16.8)
$$\lambda_s = \frac{(\sum max\, f_c + \sum max\, f_r) - (max\, F_r - max\, F_c)}{2N - (max\, F_r + max\, F_c)}$$
272

(16.9)
$$E_1 = \sum \left[\frac{N - F_r}{N} (F_r) \right]$$
275

(16.10)
$$E_2 = \sum \left[\frac{F_c - f_c}{F_c} (f_c) \right]$$
276

(16.11)
$$\tau_r = \frac{E_1 - E_2}{E_1}$$
276

(16.12)
$$\tau_r = \frac{N\sum \frac{f_c^2}{F_c} - \sum F_r^2}{N^2 - \sum F_r^2}$$
276

(16.13)
$$\gamma = \frac{C - D}{C + D}$$
280

(16.14)
$$E_1 = \tfrac{1}{2}(C + D)$$
281

(16.15)
$$E_2 = min\ (C,D)$$
281

(16.16)
$$\gamma = \frac{E_1 - E_2}{E_1}$$
281

(16.17)
$$Q = \frac{ad - bc}{ad + bc}$$
282

(16.18)
$$D = max\ |Cum\, f_o - Cum\, f_e|$$
284

(16.19)
$$K_D = max\ |Cum\, f_1 - Cum\, f_2|$$
285

(16.20)
$$U = N_1 N_2 - U'$$
288

(16.21)
$$U' = N_1 N_2 - U$$
288

(16.22) $$U = N_1 N_2 + \frac{N_1(N_1 + 1)}{2} - \Sigma R_1$$ 288

(16.23) $$U = N_1 N_2 + \frac{N_2(N_2 + 1)}{2} - \Sigma R_2$$ 288

(16.24) $$z = \frac{U - (N_1 N_2)/2}{\sqrt{N_1 N_2 (N_1 + N_2 + 1)/12}}$$ 289

(16.25) $$H = \frac{12}{N(N + 1)} \frac{\Sigma R_i^2}{N_i} - 3(N + 1)$$ 290

(16.26) $$C = 1 - \frac{(t^3 - t)}{(N^3 - N)}$$ 290

Appendix III
Tables

Acknowledgments

We would like to thank the authors and publishers listed below for permission to adapt or reproduce the following tables.

Table A: Sorenson, H. *Statistics for Students of Psychology and Education.* New York: McGraw-Hill, 1936.

Table B: Runyon, R. P., and Haber, A. *Fundamentals of Behavioral Statistics.* 3rd ed. Reading, Mass.: Addison-Wesley, 1976.

Table C: Fisher, R. A., and Yates, F. *Statistical Tables for Biological, Agricultural and Medical Research.* London: Longman Group (previously published in Edinburgh by Oliver and Boyd).

Table D: ———. *Statistical Tables for Biological, Agricultural and Medical Research.* London: Longman Group (previously published in Edinburgh by Oliver and Boyd).

Table E: Sandler, J. "A Test for the Significance of the Difference Between the Means of Correlated Measures, Based on a Simplification of Student's t." *British Journal of Psychology* (Cambridge University Press) 46(1955).

Table F: Pearson, E. S., and Hartley, H. O., eds. *Biometrika Tables for Statisticians.* New York: Cambridge, 1966.

Table G: Fisher, R. A., and Yates, F. *Statistical Tables for Biological, Agricultural and Medical Research.* London: Longman Group (previously published in Edinburgh by Oliver and Boyd).

Table H: Olds, E. G. "Distribution of Sums of Rank Differences for Small Numbers of Individuals." *Annals of Mathematical Statistics* 9(1938); and "The 5% Significance Level for Sums of Squares of Rank Differences and a Correction." *Annals of Mathematical Statistics* 20(1949).

Table I: Fisher, R. A., and Yates, F. *Statistical Tables for Biological, Agricultural and Medical Research.* London: Longman Group (previously published in Edinburgh by Oliver and Boyd).

Table J: Massey, Jr., F. J. "The Kolmogorov-Smirnov Test for Goodness of Fit." *Journal of the American Statistical Association* 46(1951).

Table K: Goodman, L.A. "Kolmogorov-Smirnov Tests for Psychological Research." *Psychological Bulletin* 51(1954); and Massey, Jr., F. J. "The Distribution of the Maximum Deviation between Two Samples Cumulative Step Functions." *Annals of Mathematical Statistics* 22(1951).

Table L: Smirnov, N. "Tables for Estimating the Goodness of Fit of Empirical Distributions." *Annals of Mathematical Statistics* 19(1948).

Table M: Mann, H. B., and Whitney, D. R. "On a Test of Whether One of Two Random Variables is Stochastically Larger than the Other." *Annals of Mathematical Statistics* 18(1948).

Table N: Kruskal, W. H., and Wallis, W. A. "Use of Ranks in One-Criterion Variance Analysis." *Journal of the American Statistical Association* 47(1952).

Table A. Squares and Square Roots

Number	Square	Square Root	Number	Square	Square Root
1	1	1.0000	41	1681	6.4031
2	4	1.4142	42	1764	6.4807
3	9	1.7321	43	1849	6.5574
4	16	2.0000	44	1936	6.6332
5	25	2.2361	45	2025	6.7082
6	36	2.4495	46	2116	6.7823
7	49	2.6458	47	2209	6.8557
8	64	2.8284	48	2304	6.9282
9	81	3.0000	49	2401	7.0000
10	100	3.1623	50	2500	7.0711
11	121	3.3166	51	2601	7.1414
12	144	3.4641	52	2704	7.2111
13	169	3.6056	53	2809	7.2801
14	196	3.7417	54	2916	7.3485
15	225	3.8730	55	3025	7.4162
16	256	4.0000	56	3136	7.4833
17	289	4.1231	57	3249	7.5498
18	324	4.2426	58	3364	7.6158
19	361	4.3589	59	3481	7.6811
20	400	4.4721	60	3600	7.7460
21	441	4.5826	61	3721	7.8102
22	484	4.6904	62	3844	7.8740
23	529	4.7958	63	3969	7.9373
24	576	4.8990	64	4096	8.0000
25	625	5.0000	65	4225	8.0623
26	676	5.0990	66	4356	8.1240
27	729	5.1962	67	4489	8.1854
28	784	5.2915	68	4624	8.2462
29	841	5.3852	69	4761	8.3066
30	900	5.4772	70	4900	8.3666
31	961	5.5678	71	5041	8.4261
32	1024	5.6569	72	5184	8.4853
33	1089	5.7446	73	5329	8.5440
34	1156	5.8310	74	5476	8.6023
35	1225	5.9161	75	5625	8.6603
36	1296	6.0000	76	5776	8.7178
37	1369	6.0828	77	5929	8.7750
38	1444	6.1644	78	6084	8.8318
39	1521	6.2450	79	6241	8.8882
40	1600	6.3246	80	6400	8.9443

Number	Square	Square Root	Number	Square	Square Root
81	6561	9.0000	121	14641	11.0000
82	6724	9.0554	122	14884	11.0454
83	6889	9.1104	123	15129	11.0905
84	7056	9.1652	124	15376	11.1355
85	7225	9.2195	125	15625	11.1803
86	7396	9.2736	126	15876	11.2250
87	7569	9.3274	127	16129	11.2694
88	7744	9.3808	128	16384	11.3137
89	7921	9.4340	129	16641	11.3578
90	8100	9.4868	130	16900	11.4018
91	8281	9.5394	131	17161	11.4455
92	8464	9.5917	132	17424	11.4891
93	8649	9.6437	133	17689	11.5326
94	8836	9.6954	134	17956	11.5758
95	9025	9.7468	135	18225	11.6190
96	9216	9.7980	136	18496	11.6619
97	9409	9.8489	137	18769	11.7047
98	9604	9.8995	138	19044	11.7473
99	9801	9.9499	139	19321	11.7898
100	10000	10.0000	140	19600	11.8322
101	10201	10.0499	141	19881	11.8743
102	10404	10.0995	142	20164	11.9164
103	10609	10.1489	143	20449	11.9583
104	10816	10.1980	144	20736	12.0000
105	11025	10.2470	145	21025	12.0416
106	11236	10.2956	146	21316	12.0830
107	11449	10.3441	147	21609	12.1244
108	11664	10.3923	148	21904	12.1655
109	11881	10.4403	149	22201	12.2066
110	12100	10.4881	150	22500	12.2474
111	12321	10.5357	151	22801	12.2882
112	12544	10.5830	152	23104	12.3288
113	12769	10.6301	153	23409	12.3693
114	12996	10.6771	154	23716	12.4097
115	13225	10.7238	155	24025	12.4499
116	13456	10.7703	156	24336	12.4900
117	13689	10.8167	157	24649	12.5300
118	13924	10.8628	158	24964	12.5698
119	14161	10.9087	159	25281	12.6095
120	14400	10.9545	160	25600	12.6491

Squares and Square Roots (*Continued*)

Number	Square	Square Root	Number	Square	Square Root
161	25921	12.6886	201	40401	14.1774
162	26244	12.7279	202	40804	14.2127
163	26569	12.7671	203	41209	14.2478
164	26896	12.8062	204	41616	14.2829
165	27225	12.8452	205	42025	14.3178
166	27556	12.8841	206	42436	14.3527
167	27889	12.9228	207	42849	14.3875
168	28224	12.9615	208	43264	14.4222
169	28561	13.0000	209	43681	14.4568
170	28900	13.0384	210	44100	14.4914
171	29241	13.0767	211	44521	14.5258
172	29584	13.1149	212	44944	14.5602
173	29929	13.1529	213	45369	14.5945
174	30276	13.1909	214	45796	14.6287
175	30625	13.2288	215	46225	14.6629
176	30976	13.2665	216	46656	14.6969
177	31329	13.3041	217	47089	14.7309
178	31684	13.3417	218	47524	14.7648
179	32041	13.3791	219	47961	14.7986
180	32400	13.4164	220	48400	14.8324
181	32761	13.4536	221	48841	14.8661
182	33124	13.4907	222	49284	14.8997
183	33489	13.5277	223	49729	14.9332
184	33856	13.5647	224	50176	14.9666
185	34225	13.6015	225	50625	15.0000
186	34596	13.6382	226	51076	15.0333
187	34969	13.6748	227	51529	15.0665
188	35344	13.7113	228	51984	15.0997
189	35721	13.7477	229	52441	15.1327
190	36100	13.7840	230	52900	15.1658
191	36481	13.8203	231	53361	15.1987
192	36864	13.8564	232	53824	15.2315
193	37249	13.8924	233	54289	15.2643
194	37636	13.9284	234	54756	15.2971
195	38025	13.9642	235	55225	15.3297
196	38416	14.0000	236	55696	15.3623
197	38809	14.0357	237	56169	15.3948
198	39204	14.0712	238	56644	15.4272
199	39601	14.1067	239	57121	15.4596
200	40000	14.1421	240	57600	15.4919

Squares and Square Roots (*Continued*)

Number	Square	Square Root	Number	Square	Square Root
241	58081	15.5242	281	78961	16.7631
242	58564	15.5563	282	79524	16.7929
243	59049	15.5885	283	80089	16.8226
244	59536	15.6205	284	80656	16.8523
245	60025	15.6525	285	81225	16.8819
246	60516	15.6844	286	81796	16.9115
247	61009	15.7162	287	82369	16.9411
248	61504	15.7480	288	82944	16.9706
249	62001	15.7797	289	83521	17.0000
250	62500	15.8114	290	84100	17.0294
251	63001	15.8430	291	84681	17.0587
252	63504	15.8745	292	85264	17.0880
253	64009	15.9060	293	85849	17.1172
254	64516	15.9374	294	86436	17.1464
255	65025	15.9687	295	87025	17.1756
256	65536	16.0000	296	87616	17.2047
257	66049	16.0312	297	88209	17.2337
258	66564	16.0624	298	88804	17.2627
259	67081	16.0935	299	89401	17.2916
260	67600	16.1245	300	90000	17.3205
261	68121	16.1555	301	90601	17.3494
262	68644	16.1864	302	91204	17.3781
263	69169	16.2173	303	91809	17.4069
264	69696	16.2481	304	92416	17.4356
265	70225	16.2788	305	93025	17.4642
266	70756	16.3095	306	93636	17.4929
267	71289	16.3401	307	94249	17.5214
268	71824	16.3707	308	94864	17.5499
269	72361	16.4012	309	95481	17.5784
270	72900	16.4317	310	96100	17.6068
271	73441	16.4621	311	96721	17.6352
272	73984	16.4924	312	97344	17.6635
273	74529	16.5227	313	97969	17.6918
274	75076	16.5529	314	98596	17.7200
275	75625	16.5831	315	99225	17.7482
276	76176	16.6132	316	99856	17.7764
277	76729	16.6433	317	100489	17.8045
278	77284	16.6733	318	101124	17.8326
279	77841	16.7033	319	101761	17.8606
280	78400	16.7332	320	102400	17.8885

Number	Square	Square Root	Number	Square	Square Root
321	103041	17.9165	361	130321	19.0000
322	103684	17.9444	362	131044	19.0263
323	104329	17.9722	363	131769	19.0526
324	104976	18.0000	364	132496	19.0788
325	105625	18.0278	365	133225	19.1050
326	106276	18.0555	366	133956	19.1311
327	106929	18.0831	367	134689	19.1572
328	107584	18.1108	368	135424	19.1833
329	108241	18.1384	369	136161	19.2094
330	108900	18.1659	370	136900	19.2354
331	109561	18.1934	371	137641	19.2614
332	110224	18.2209	372	138384	19.2873
333	110889	18.2483	373	139129	19.3132
334	111556	18.2757	374	139876	19.3391
335	112225	18.3030	375	140625	19.3649
336	112896	18.3303	376	141376	19.3907
337	113569	18.3576	377	142129	19.4165
338	114244	18.3848	378	142884	19.4422
339	114921	18.4120	379	143641	19.4679
340	115600	18.4391	380	144400	19.4936
341	116281	18.4662	381	145161	19.5192
342	116964	18.4932	382	145924	19.5448
343	117649	18.5203	383	146689	19.5704
344	118336	18.5472	384	147456	19.5959
345	119025	18.5742	385	148225	19.6214
346	119716	18.6011	386	148996	19.6469
347	120409	18.6279	387	149769	19.6723
348	121104	18.6548	388	150544	19.6977
349	121801	18.6815	389	151321	19.7231
350	122500	18.7083	390	152100	19.7484
351	123201	18.7350	391	152881	19.7737
352	123904	18.7617	392	153664	19.7990
353	124609	18.7883	393	154449	19.8242
354	125316	18.8149	394	155236	19.8494
355	126025	18.8414	395	156025	19.8746
356	126736	18.8680	396	156816	19.8997
357	127449	18.8944	397	157609	19.9249
358	128164	18.9209	398	158404	19.9499
359	128881	18.9473	399	159201	19.9750
360	129600	18.9737	400	160000	20.0000

Squares and Square Roots (*Continued*)

Number	Square	Square Root	Number	Square	Square Root
401	160801	20.0250	441	194481	21.0000
402	161604	20.0499	442	195364	21.0238
403	162409	20.0749	443	196249	21.0476
404	163216	20.0998	444	197136	21.0713
405	164025	20.1246	445	198025	21.0950
406	164836	20.1494	446	198916	21.1187
407	165649	20.1742	447	199809	21.1424
408	166464	20.1990	448	200704	21.1660
409	167281	20.2237	449	201601	21.1896
410	168100	20.2485	450	202500	21.2132
411	168921	20.2731	451	203401	21.2368
412	169744	20.2978	452	204304	21.2603
413	170569	20.3224	453	205209	21.2838
414	171396	20.3470	454	206116	21.3073
415	172225	20.3715	455	207025	21.3307
416	173056	20.3961	456	207936	21.3542
417	173889	20.4206	457	208849	21.3776
418	174724	20.4450	458	209764	21.4009
419	175561	20.4695	459	210681	21.4243
420	176400	20.4939	460	211600	21.4476
421	177241	20.5183	461	212521	21.4709
422	178084	20.5426	462	213444	21.4942
423	178929	20.5670	463	214369	21.5174
424	179776	20.5913	464	215296	21.5407
425	180625	20.6155	465	216225	21.5639
426	181476	20.6398	466	217156	21.5870
427	182329	20.6640	467	218089	21.6102
428	183184	20.6882	468	219024	21.6333
429	184041	20.7123	469	219961	21.6564
430	184900	20.7364	470	220900	21.6795
431	185761	20.7605	471	221841	21.7025
432	186624	20.7846	472	222784	21.7256
433	187489	20.8087	473	223729	21.7486
434	188356	20.8327	474	224676	21.7715
435	189225	20.8567	475	225625	21.7945
436	190096	20.8806	476	226576	21.8174
437	190969	20.9045	477	227529	21.8403
438	191844	20.9284	478	228484	21.8632
439	192721	20.9523	479	229441	21.8861
440	193600	20.9762	480	230400	21.9089

Squares and Square Roots (*Continued*)

Number	Square	Square Root	Number	Square	Square Root
481	231361	21.9317	521	271441	22.8254
482	232324	21.9545	522	272484	22.8473
483	233289	21.9773	523	273529	22.8692
484	234256	22.0000	524	274576	22.8910
485	235225	22.0227	525	275625	22.9129
486	236196	22.0454	526	276676	22.9347
487	237169	22.0681	527	277729	22.9565
488	238144	22.0907	528	278784	22.9783
489	239121	22.1133	529	279841	23.0000
490	240100	22.1359	530	280900	23.0217
491	241081	22.1585	531	281961	23.0434
492	242064	22.1811	532	283024	23.0651
493	243049	22.2036	533	284089	23.0868
494	244036	22.2261	534	285156	23.1084
495	245025	22.2486	535	286225	23.1301
496	246016	22.2711	536	287296	23.1517
497	247009	22.2935	537	288369	23.1733
498	248004	22.3159	538	289444	23.1948
499	249001	22.3383	539	290521	23.2164
500	250000	22.3607	540	291600	23.2379
501	251001	22.3830	541	292681	23.2594
502	252004	22.4054	542	293764	23.2809
503	253009	22.4277	543	294849	23.3024
504	254016	22.4499	544	295936	23.3238
505	255025	22.4722	545	297025	23.3452
506	256036	22.4944	546	298116	23.3666
507	257049	22.5167	547	299209	23.3880
508	258064	22.5389	548	300304	23.4094
509	259081	22.5610	549	301401	23.4307
510	260100	22.5832	550	302500	23.4521
511	261121	22.6053	551	303601	23.4734
512	262144	22.6274	552	304704	23.4947
513	263169	22.6495	553	305809	23.5160
514	264196	22.6716	554	306916	23.5372
515	265225	22.6936	555	308025	23.5584
516	266256	22.7156	556	309136	23.5797
517	267289	22.7376	557	310249	23.6008
518	268324	22.7596	558	311364	23.6220
519	269361	22.7816	559	312481	23.6432
520	270400	22.8035	560	313600	23.6643

Squares and Square Roots (Continued)

Number	Square	Square Root	Number	Square	Square Root
561	314721	23.6854	601	361201	24.5153
562	315844	23.7065	602	362404	24.5357
563	316969	23.7276	603	363609	24.5561
564	318096	23.7487	604	364816	24.5764
565	319225	23.7697	605	366025	24.5967
566	320356	23.7908	606	367236	24.6171
567	321489	23.8118	607	368449	24.6374
568	322624	23.8328	608	369664	24.6577
569	323761	23.8537	609	370881	24.6779
570	324900	23.8747	610	372100	24.6982
571	326041	23.8956	611	373321	24.7184
572	327184	23.9165	612	374544	24.7385
573	328329	23.9374	613	375769	24.7588
574	329476	23.9583	614	376996	24.7790
575	330625	23.9792	615	378225	24.7992
576	331776	24.0000	616	379456	24.8193
577	332929	24.0208	617	380689	24.8395
578	334084	24.0416	618	381924	24.8596
579	335241	24.0624	619	383161	24.8797
580	336400	24.0832	620	384400	24.8998
581	337561	24.1039	621	385641	24.9199
582	338724	24.1247	622	386884	24.9399
583	339889	24.1454	623	388129	24.9600
584	341056	24.1661	624	389376	24.9800
585	342225	24.1868	625	390625	25.0000
586	343396	24.2074	626	391876	25.0200
587	344569	24.2281	627	393129	25.0400
588	345744	24.2487	628	394384	25.0599
589	346921	24.2693	629	395641	25.0799
590	348100	24.2899	630	396900	25.0998
591	349281	24.3105	631	398161	25.1197
592	350464	24.3311	632	399424	25.1396
593	351649	24.3516	633	400689	25.1595
594	352836	24.3721	634	401956	25.1794
595	354025	24.3926	635	403225	25.1992
596	355216	24.4131	636	404496	25.2190
597	356409	24.4336	637	405769	25.2389
598	357604	24.4540	638	407044	25.2587
599	358801	24.4745	639	408321	25.2784
600	360000	24.4949	640	409600	25.2982

Squares and Square Roots (Continued)

Number	Square	Square Root	Number	Square	Square Root
641	410881	25.3180	681	463761	26.0960
642	412164	25.3377	682	465124	26.1151
643	413449	25.3574	683	466489	26.1343
644	414736	25.3772	684	467856	26.1534
645	416025	25.3969	685	469225	26.1725
646	417316	25.4165	686	470596	26.1916
647	418609	25.4362	687	471969	26.2107
648	419904	25.4558	688	473344	26.2298
649	421201	25.4755	689	474721	26.2488
650	422500	25.4951	690	476100	26.2679
651	423801	25.5147	691	477481	26.2869
652	425104	25.5343	692	478864	26.3059
653	426409	25.5539	693	480249	26.3249
654	427716	25.5734	694	481636	26.3439
655	429025	25.5930	695	483025	26.3629
656	430336	25.6125	696	484416	26.3818
657	431649	25.6320	697	485809	26.4008
658	432964	25.6515	698	487204	26.4197
659	434281	25.6710	699	488601	26.4386
660	435600	25.6905	700	490000	26.4575
661	436921	25.7099	701	491401	26.4764
662	438244	25.7294	702	492804	26.4953
663	439569	25.7488	703	494209	26.5141
664	440896	25.7682	704	495616	26.5330
665	442225	25.7876	705	497025	26.5518
666	443556	25.8070	706	498436	26.5707
667	444889	25.8263	707	499849	26.5895
668	446224	25.8457	708	501264	26.6083
669	447561	25.8650	709	502681	26.6271
670	448900	25.8844	710	504100	26.6458
671	450241	25.9037	711	505521	26.6646
672	451584	25.9230	712	506944	26.6833
673	452929	25.9422	713	508369	26.7021
674	454276	25.9615	714	509796	26.7208
675	455625	25.9808	715	511225	26.7395
676	456976	26.0000	716	512656	26.7582
677	458329	26.0192	717	514089	26.7769
678	459684	26.0384	718	515524	26.7955
679	461041	26.0576	719	516961	26.8142
680	462400	26.0768	720	518400	26.8328

Number	Square	Square Root	Number	Square	Square Root
721	519841	26.8514	761	579121	27.5862
722	521284	26.8701	762	580644	27.6043
723	522729	26.8887	763	582169	27.6225
724	524176	26.9072	764	583696	27.6405
725	525625	26.9258	765	585225	27.6586
726	527076	26.9444	766	586756	27.6767
727	528529	26.9629	767	588289	27.6948
728	529984	26.9815	768	589824	27.7128
729	531441	27.0000	769	591361	27.7308
730	532900	27.0185	770	592900	27.7489
731	534361	27.0370	771	594441	27.7669
732	535824	27.0555	772	595984	27.7849
733	537289	27.0740	773	597529	27.8029
734	538756	27.0924	774	599076	27.8209
735	540225	27.1109	775	600625	27.8388
736	541696	27.1293	776	602176	27.8568
737	543169	27.1477	777	603729	27.8747
738	544644	27.1662	778	605284	27.8927
739	546121	27.1846	779	606841	27.9106
740	547600	27.2029	780	608400	27.9285
741	549081	27.2213	781	609961	27.9464
742	550564	27.2397	782	611524	27.9643
743	552049	27.2580	783	613089	27.9821
744	553536	27.2764	784	614656	28.0000
745	555025	27.2947	785	616225	28.0179
746	556516	27.3130	786	617796	28.0357
747	558009	27.3313	787	619369	28.0535
748	559504	27.3496	788	620944	28.0713
749	561001	27.3679	789	622521	28.0891
750	562500	27.3861	790	624100	28.1069
751	564001	27.4044	791	625681	28.1247
752	565504	27.4226	792	627264	28.1425
753	567009	27.4408	793	628849	28.1603
754	568516	27.4591	794	630436	28.1780
755	570025	27.4773	795	632025	28.1957
756	571536	27.4955	796	633616	28.2135
757	573049	27.5136	797	635209	28.2312
758	574564	27.5318	798	636804	28.2489
759	576081	27.5500	799	638401	28.2666
760	577600	27.5681	800	640000	28.2843

330

Squares and Square Roots (*Continued*)

Number	Square	Square Root	Number	Square	Square Root
801	641601	28.3019	841	707281	29.0000
802	643204	28.3196	842	708964	29.0172
803	644809	28.3373	843	710649	29.0345
804	646416	28.3549	844	712336	29.0517
805	648025	28.3725	845	714025	29.0689
806	649636	28.3901	846	715716	29.0861
807	651249	28.4077	847	717409	29.1033
808	652864	28.4253	848	719104	29.1204
809	654481	28.4429	849	720801	29.1376
810	656100	28.4605	850	722500	29.1548
811	657721	28.4781	851	724201	29.1719
812	659344	28.4956	852	725904	29.1890
813	660969	28.5132	853	727609	29.2062
814	662596	28.5307	854	729316	29.2233
815	664225	28.5482	855	731025	29.2404
816	665856	28.5657	856	732736	29.2575
817	667489	28.5832	857	734449	29.2746
818	669124	28.6007	858	736164	29.2916
819	670761	28.6182	859	737881	29.3087
820	672400	28.6356	860	739600	29.3258
821	674041	28.6531	861	741321	29.3428
822	675684	28.6705	862	743044	29.3598
823	677329	28.6880	863	744769	29.3769
824	678976	28.7054	864	746496	29.3939
825	680625	28.7228	865	748225	29.4109
826	682276	28.7402	866	749956	29.4279
827	683929	28.7576	867	751689	29.4449
828	685584	28.7750	868	753424	29.4618
829	687241	28.7924	869	755161	29.4788
830	688900	28.8097	870	756900	29.4958
831	690561	28.8271	871	758641	29.5127
832	692224	28.8444	872	760384	29.5296
833	693889	28.8617	873	762129	29.5466
834	695556	28.8791	874	763876	29.5635
835	697225	28.8964	875	765625	29.5804
836	698896	28.9137	876	767376	29.5973
837	700569	28.9310	877	769129	29.6142
838	702244	28.9482	878	770884	29.6311
839	703921	28.9655	879	772641	29.6479
840	705600	28.9828	880	774400	29.6648

Number	Square	Square Root	Number	Square	Square Root
881	776161	29.6816	921	848241	30.3480
882	777924	29.6985	922	850084	30.3645
883	779689	29.7153	923	851929	30.3809
884	781456	29.7321	924	853776	30.3974
885	783225	29.7489	925	855625	30.4138
886	784996	29.7658	926	857476	30.4302
887	786769	29.7825	927	859329	30.4467
888	788544	29.7993	928	861184	30.4631
889	790321	29.8161	929	863041	30.4795
890	792100	29.8329	930	864900	30.4959
891	793881	29.8496	931	866761	30.5123
892	795664	29.8664	932	868624	30.5287
893	797449	29.8831	933	870489	30.5450
894	799236	29.8998	934	872356	30.5614
895	801025	29.9166	935	874225	30.5778
896	802816	29.9333	936	876096	30.5941
897	804609	29.9500	937	877969	30.6105
898	806404	29.9666	938	879844	30.6268
899	808201	29.9833	939	881721	30.6431
900	810000	30.0000	940	883600	30.6594
901	811801	30.0167	941	885481	30.6757
902	813604	30.0333	942	887364	30.6920
903	815409	30.0500	943	889249	30.7083
904	817216	30.0666	944	891136	30.7246
905	819025	30.0832	945	893025	30.7409
906	820836	30.0998	946	894916	30.7571
907	822649	30.1164	947	896809	30.7734
908	824464	30.1330	948	898704	30.7896
909	826281	30.1496	949	900601	30.8058
910	828100	30.1662	950	902500	30.8221
911	829921	30.1828	951	904401	30.8383
912	831744	30.1993	952	906304	30.8545
913	833569	30.2159	953	908209	30.8707
914	835396	30.2324	954	910116	30.8869
915	837225	30.2490	955	912025	30.9031
916	839056	30.2655	956	913936	30.9192
917	840889	30.2820	957	915849	30.9354
918	842724	30.2985	958	917764	30.9516
919	844561	30.3150	959	919681	30.9677
920	846400	30.3315	960	921600	30.9839

Squares and Square Roots (*Continued*)

Number	Square	Square Root	Number	Square	Square Root
961	923521	31.0000	981	962361	31.3209
962	925444	31.0161	982	964324	31.3369
963	927369	31.0322	983	966289	31.3528
964	929296	31.0483	984	968256	31.3688
965	931225	31.0644	985	970225	31.3847
966	933156	31.0805	986	972196	31.4006
967	935089	31.0966	987	974169	31.4166
968	937024	31.1127	988	976144	31.4325
969	938961	31.1288	989	978121	31.4484
970	940900	31.1448	990	980100	31.4643
971	942841	31.1609	991	982081	31.4802
972	944784	31.1769	992	984064	31.4960
973	946729	31.1929	993	986049	31.5119
974	948676	31.2090	994	988036	31.5278
975	950625	31.2250	995	990025	31.5436
976	952576	31.2410	996	992016	31.5595
977	954529	31.2570	997	994009	31.5753
978	956484	31.2730	998	996004	31.5911
979	958441	31.2890	999	998001	31.6070
980	960400	31.3050	1000	1000000	31.6228

Table B. Critical Values of z (Proportion of Area Under the Normal Curve)

(A) z	(B) Area Between Mean and z	(C) Area Beyond z	(A) z	(B) Area Between Mean and z	(C) Area Beyond z
0.00	.0000	.5000	0.40	.1554	.3446
0.01	.0040	.4960	0.41	.1591	.3409
0.02	.0080	.4920	0.42	.1628	.3372
0.03	.0120	.4880	0.43	.1664	.3336
0.04	.0160	.4840	0.44	.1700	.3300
0.05	.0199	.4801	0.45	.1736	.3264
0.06	.0239	.4761	0.46	.1772	.3228
0.07	.0279	.4721	0.47	.1808	.3192
0.08	.0319	.4681	0.48	.1844	.3156
0.09	.0359	.4641	0.49	.1879	.3121
0.10	.0398	.4602	0.50	.1915	.3085
0.11	.0438	.4562	0.51	.1950	.3050
0.12	.0478	.4522	0.52	.1985	.3015
0.13	.0517	.4483	0.53	.2019	.2981
0.14	.0557	.4443	0.54	.2054	.2946
0.15	.0596	.4404	0.55	.2088	.2912
0.16	.0636	.4364	0.56	.2123	.2877
0.17	.0675	.4325	0.57	.2157	.2843
0.18	.0714	.4286	0.58	.2190	.2810
0.19	.0753	.4247	0.59	.2224	.2776
0.20	.0793	.4207	0.60	.2257	.2743
0.21	.0832	.4168	0.61	.2291	.2709
0.22	.0871	.4129	0.62	.2324	.2676
0.23	.0910	.4090	0.63	.2357	.2643
0.24	.0948	.4052	0.64	.2389	.2611
0.25	.0987	.4013	0.65	.2422	.2578
0.26	.1026	.3974	0.66	.2454	.2546
0.27	.1064	.3936	0.67	.2486	.2514
0.28	.1103	.3897	0.68	.2517	.2483
0.29	.1141	.3859	0.69	.2549	.2451
0.30	.1179	.3821	0.70	.2580	.2420
0.31	.1217	.3783	0.71	.2611	.2389
0.32	.1255	.3745	0.72	.2642	.2358
0.33	.1293	.3707	0.73	.2673	.2327
0.34	.1331	.3669	0.74	.2704	.2296
0.35	.1368	.3632	0.75	.2734	.2266
0.36	.1406	.3594	0.76	.2764	.2236
0.37	.1443	.3557	0.77	.2794	.2206
0.38	.1480	.3520	0.78	.2823	.2177
0.39	.1517	.3483	0.79	.2852	.2148

(A) z	(B) Area Between Mean and z	(C) Area Beyond z	(A) z	(B) Area Between Mean and z	(C) Area Beyond z
0.80	.2881	.2119	1.20	.3849	.1151
0.81	.2910	.2090	1.21	.3869	.1131
0.82	.2939	.2061	1.22	.3888	.1112
0.83	.2967	.2033	1.23	.3907	.1093
0.84	.2995	.2005	1.24	.3925	.1075
0.85	.3023	.1977	1.25	.3944	.1056
0.86	.3051	.1949	1.26	.3962	.1038
0.87	.3078	.1922	1.27	.3980	.1020
0.88	.3106	.1894	1.28	.3997	.1003
0.89	.3133	.1867	1.29	.4015	.0985
0.90	.3159	.1841	1.30	.4032	.0968
0.91	.3186	.1814	1.31	.4049	.0951
0.92	.3212	.1788	1.32	.4066	.0934
0.93	.3238	.1762	1.33	.4082	.0918
0.94	.3264	.1736	1.34	.4099	.0901
0.95	.3289	.1711	1.35	.4115	.0885
0.96	.3315	.1685	1.36	.4131	.0869
0.97	.3340	.1660	1.37	.4147	.0853
0.98	.3365	.1635	1.38	.4162	.0838
0.99	.3389	.1611	1.39	.4177	.0823
1.00	.3413	.1587	1.40	.4192	.0808
1.01	.3438	.1562	1.41	.4207	.0793
1.02	.3461	.1539	1.42	.4222	.0778
1.03	.3485	.1515	1.43	.4236	.0764
1.04	.3508	.1492	1.44	.4251	.0749
1.05	.3531	.1469	1.45	.4265	.0735
1.06	.3554	.1446	1.46	.4279	.0721
1.07	.3577	.1423	1.47	.4292	.0708
1.08	.3599	.1401	1.48	.4306	.0694
1.09	.3621	.1379	1.49	.4319	.0681
1.10	.3643	.1357	1.50	.4332	.0668
1.11	.3665	.1335	1.51	.4345	.0655
1.12	.3686	.1314	1.52	.4357	.0643
1.13	.3708	.1292	1.53	.4370	.0630
1.14	.3729	.1271	1.54	.4382	.0618
1.15	.3749	.1251	1.55	.4394	.0606
1.16	.3770	.1230	1.56	.4406	.0594
1.17	.3790	.1210	1.57	.4418	.0582
1.18	.3810	.1190	1.58	.4429	.0571
1.19	.3830	.1170	1.59	.4441	.0559

Critical Values of z (Continued)

(A) z	(B) Area Between Mean and z	(C) Area Beyond z	(A) z	(B) Area Between Mean and z	(C) Area Beyond z
1.60	.4452	.0548	2.00	.4772	.0228
1.61	.4463	.0537	2.01	.4778	.0222
1.62	.4474	.0526	2.02	.4783	.0217
1.63	.4484	.0516	2.03	.4788	.0212
1.64	.4495	.0505	2.04	.4793	.0207
1.65	.4505	.0495	2.05	.4798	.0202
1.66	.4515	.0485	2.06	.4803	.0197
1.67	.4525	.0475	2.07	.4808	.0192
1.68	.4535	.0465	2.08	.4812	.0188
1.69	.4545	.0455	2.09	.4817	.0183
1.70	.4554	.0446	2.10	.4821	.0179
1.71	.4564	.0436	2.11	.4826	.0174
1.72	.4573	.0427	2.12	.4830	.0170
1.73	.4582	.0418	2.13	.4834	.0166
1.74	.4591	.0409	2.14	.4838	.0162
1.75	.4599	.0401	2.15	.4842	.0158
1.76	.4608	.0392	2.16	.4846	.0154
1.77	.4616	.0384	2.17	.4850	.0150
1.78	.4625	.0375	2.18	.4854	.0146
1.79	.4633	.0367	2.19	.4857	.0143
1.80	.4641	.0359	2.20	.4861	.0139
1.81	.4649	.0351	2.21	.4864	.0136
1.82	.4656	.0344	2.22	.4868	.0132
1.83	.4664	.0336	2.23	.4871	.0129
· 1.84	.4671	.0329	2.24	.4875	.0125
1.85	.4678	.0322	2.25	.4878	.0122
1.86	.4686	.0314	2.26	.4881	.0119
1.87	.4693	.0307	2.27	.4884	.0116
1.88	.4699	.0301	2.28	.4887	.0113
1.89	.4706	.0294	2.29	.4890	.0110
1.90	.4713	.0287	2.30	.4893	.0107
1.91	.4719	.0281	2.31	.4896	.0104
1.92	.4726	.0274	2.32	.4898	.0102
1.93	.4732	.0268	2.33	.4901	.0099
1.94	.4738	.0262	2.34	.4904	.0096
1.95	.4744	.0256	2.35	.4906	.0094
1.96	.4750	.0250	2.36	.4909	.0091
1.97	.4756	.0244	2.37	.4911	.0089
1.98	.4761	.0239	2.38	.4913	.0087
1.99	.4767	.0233	2.39	.4916	.0084

Critical Values of z (Continued)

(A) z	(B) Area Between Mean and z	(C) Area Beyond z	(A) z	(B) Area Between Mean and z	(C) Area Beyond z
2.40	.4918	.0082	2.80	.4974	.0026
2.41	.4920	.0080	2.81	.4975	.0025
2.42	.4922	.0078	2.82	.4976	.0024
2.43	.4925	.0075	2.83	.4977	.0023
2.44	.4927	.0073	2.84	.4977	.0023
2.45	.4929	.0071	2.85	.4978	.0022
2.46	.4931	.0069	2.86	.4979	.0021
2.47	.4932	.0068	2.87	.4979	.0021
2.48	.4934	.0066	2.88	.4980	.0020
2.49	.4936	.0064	2.89	.4981	.0019
2.50	.4938	.0062	2.90	.4981	.0019
2.51	.4940	.0060	2.91	.4982	.0018
2.52	.4941	.0059	2.92	.4982	.0018
2.53	.4943	.0057	2.93	.4983	.0017
2.54	.4945	.0055	2.94	.4984	.0016
2.55	.4946	.0054	2.95	.4984	.0016
2.56	.4948	.0052	2.96	.4985	.0015
2.57	.4949	.0051	2.97	.4985	.0015
2.58	.4951	.0049	2.98	.4986	.0014
2.59	.4952	.0048	2.99	.4986	.0014
2.60	.4953	.0047	3.00	.4987	.0013
2.61	.4955	.0045	3.01	.4987	.0013
2.62	.4956	.0044	3.02	.4987	.0013
2.63	.4957	.0043	3.03	.4988	.0012
2.64	.4959	.0041	3.04	.4988	.0012
2.65	.4960	.0040	3.05	.4989	.0011
2.66	.4961	.0039	3.06	.4989	.0011
2.67	.4962	.0038	3.07	.4989	.0011
2.68	.4963	.0037	3.08	.4990	.0010
2.69	.4964	.0036	3.09	.4990	.0010
2.70	.4965	.0035	3.10	.4990	.0010
2.71	.4966	.0034	3.11	.4991	.0009
2.72	.4967	.0033	3.12	.4991	.0009
2.73	.4968	.0032	3.13	.4991	.0009
2.74	.4969	.0031	3.14	.4992	.0008
2.75	.4970	.0030	3.15	.4992	.0008
2.76	.4971	.0029	3.16	.4992	.0008
2.77	.4972	.0028	3.17	.4992	.0008
2.78	.4973	.0027	3.18	.4993	.0007
2.79	.4974	.0026	3.19	.4993	.0007

Critical Values of z (Continued)

(A) z	(B) Area Between Mean and z	(C) Area Beyond z	(A) z	(B) Area Between Mean and z	(C) Area Beyond z
3.20	.4993	.0007	3.40	.4997	.0003
3.21	.4993	.0007	3.45	.4997	.0003
3.22	.4994	.0006	3.50	.4998	.0002
3.23	.4994	.0006	3.60	.4998	.0002
3.24	.4994	.0006	3.70	.4999	.0001
3.25	.4994	.0006	3.80	.4999	.0001
3.30	.4995	.0005	3.90	.49995	.00005
3.35	.4996	.0004	4.00	.49997	.00003

Table C.   Critical Values of t (Student t Test)

df	LEVEL OF SIGNIFICANCE FOR ONE-TAILED TEST					
	.10	.05	.025	.01	.005	.0005
	LEVEL OF SIGNIFICANCE FOR TWO-TAILED TEST					
df	.20	.10	.05	.02	.01	.001
1	3.078	6.314	12.706	31.821	63.657	636.619
2	1.886	2.920	4.303	6.965	9.925	31.598
3	1.638	2.353	3.182	4.541	5.841	12.941
4	1.533	2.132	2.776	3.747	4.604	8.610
5	1.476	2.015	2.571	3.365	4.032	6.859
6	1.440	1.943	2.447	3.143	3.707	5.959
7	1.415	1.895	2.365	2.998	3.499	5.405
8	1.397	1.860	2.306	2.896	3.355	5.041
9	1.383	1.833	2.262	2.821	3.250	4.781
10	1.372	1.812	2.228	2.764	3.169	4.587
11	1.363	1.796	2.201	2.718	3.106	4.437
12	1.356	1.782	2.179	2.681	3.055	4.318
13	1.350	1.771	2.160	2.650	3.012	4.221
14	1.345	1.761	2.145	2.624	2.977	4.140
15	1.341	1.753	2.131	2.602	2.947	4.073
16	1.337	1.746	2.120	2.583	2.921	4.015
17	1.333	1.740	2.110	2.567	2.898	3.965
18	1.330	1.734	2.101	2.552	2.878	3.922
19	1.328	1.729	2.093	2.539	2.861	3.883
20	1.325	1.725	2.086	2.528	2.845	3.850
21	1.323	1.721	2.080	2.518	2.831	3.819
22	1.321	1.717	2.074	2.508	2.819	3.792
23	1.319	1.714	2.069	2.500	2.807	3.767
24	1.318	1.711	2.064	2.492	2.797	3.745
25	1.316	1.708	2.060	2.485	2.787	3.725
26	1.315	1.706	2.056	2.479	2.779	3.707
27	1.314	1.703	2.052	2.473	2.771	3.690
28	1.313	1.701	2.048	2.467	2.763	3.674
29	1.311	1.699	2.045	2.462	2.756	3.659
30	1.310	1.697	2.042	2.457	2.750	3.646
40	1.303	1.684	2.021	2.423	2.704	3.551
60	1.296	1.671	2.000	2.390	2.660	3.460
120	1.289	1.658	1.980	2.358	2.617	3.373
∞	1.282	1.645	1.960	2.326	2.576	3.291

Table D. Critical Values of F

The obtained F is significant at a given level if it is equal to or *greater than* the value shown in the table. 0.05 (light row) and 0.01 (dark row) points for the distribution of F

DEGREES OF FREEDOM FOR GREATER MEAN SQUARE (light = 0.05 / dark = 0.01)

DEGREES OF FREEDOM FOR LESSER MEAN SQUARE (row label at left)

df	1	2	3	4	5	6	7	8	9	10	11	12	14	16	20	24	30	40	50	75	100	200	500	∞
1	161 / 4052	200 / 4999	216 / 5403	225 / 5625	230 / 5764	234 / 5859	237 / 5928	239 / 5981	241 / 6022	242 / 6056	243 / 6082	244 / 6106	245 / 6142	246 / 6169	248 / 6208	249 / 6234	250 / 6258	251 / 6286	252 / 6302	253 / 6323	253 / 6334	254 / 6352	254 / 6361	254 / 6366
2	18.51 / 98.49	19.00 / 99.01	19.16 / 99.17	19.25 / 99.25	19.30 / 99.30	19.33 / 99.33	19.36 / 99.34	19.37 / 99.36	19.38 / 99.38	19.39 / 99.40	19.40 / 99.41	19.41 / 99.42	19.42 / 99.43	19.43 / 99.44	19.44 / 99.45	19.45 / 99.46	19.46 / 99.47	19.47 / 99.48	19.47 / 99.48	19.48 / 99.49	19.49 / 99.49	19.49 / 99.49	19.50 / 99.50	19.50 / 99.50
3	10.13 / 34.12	9.55 / 30.81	9.28 / 29.46	9.12 / 28.71	9.01 / 28.24	8.94 / 27.91	8.88 / 27.67	8.84 / 27.49	8.81 / 27.34	8.78 / 27.23	8.76 / 27.13	8.74 / 27.05	8.71 / 26.92	8.69 / 26.83	8.66 / 26.69	8.64 / 26.60	8.62 / 26.50	8.60 / 26.41	8.58 / 26.30	8.57 / 26.27	8.56 / 26.23	8.54 / 26.18	8.54 / 26.14	8.53 / 26.12
4	7.71 / 21.20	6.94 / 18.00	6.59 / 16.69	6.39 / 15.98	6.26 / 15.52	6.16 / 15.21	6.09 / 14.98	6.04 / 14.80	6.00 / 14.66	5.96 / 14.54	5.93 / 14.45	5.91 / 14.37	5.87 / 14.24	5.84 / 14.15	5.80 / 14.02	5.77 / 13.93	5.74 / 13.83	5.71 / 13.74	5.70 / 13.69	5.68 / 13.61	5.66 / 13.57	5.65 / 13.52	5.64 / 13.48	5.63 / 13.46
5	6.61 / 16.26	5.79 / 13.27	5.41 / 12.06	5.19 / 11.39	5.05 / 10.97	4.95 / 10.67	4.88 / 10.45	4.82 / 10.27	4.78 / 10.15	4.74 / 10.05	4.70 / 9.96	4.68 / 9.89	4.64 / 9.77	4.60 / 9.68	4.56 / 9.55	4.53 / 9.47	4.50 / 9.38	4.46 / 9.29	4.44 / 9.24	4.42 / 9.17	4.40 / 9.13	4.38 / 9.07	4.37 / 9.04	4.36 / 9.02
6	5.99 / 13.74	5.14 / 10.92	4.76 / 9.78	4.53 / 9.15	4.39 / 8.75	4.28 / 8.47	4.21 / 8.26	4.15 / 8.10	4.10 / 7.98	4.06 / 7.87	4.03 / 7.79	4.00 / 7.72	3.96 / 7.60	3.92 / 7.52	3.87 / 7.39	3.84 / 7.31	3.81 / 7.23	3.77 / 7.14	3.75 / 7.09	3.72 / 7.02	3.71 / 6.99	3.69 / 6.94	3.68 / 6.90	3.67 / 6.88
7	5.59 / 12.25	4.74 / 9.55	4.35 / 8.45	4.12 / 7.85	3.97 / 7.46	3.87 / 7.19	3.79 / 7.00	3.73 / 6.84	3.68 / 6.71	3.63 / 6.62	3.60 / 6.54	3.57 / 6.47	3.52 / 6.35	3.49 / 6.27	3.44 / 6.15	3.41 / 6.07	3.38 / 5.98	3.34 / 5.90	3.32 / 5.85	3.29 / 5.78	3.28 / 5.75	3.25 / 5.70	3.24 / 5.67	3.23 / 5.65
8	5.32 / 11.26	4.46 / 8.65	4.07 / 7.59	3.84 / 7.01	3.69 / 6.63	3.58 / 6.37	3.50 / 6.19	3.44 / 6.03	3.39 / 5.91	3.34 / 5.82	3.31 / 5.74	3.28 / 5.67	3.23 / 5.56	3.20 / 5.48	3.15 / 5.36	3.12 / 5.28	3.08 / 5.20	3.05 / 5.11	3.03 / 5.06	3.00 / 5.00	2.98 / 4.96	2.96 / 4.91	2.94 / 4.88	2.93 / 4.86
9	5.12 / 10.56	4.26 / 8.02	3.86 / 6.99	3.63 / 6.42	3.48 / 6.06	3.37 / 5.80	3.29 / 5.62	3.23 / 5.47	3.18 / 5.35	3.13 / 5.26	3.10 / 5.18	3.07 / 5.11	3.02 / 5.00	2.98 / 4.92	2.93 / 4.80	2.90 / 4.73	2.86 / 4.64	2.82 / 4.56	2.80 / 4.51	2.77 / 4.45	2.76 / 4.41	2.73 / 4.36	2.72 / 4.33	2.71 / 4.31
10	4.96 / 10.04	4.10 / 7.56	3.71 / 6.55	3.48 / 5.99	3.33 / 5.64	3.22 / 5.39	3.14 / 5.21	3.07 / 5.06	3.02 / 4.95	2.97 / 4.85	2.94 / 4.78	2.91 / 4.71	2.86 / 4.60	2.82 / 4.52	2.77 / 4.41	2.74 / 4.33	2.70 / 4.25	2.67 / 4.17	2.64 / 4.12	2.61 / 4.05	2.59 / 4.01	2.56 / 3.96	2.55 / 3.93	2.54 / 3.91
11	4.84 / 9.65	3.98 / 7.20	3.59 / 6.22	3.36 / 5.67	3.20 / 5.32	3.09 / 5.07	3.01 / 4.88	2.95 / 4.74	2.90 / 4.63	2.86 / 4.54	2.82 / 4.46	2.79 / 4.40	2.74 / 4.29	2.70 / 4.21	2.65 / 4.10	2.61 / 4.02	2.57 / 3.94	2.53 / 3.86	2.50 / 3.80	2.47 / 3.74	2.45 / 3.70	2.42 / 3.66	2.41 / 3.62	2.40 / 3.60
12	4.75 / 9.33	3.88 / 6.93	3.49 / 5.95	3.26 / 5.41	3.11 / 5.06	3.00 / 4.82	2.92 / 4.65	2.85 / 4.50	2.80 / 4.39	2.76 / 4.30	2.72 / 4.22	2.69 / 4.16	2.64 / 4.05	2.60 / 3.98	2.54 / 3.86	2.50 / 3.78	2.46 / 3.70	2.42 / 3.61	2.40 / 3.56	2.36 / 3.49	2.35 / 3.46	2.32 / 3.41	2.31 / 3.38	2.30 / 3.36
13	4.67 / 9.07	3.80 / 6.70	3.41 / 5.74	3.18 / 5.20	3.02 / 4.86	2.92 / 4.62	2.84 / 4.44	2.77 / 4.30	2.72 / 4.19	2.67 / 4.10	2.63 / 4.02	2.60 / 3.96	2.55 / 3.85	2.51 / 3.78	2.46 / 3.67	2.42 / 3.59	2.38 / 3.51	2.34 / 3.42	2.32 / 3.37	2.28 / 3.30	2.26 / 3.27	2.24 / 3.21	2.22 / 3.18	2.21 / 3.16
14	4.60 / 8.86	3.74 / 6.51	3.34 / 5.56	3.11 / 5.03	2.96 / 4.69	2.85 / 4.46	2.77 / 4.28	2.70 / 4.14	2.65 / 4.03	2.60 / 3.94	2.56 / 3.86	2.53 / 3.80	2.48 / 3.70	2.44 / 3.62	2.39 / 3.51	2.35 / 3.43	2.31 / 3.34	2.27 / 3.26	2.24 / 3.21	2.21 / 3.14	2.19 / 3.11	2.16 / 3.06	2.14 / 3.02	2.13 / 3.00
15	4.54 / 8.68	3.68 / 6.36	3.29 / 5.42	3.06 / 4.89	2.90 / 4.56	2.79 / 4.32	2.70 / 4.14	2.64 / 4.00	2.59 / 3.89	2.55 / 3.80	2.51 / 3.73	2.48 / 3.67	2.43 / 3.56	2.39 / 3.48	2.33 / 3.36	2.29 / 3.29	2.25 / 3.20	2.21 / 3.12	2.18 / 3.07	2.15 / 3.00	2.12 / 2.97	2.10 / 2.92	2.08 / 2.89	2.07 / 2.87

Critical Values of *F* (Continued)

DEGREES OF FREEDOM FOR GREATER MEAN SQUARE

DEGREES OF FREEDOM FOR LESSER MEAN SQUARE

	1	2	3	4	5	6	7	8	9	10	11	12	14	16	20	24	30	40	50	75	100	200	500	∞
16	4.49 / 8.53	3.63 / 6.23	3.24 / 5.29	3.01 / 4.77	2.85 / 4.44	2.74 / 4.20	2.66 / 4.03	2.59 / 3.89	2.54 / 3.78	2.49 / 3.69	2.45 / 3.61	2.42 / 3.55	2.37 / 3.45	2.33 / 3.37	2.28 / 3.25	2.24 / 3.18	2.20 / 3.10	2.16 / 3.01	2.13 / 2.96	2.09 / 2.89	2.07 / 2.86	2.04 / 2.80	2.02 / 2.77	2.01 / 2.75
17	4.45 / 8.40	3.59 / 6.11	3.20 / 5.18	2.96 / 4.67	2.81 / 4.34	2.70 / 4.10	2.62 / 3.93	2.55 / 3.79	2.50 / 3.68	2.45 / 3.59	2.41 / 3.52	2.38 / 3.45	2.33 / 3.35	2.29 / 3.27	2.23 / 3.16	2.19 / 3.08	2.15 / 3.00	2.11 / 2.92	2.08 / 2.86	2.04 / 2.79	2.02 / 2.76	1.99 / 2.70	1.97 / 2.67	1.96 / 2.65
18	4.41 / 8.28	3.55 / 6.01	3.16 / 5.09	2.93 / 4.58	2.77 / 4.25	2.66 / 4.01	2.58 / 3.85	2.51 / 3.71	2.46 / 3.60	2.41 / 3.51	2.37 / 3.44	2.34 / 3.37	2.29 / 3.27	2.25 / 3.19	2.19 / 3.07	2.15 / 3.00	2.11 / 2.91	2.07 / 2.83	2.04 / 2.78	2.00 / 2.71	1.98 / 2.68	1.95 / 2.62	1.93 / 2.59	1.92 / 2.57
19	4.38 / 8.18	3.52 / 5.93	3.13 / 5.01	2.90 / 4.50	2.74 / 4.17	2.63 / 3.94	2.55 / 3.77	2.48 / 3.63	2.43 / 3.52	2.38 / 3.43	2.34 / 3.36	2.31 / 3.30	2.26 / 3.19	2.21 / 3.12	2.15 / 3.00	2.11 / 2.92	2.07 / 2.84	2.02 / 2.76	2.00 / 2.70	1.96 / 2.63	1.94 / 2.60	1.91 / 2.54	1.90 / 2.51	1.88 / 2.49
20	4.35 / 8.10	3.49 / 5.85	3.10 / 4.94	2.87 / 4.43	2.71 / 4.10	2.60 / 3.87	2.52 / 3.71	2.45 / 3.56	2.40 / 3.45	2.35 / 3.37	2.31 / 3.30	2.28 / 3.23	2.23 / 3.13	2.18 / 3.05	2.12 / 2.94	2.08 / 2.86	2.04 / 2.77	1.99 / 2.69	1.96 / 2.63	1.92 / 2.56	1.90 / 2.53	1.87 / 2.47	1.85 / 2.44	1.84 / 2.42
21	4.32 / 8.02	3.47 / 5.78	3.07 / 4.87	2.84 / 4.37	2.68 / 4.04	2.57 / 3.81	2.49 / 3.65	2.42 / 3.51	2.37 / 3.40	2.32 / 3.31	2.28 / 3.24	2.25 / 3.17	2.20 / 3.07	2.15 / 2.99	2.09 / 2.88	2.05 / 2.80	2.00 / 2.72	1.96 / 2.63	1.93 / 2.58	1.80 / 2.51	1.87 / 2.47	1.84 / 2.42	1.82 / 2.38	1.81 / 2.36
22	4.30 / 7.94	3.44 / 5.72	3.05 / 4.82	2.82 / 4.31	2.66 / 3.99	2.55 / 3.76	2.47 / 3.59	2.40 / 3.45	2.35 / 3.35	2.30 / 3.26	2.26 / 3.18	2.23 / 3.12	2.18 / 3.02	2.13 / 2.94	2.07 / 2.83	2.03 / 2.75	1.98 / 2.67	1.93 / 2.58	1.91 / 2.53	1.87 / 2.46	1.84 / 2.42	1.81 / 2.37	1.80 / 2.33	1.78 / 2.31
23	4.28 / 7.88	3.42 / 5.66	3.03 / 4.76	2.80 / 4.26	2.64 / 3.94	2.53 / 3.71	2.45 / 3.54	2.38 / 3.41	2.32 / 3.30	2.28 / 3.21	2.24 / 3.14	2.20 / 3.07	2.14 / 2.97	2.10 / 2.89	2.04 / 2.78	2.00 / 2.70	1.96 / 2.62	1.91 / 2.53	1.88 / 2.48	1.84 / 2.41	1.82 / 2.37	1.79 / 2.32	1.77 / 2.28	1.76 / 2.26
24	4.26 / 7.82	3.40 / 5.61	3.01 / 4.72	2.78 / 4.22	2.62 / 3.90	2.51 / 3.67	2.43 / 3.50	2.36 / 3.36	2.30 / 3.25	2.26 / 3.17	2.22 / 3.09	2.18 / 3.03	2.13 / 2.93	2.09 / 2.85	2.02 / 2.74	1.98 / 2.66	1.94 / 2.58	1.89 / 2.49	1.86 / 2.44	1.82 / 2.36	1.80 / 2.33	1.76 / 2.27	1.74 / 2.23	1.73 / 2.21
25	4.24 / 7.77	3.38 / 5.57	2.99 / 4.68	2.76 / 4.18	2.60 / 3.86	2.49 / 3.63	2.41 / 3.46	2.34 / 3.32	2.28 / 3.21	2.24 / 3.13	2.20 / 3.05	2.16 / 2.99	2.11 / 2.89	2.06 / 2.81	2.00 / 2.70	1.96 / 2.62	1.92 / 2.54	1.87 / 2.45	1.84 / 2.40	1.80 / 2.32	1.77 / 2.29	1.74 / 2.23	1.72 / 2.19	1.71 / 2.17
26	4.22 / 7.72	3.37 / 5.53	2.89 / 4.64	2.74 / 4.14	2.59 / 3.82	2.47 / 3.59	2.39 / 3.42	2.32 / 3.29	2.27 / 3.17	2.22 / 3.09	2.18 / 3.02	2.15 / 2.96	2.10 / 2.86	2.05 / 2.77	1.99 / 2.66	1.95 / 2.58	1.90 / 2.50	1.85 / 2.41	1.82 / 2.36	1.78 / 2.28	1.76 / 2.25	1.72 / 2.19	1.70 / 2.15	1.69 / 2.13
27	4.21 / 7.68	3.35 / 5.49	2.96 / 4.60	2.73 / 4.11	2.57 / 3.79	2.46 / 3.56	2.37 / 3.39	2.30 / 3.26	2.25 / 3.14	2.20 / 3.06	2.16 / 2.98	2.13 / 2.93	2.08 / 2.83	2.03 / 2.74	1.97 / 2.63	1.93 / 2.55	1.88 / 2.47	1.84 / 2.38	1.80 / 2.33	1.76 / 2.25	1.74 / 2.21	1.71 / 2.16	1.68 / 2.12	1.67 / 2.10
28	4.20 / 7.64	3.34 / 5.45	2.95 / 4.57	2.71 / 4.07	2.56 / 3.76	2.44 / 3.53	2.36 / 3.36	2.29 / 3.23	2.24 / 3.11	2.19 / 3.03	2.15 / 2.95	2.12 / 2.90	2.06 / 2.80	2.02 / 2.71	1.95 / 2.60	1.91 / 2.52	1.87 / 2.44	1.81 / 2.35	1.78 / 2.30	1.75 / 2.22	1.72 / 2.18	1.69 / 2.13	1.67 / 2.09	1.65 / 2.06
29	4.18 / 7.60	3.33 / 5.52	2.93 / 4.54	2.70 / 4.04	2.54 / 3.73	2.43 / 3.50	2.35 / 3.33	2.28 / 3.20	2.22 / 3.08	2.18 / 3.00	2.14 / 2.92	2.10 / 2.87	2.05 / 2.77	2.00 / 2.68	1.94 / 2.57	1.90 / 2.49	1.85 / 2.41	1.80 / 2.32	1.77 / 2.27	1.73 / 2.19	1.71 / 2.15	1.68 / 2.10	1.65 / 2.06	1.64 / 2.03
30	4.17 / 7.56	3.32 / 5.39	2.92 / 4.51	2.69 / 4.02	2.53 / 3.70	2.42 / 3.47	2.34 / 3.30	2.27 / 3.17	2.21 / 3.06	2.16 / 2.98	2.12 / 2.90	2.09 / 2.84	2.04 / 2.74	1.99 / 2.66	1.93 / 2.55	1.89 / 2.47	1.84 / 2.38	1.79 / 2.29	1.76 / 2.24	1.72 / 2.16	1.69 / 2.13	1.66 / 2.07	1.64 / 2.03	1.62 / 2.01

Critical Values of F (Continued)

The obtained F is significant at a given level if it is equal to or *greater than* the value shown in the table.

0.05 (light row) and 0.01 (dark row) points for the distribution of F

DEGREES OF FREEDOM FOR GREATER MEAN SQUARE

DEGREES OF FREEDOM FOR LESSER MEAN SQUARE

	1	2	3	4	5	6	7	8	9	10	11	12	14	16	20	24	30	40	50	75	100	200	500	∞
32	4.15 / 7.50	3.30 / 5.34	2.90 / 4.46	2.67 / 3.97	2.51 / 3.66	2.40 / 3.42	2.32 / 3.25	2.25 / 3.12	2.19 / 3.01	2.14 / 2.94	2.10 / 2.86	2.07 / 2.80	2.02 / 2.70	1.97 / 2.62	1.91 / 2.51	1.86 / 2.42	1.82 / 2.34	1.76 / 2.25	1.74 / 2.20	1.69 / 2.12	1.67 / 2.08	1.64 / 2.02	1.61 / 1.98	1.59 / 1.96
34	4.13 / 7.44	3.28 / 5.29	2.88 / 4.42	2.65 / 3.93	2.49 / 3.61	2.38 / 3.38	2.30 / 3.21	2.23 / 3.08	2.17 / 2.97	2.12 / 2.89	2.08 / 2.82	2.05 / 2.76	2.00 / 2.66	1.95 / 2.58	1.89 / 2.47	1.84 / 2.38	1.80 / 2.30	1.74 / 2.21	1.71 / 2.15	1.67 / 2.08	1.64 / 2.04	1.61 / 1.98	1.59 / 1.94	1.57 / 1.91
36	4.11 / 7.39	3.26 / 5.25	2.86 / 4.38	2.63 / 3.89	2.48 / 3.58	2.36 / 3.35	2.28 / 3.18	2.21 / 3.04	2.15 / 2.94	2.10 / 2.86	2.06 / 2.78	2.03 / 2.72	1.98 / 2.62	1.93 / 2.54	1.87 / 2.43	1.82 / 2.35	1.78 / 2.26	1.72 / 2.17	1.69 / 2.12	1.65 / 2.04	1.62 / 2.00	1.59 / 1.94	1.56 / 1.90	1.55 / 1.87
38	4.10 / 7.35	3.25 / 5.21	2.85 / 4.34	2.62 / 3.86	2.46 / 3.54	2.35 / 3.32	2.26 / 3.15	2.19 / 3.02	2.14 / 2.91	2.09 / 2.82	2.05 / 2.75	2.02 / 2.69	1.96 / 2.59	1.92 / 2.51	1.85 / 2.40	1.80 / 2.32	1.76 / 2.22	1.71 / 2.14	1.67 / 2.08	1.63 / 2.00	1.60 / 1.97	1.57 / 1.90	1.54 / 1.86	1.53 / 1.84
40	4.08 / 7.31	3.23 / 5.18	2.84 / 4.31	2.61 / 3.83	2.45 / 3.51	2.34 / 3.29	2.25 / 3.12	2.18 / 2.99	2.12 / 2.88	2.07 / 2.80	2.04 / 2.73	2.00 / 2.66	1.95 / 2.56	1.90 / 2.49	1.84 / 2.37	1.79 / 2.29	1.74 / 2.20	1.69 / 2.11	1.66 / 2.05	1.61 / 1.97	1.59 / 1.94	1.55 / 1.88	1.53 / 1.84	1.51 / 1.81
42	4.07 / 7.27	3.22 / 5.15	2.83 / 4.29	2.59 / 3.80	2.44 / 3.49	2.32 / 3.26	2.24 / 3.10	2.17 / 2.96	2.11 / 2.86	2.06 / 2.77	2.02 / 2.70	1.99 / 2.64	1.94 / 2.54	1.89 / 2.46	1.82 / 2.35	1.78 / 2.26	1.73 / 2.17	1.68 / 2.08	1.64 / 2.02	1.60 / 1.94	1.57 / 1.91	1.54 / 1.85	1.51 / 1.80	1.49 / 1.78
44	4.06 / 7.24	3.21 / 5.12	2.82 / 4.26	2.58 / 3.78	2.43 / 3.46	2.31 / 3.24	2.23 / 3.07	2.16 / 2.94	2.10 / 2.84	2.05 / 2.75	2.01 / 2.68	1.98 / 2.62	1.92 / 2.52	1.88 / 2.44	1.81 / 2.32	1.76 / 2.24	1.72 / 2.15	1.66 / 2.06	1.63 / 2.00	1.58 / 1.92	1.56 / 1.88	1.52 / 1.82	1.50 / 1.78	1.48 / 1.75
46	4.05 / 7.21	3.20 / 5.10	2.81 / 4.24	2.57 / 3.76	2.42 / 3.44	2.30 / 3.22	2.22 / 3.05	2.14 / 2.92	2.09 / 2.82	2.04 / 2.73	2.00 / 2.66	1.97 / 2.60	1.91 / 2.50	1.87 / 2.42	1.80 / 2.30	1.75 / 2.22	1.71 / 2.13	1.65 / 2.04	1.62 / 1.98	1.57 / 1.90	1.54 / 1.86	1.51 / 1.80	1.48 / 1.76	1.46 / 1.72
48	4.04 / 7.19	3.19 / 5.08	2.80 / 4.22	2.56 / 3.74	2.41 / 3.42	2.30 / 3.20	2.21 / 3.04	2.14 / 2.90	2.08 / 2.80	2.03 / 2.71	1.99 / 2.64	1.96 / 2.58	1.90 / 2.48	1.86 / 2.40	1.79 / 2.28	1.74 / 2.20	1.70 / 2.11	1.64 / 2.02	1.61 / 1.96	1.56 / 1.88	1.53 / 1.84	1.50 / 1.78	1.47 / 1.73	1.45 / 1.70
50	4.03 / 7.17	3.18 / 5.06	2.79 / 4.20	2.56 / 3.72	2.40 / 3.41	2.29 / 3.18	2.20 / 3.02	2.13 / 2.88	2.07 / 2.78	2.02 / 2.70	1.98 / 2.62	1.95 / 2.56	1.90 / 2.46	1.85 / 2.39	1.78 / 2.26	1.74 / 2.18	1.69 / 2.10	1.63 / 2.00	1.60 / 1.94	1.55 / 1.86	1.52 / 1.82	1.48 / 1.76	1.46 / 1.71	1.44 / 1.68
55	4.02 / 7.12	3.17 / 5.01	2.78 / 4.16	2.54 / 3.68	2.38 / 3.37	2.27 / 3.15	2.18 / 2.98	2.11 / 2.85	2.05 / 2.75	2.00 / 2.66	1.97 / 2.59	1.93 / 2.53	1.88 / 2.43	1.83 / 2.35	1.76 / 2.23	1.72 / 2.15	1.67 / 2.06	1.61 / 1.96	1.58 / 1.90	1.52 / 1.82	1.50 / 1.78	1.46 / 1.71	1.43 / 1.66	1.41 / 1.64
60	4.00 / 7.08	3.15 / 4.98	2.76 / 4.13	2.52 / 3.65	2.37 / 3.34	2.25 / 3.12	2.17 / 2.95	2.10 / 2.82	2.04 / 2.72	1.99 / 2.63	1.95 / 2.56	1.92 / 2.50	1.86 / 2.40	1.81 / 2.32	1.75 / 2.20	1.70 / 2.12	1.65 / 2.03	1.59 / 1.93	1.56 / 1.87	1.50 / 1.79	1.48 / 1.74	1.44 / 1.68	1.41 / 1.63	1.39 / 1.60
65	3.99 / 7.04	3.14 / 4.95	2.75 / 4.10	2.51 / 3.62	2.36 / 3.31	2.24 / 3.09	2.15 / 2.93	2.08 / 2.79	2.02 / 2.70	1.98 / 2.61	1.94 / 2.54	1.90 / 2.47	1.85 / 2.37	1.80 / 2.30	1.73 / 2.18	1.68 / 2.09	1.63 / 2.00	1.57 / 1.90	1.54 / 1.84	1.49 / 1.76	1.46 / 1.71	1.42 / 1.64	1.39 / 1.60	1.37 / 1.56
70	3.98 / 7.01	3.13 / 4.92	2.74 / 4.08	2.50 / 3.60	2.35 / 3.29	2.22 / 3.07	2.14 / 2.91	2.07 / 2.77	2.01 / 2.67	1.97 / 2.59	1.93 / 2.51	1.89 / 2.45	1.84 / 2.35	1.79 / 2.28	1.72 / 2.15	1.67 / 2.07	1.62 / 1.98	1.56 / 1.88	1.53 / 1.82	1.47 / 1.74	1.45 / 1.69	1.40 / 1.62	1.37 / 1.56	1.35 / 1.53
80	3.96 / 6.96	3.11 / 4.88	2.72 / 4.04	2.48 / 3.56	2.33 / 3.25	2.21 / 3.04	2.12 / 2.87	2.05 / 2.74	1.99 / 2.64	1.95 / 2.55	1.91 / 2.48	1.88 / 2.41	1.82 / 2.32	1.77 / 2.24	1.70 / 2.11	1.65 / 2.03	1.60 / 1.94	1.54 / 1.84	1.51 / 1.78	1.45 / 1.70	1.42 / 1.65	1.38 / 1.57	1.35 / 1.52	1.32 / 1.49

Critical Values of F (Continued)

	1	2	3	4	5	6	7	8	9	10	11	12	14	16	20	24	30	40	50	75	100	200	500	∞
100	3.94 **6.90**	3.09 **4.82**	2.70 **3.98**	2.46 **3.51**	2.30 **3.20**	2.19 **2.99**	2.10 **2.82**	2.03 **2.69**	1.97 **2.59**	1.92 **2.51**	1.88 **2.43**	1.85 **2.36**	1.79 **2.26**	1.75 **2.19**	1.68 **2.06**	1.63 **1.98**	1.57 **1.89**	1.51 **1.79**	1.48 **1.73**	1.42 **1.64**	1.39 **1.59**	1.34 **1.51**	1.30 **1.46**	1.28 **1.43**
125	3.92 **6.84**	3.07 **4.78**	2.68 **3.94**	2.44 **3.47**	2.29 **3.17**	2.17 **2.95**	2.08 **2.79**	2.01 **2.65**	1.95 **2.56**	1.90 **2.47**	1.86 **2.40**	1.83 **2.33**	1.77 **2.23**	1.72 **2.15**	1.65 **2.03**	1.60 **1.94**	1.55 **1.85**	1.49 **1.75**	1.45 **1.68**	1.39 **1.59**	1.36 **1.54**	1.31 **1.46**	1.27 **1.40**	1.25 **1.37**
150	3.91 **6.81**	3.06 **4.75**	2.67 **3.91**	2.43 **3.44**	2.27 **3.13**	2.16 **2.92**	2.07 **2.76**	2.00 **2.62**	1.94 **2.53**	1.89 **2.44**	1.85 **2.37**	1.82 **2.30**	1.76 **2.20**	1.71 **2.12**	1.64 **2.00**	1.59 **1.91**	1.54 **1.83**	1.47 **1.72**	1.44 **1.66**	1.37 **1.56**	1.34 **1.51**	1.29 **1.43**	1.25 **1.37**	1.22 **1.33**
200	3.89 **6.76**	3.04 **4.71**	2.65 **3.88**	2.41 **3.41**	2.26 **3.11**	2.14 **2.90**	2.05 **2.73**	1.98 **2.60**	1.92 **2.50**	1.87 **2.41**	1.83 **2.34**	1.80 **2.28**	1.74 **2.17**	1.69 **2.09**	1.62 **1.97**	1.57 **1.88**	1.52 **1.79**	1.45 **1.69**	1.42 **1.62**	1.35 **1.53**	1.32 **1.48**	1.26 **1.39**	1.22 **1.33**	1.19 **1.28**
400	3.86 **6.70**	3.02 **4.66**	2.62 **3.83**	2.39 **3.36**	2.23 **3.06**	2.12 **2.85**	2.03 **2.69**	1.96 **2.55**	1.90 **2.46**	1.85 **2.37**	1.81 **2.29**	1.78 **2.23**	1.72 **2.12**	1.67 **2.04**	1.60 **1.92**	1.54 **1.84**	1.49 **1.74**	1.42 **1.64**	1.38 **1.57**	1.32 **1.47**	1.28 **1.42**	1.22 **1.32**	1.16 **1.24**	1.13 **1.19**
1000	3.85 **6.66**	3.00 **4.62**	2.61 **3.80**	2.38 **3.34**	2.22 **3.04**	2.10 **2.82**	2.02 **2.66**	1.95 **2.53**	1.89 **2.43**	1.84 **2.34**	1.80 **2.26**	1.76 **2.20**	1.70 **2.09**	1.65 **2.01**	1.58 **1.89**	1.53 **1.81**	1.47 **1.71**	1.41 **1.61**	1.36 **1.54**	1.30 **1.44**	1.26 **1.38**	1.19 **1.28**	1.13 **1.19**	1.08 **1.11**
∞	3.84 **6.64**	2.99 **4.60**	2.60 **3.78**	2.37 **3.32**	2.21 **3.02**	2.09 **2.80**	2.01 **2.64**	1.94 **2.51**	1.88 **2.41**	1.83 **2.32**	1.79 **2.24**	1.75 **2.18**	1.69 **2.07**	1.64 **1.99**	1.57 **1.87**	1.52 **1.79**	1.46 **1.69**	1.40 **1.59**	1.35 **1.52**	1.28 **1.41**	1.24 **1.36**	1.17 **1.25**	1.11 **1.15**	1.00 **1.00**

Degrees of Freedom for Lesser Mean Square

Table E. Critical Values of A

	LEVEL OF SIGNIFICANCE FOR ONE-TAILED TEST				
	.05	.025	.01	.005	.0005
	LEVEL OF SIGNIFICANCE FOR TWO-TAILED TEST				
df	.10	.05	.02	.01	.001
1	0.512	0.503	0.500	0.500	0.500
2	0.412	0.369	0.347	0.340	0.334
3	0.385	0.324	0.286	0.272	0.254
4	0.376	0.304	0.257	0.238	0.211
5	0.372	0.293	0.240	0.218	0.184
6	0.370	0.286	0.230	0.205	0.167
7	0.369	0.281	0.222	0.196	0.155
8	0.368	0.278	0.217	0.190	0.146
9	0.368	0.276	0.213	0.185	0.139
10	0.368	0.274	0.210	0.181	0.134
11	0.368	0.273	0.207	0.178	0.130
12	0.368	0.271	0.205	0.176	0.126
13	0.368	0.270	0.204	0.174	0.124
14	0.368	0.270	0.202	0.172	0.121
15	0.368	0.269	0.201	0.170	0.119
16	0.368	0.268	0.200	0.169	0.117
17	0.368	0.268	0.199	0.168	0.116
18	0.368	0.267	0.198	0.167	0.114
19	0.368	0.267	0.197	0.166	0.113
20	0.368	0.266	0.197	0.165	0.112
21	0.368	0.266	0.196	0.165	0.111
22	0.368	0.266	0.196	0.164	0.110
23	0.368	0.266	0.195	0.163	0.109
24	0.368	0.265	0.195	0.163	0.108
25	0.368	0.265	0.194	0.162	0.108
26	0.368	0.265	0.194	0.162	0.107
27	0.368	0.265	0.193	0.161	0.107
28	0.368	0.265	0.193	0.161	0.106
29	0.368	0.264	0.193	0.161	0.106
30	0.368	0.264	0.193	0.160	0.105
40	0.368	0.263	0.191	0.158	0.102
60	0.369	0.262	0.189	0.155	0.099
120	0.369	0.261	0.187	0.153	0.095
∞	0.370	0.260	0.185	0.151	0.092

Table F. Percentage Points of the Studentized Range

df	α	k = NUMBER OF MEANS OR NUMBER OF STEPS BETWEEN ORDERED MEANS									
		2	3	4	5	6	7	8	9	10	11
5	.05	3.64	4.60	5.22	5.67	6.03	6.33	6.58	6.80	6.99	7.17
	.01	5.70	6.98	7.80	8.42	8.91	9.32	9.67	9.97	10.24	10.48
6	.05	3.46	4.34	4.90	5.30	5.63	5.90	6.12	6.32	6.49	6.65
	.01	5.24	6.33	7.03	7.56	7.97	8.32	8.61	8.87	9.10	9.30
7	.05	3.34	4.16	4.68	5.06	5.36	5.61	5.82	6.00	6.16	6.30
	.01	4.95	5.92	6.54	7.01	7.37	7.68	7.94	8.17	8.37	8.55
8	.05	3.26	4.04	4.53	4.89	5.17	5.40	5.60	5.77	5.92	6.05
	.01	4.75	5.64	6.20	6.62	6.96	7.24	7.47	7.68	7.86	8.03
9	.05	3.20	3.95	4.41	4.76	5.02	5.24	5.43	5.59	5.74	5.87
	.01	4.60	5.43	5.96	6.35	6.66	6.91	7.13	7.33	7.49	7.65
10	.05	3.15	3.88	4.33	4.65	4.91	5.12	5.30	5.46	5.60	5.72
	.01	4.48	5.27	5.77	6.14	6.43	6.67	6.87	7.05	7.21	7.36
11	.05	3.11	3.82	4.26	4.57	4.82	5.03	5.20	5.35	5.49	5.61
	.01	4.39	5.15	5.62	5.97	6.25	6.48	6.67	6.84	6.99	7.13
12	.05	3.08	3.77	4.20	4.51	4.75	4.95	5.12	5.27	5.39	5.51
	.01	4.32	5.05	5.50	5.84	6.10	6.32	6.51	6.67	6.81	6.94
13	.05	3.06	3.73	4.15	4.45	4.69	4.88	5.05	5.19	5.32	5.43
	.01	4.26	4.96	5.40	5.73	5.98	6.19	6.37	6.53	6.67	6.79
14	.05	3.03	3.70	4.11	4.41	4.64	4.83	4.99	5.13	5.25	5.36
	.01	4.21	4.89	5.32	5.63	5.88	6.08	6.26	6.41	6.54	6.66
15	.05	3.01	3.67	4.08	4.37	4.59	4.78	4.94	5.08	5.20	5.31
	.01	4.17	4.84	5.25	5.56	5.80	5.99	6.16	6.31	6.44	6.55
16	.05	3.00	3.65	4.05	4.33	4.56	4.74	4.90	5.03	5.15	5.26
	.01	4.13	4.79	5.19	5.49	5.72	5.92	6.08	6.22	6.35	6.46
17	.05	2.98	3.63	4.02	4.30	4.52	4.70	4.86	4.99	5.11	5.21
	.01	4.10	4.74	5.14	5.43	5.66	5.85	6.01	6.15	6.27	6.38
18	.05	2.97	3.61	4.00	4.28	4.49	4.67	4.82	4.96	5.07	5.17
	.01	4.07	4.70	5.09	5.38	5.60	5.79	5.94	6.08	6.20	6.31
19	.05	2.96	3.59	3.98	4.25	4.47	4.65	4.79	4.92	5.04	5.14
	.01	4.05	4.67	5.05	5.33	5.55	5.73	5.89	6.02	6.14	6.25
20	.05	2.95	3.58	3.96	4.23	4.45	4.62	4.77	4.90	5.01	5.11
	.01	4.02	4.64	5.02	5.29	5.51	5.69	5.84	5.97	6.09	6.19
24	.05	2.92	3.53	3.90	4.17	4.37	4.54	4.68	4.81	4.92	5.01
	.01	3.96	4.55	4.91	5.17	5.37	5.54	5.69	5.81	5.92	6.02
30	.05	2.89	3.49	3.85	4.10	4.30	4.46	4.60	4.72	4.82	4.92
	.01	3.89	4.45	4.80	5.05	5.24	5.40	5.54	5.65	5.76	5.85
40	.05	2.86	3.44	3.79	4.04	4.23	4.39	4.52	4.63	4.73	4.82
	.01	3.82	4.37	4.70	4.93	5.11	5.26	5.39	5.50	5.60	5.69
60	.05	2.83	3.40	3.74	3.98	4.16	4.31	4.44	4.55	4.65	4.73
	.01	3.76	4.28	4.59	4.82	4.99	5.13	5.25	5.36	5.45	5.53
120	.05	2.80	3.36	3.68	3.92	4.10	4.24	4.36	4.47	4.56	4.64
	.01	3.70	4.20	4.50	4.71	4.87	5.01	5.12	5.21	5.30	5.37
∞	.05	2.77	3.31	3.63	3.86	4.03	4.17	4.29	4.39	4.47	4.55
	.01	3.64	4.12	4.40	4.60	4.76	4.88	4.99	5.08	5.16	5.23

Table G. Critical Values of *r* (Pearson Product-Moment Correlation Coefficient)

df	.1	.05	.02	.01	.001
1	.98769	.99692	.999507	.999877	.9999988
2	.9000	.9500	.9800	.9900	.99900
3	.8054	.8783	.9343	.9587	.99116
4	.7293	.8114	.8822	.9172	.97406
5	.6694	.7545	.8329	.8745	.9507
6	.6215	.7067	.7887	.8343	.9249
7	.5822	.6664	.7498	.7977	.8982
8	.5494	.6319	.7155	.7646	.8721
9	.5214	.6021	.6851	.7348	.8471
10	.4973	.5760	.6581	.7079	.8233
11	.4762	.5529	.6339	.6835	.8010
12	.4575	.5324	.6120	.6614	.7800
13	.4409	.5139	.5923	.6411	.7603
14	.4259	.4973	.5742	.6226	.7420
15	.4124	.4821	.5577	.6055	.7246
16	.4000	.4683	.5425	.5897	.7084
17	.3887	.4555	.5285	.5751	.6932
18	.3783	.4438	.5155	.5614	.6787
19	.3687	.4329	.5034	.5487	.6652
20	.3598	.4227	.4921	.5368	.6524
25	.3233	.3809	.4451	.4869	.5974
30	.2960	.3494	.4093	.4487	.5541
35	.2746	.3246	.3810	.4182	.5189
40	.2573	.3044	.3578	.3932	.4896
45	.2428	.2875	.3384	.3721	.4648
50	.2306	.2732	.3218	.3541	.4433
60	.2108	.2500	.2948	.3248	.4078
70	.1954	.2319	.2737	.3017	.3799
80	.1829	.2172	.2565	.2830	.3568
90	.1726	.2050	.2422	.2673	.3375
100	.1638	.1946	.2301	.2540	.3211

Table H. Critical Values of r_s (Spearman Rank-Order Correlation Coefficient)

df	LEVEL OF SIGNIFICANCE FOR ONE-TAILED TEST			
	.05	.025	.01	.005
	LEVEL OF SIGNIFICANCE FOR TWO-TAILED TEST			
	.10	.05	.02	.01
5	.900	1.000	1.000	—
6	.829	.886	.943	1.000
7	.714	.786	.893	.929
8	.643	.738	.833	.881
9	.600	.683	.783	.833
10	.564	.648	.746	.794
12	.506	.591	.712	.777
14	.456	.544	.645	.715
16	.425	.506	.601	.665
18	.399	.475	.564	.625
20	.377	.450	.534	.591
22	.359	.428	.508	.562
24	.343	.409	.485	.537
26	.329	.392	.465	.515
28	.317	.377	.448	.496
30	.306	.364	.432	.478

347

Table I. Critical Values of χ^2

df	.20	.10	.05	.02	.01	.001
1	1.64	2.71	3.84	5.41	6.64	10.83
2	3.22	4.60	5.99	7.82	9.21	13.82
3	4.64	6.25	7.82	9.84	11.34	16.27
4	5.99	7.78	9.49	11.67	13.28	18.46
5	7.29	9.24	11.07	13.39	15.09	20.52
6	8.56	10.64	12.59	15.03	16.81	22.46
7	9.80	12.02	14.07	16.62	18.48	24.32
8	11.03	13.36	15.51	18.17	20.09	26.12
9	12.24	14.68	16.92	19.68	21.67	27.88
10	13.44	15.99	18.31	21.16	23.21	29.59
11	14.63	17.28	19.68	22.62	24.72	31.26
12	15.81	18.55	21.03	24.05	26.22	32.91
13	16.98	19.81	22.36	25.47	27.69	34.53
14	18.15	21.06	23.68	26.87	29.14	36.12
15	19.31	22.31	25.00	28.26	30.58	37.70
16	20.46	23.54	26.30	29.63	32.00	39.29
17	21.62	24.77	27.59	31.00	33.41	40.75
18	22.76	25.99	28.87	32.35	34.80	42.31
19	23.90	27.20	30.14	33.69	36.19	43.82
20	25.04	28.41	31.41	35.02	37.57	45.32
21	26.17	29.62	32.67	36.34	38.93	46.80
22	27.30	30.81	33.92	37.66	40.29	48.27
23	28.43	32.01	35.17	38.97	41.64	49.73
24	29.55	33.20	36.42	40.27	42.98	51.18
25	30.68	34.38	37.65	41.57	44.31	52.62
26	31.80	35.56	38.88	42.86	45.64	54.05
27	32.91	36.74	40.11	44.14	46.96	55.48
28	34.03	37.92	41.34	45.42	48.28	56.89
29	35.14	39.09	42.69	46.69	49.59	58.30
30	36.25	40.26	43.77	47.96	50.89	59.70
32	38.47	42.59	46.19	50.49	53.49	62.49
34	40.68	44.90	48.60	53.00	56.06	65.25
36	42.88	47.21	51.00	55.49	58.62	67.99
38	45.08	49.51	53.38	57.97	61.16	70.70
40	47.27	51.81	55.76	60.44	63.69	73.40
44	51.64	56.37	60.48	65.34	68.71	78.75
48	55.99	60.91	65.17	70.20	73.68	84.04
52	60.33	65.42	69.83	75.02	78.62	89.27
56	64.66	69.92	74.47	79.82	83.51	94.46
60	68.97	74.40	79.08	84.58	88.38	99.61

Table J. **Critical Values of _D_ (Kolmogorov-Smirnov One-Sample Test)**

N	LEVEL OF SIGNIFICANCE FOR TWO-TAILED TEST				
	.20	.15	.10	.05	.01
1	.900	.925	.950	.975	.995
2	.684	.726	.776	.842	.929
3	.565	.597	.642	.708	.828
4	.494	.525	.564	.624	.733
5	.446	.474	.510	.565	.669
6	.410	.436	.470	.521	.618
7	.381	.405	.438	.486	.577
8	.358	.381	.411	.457	.543
9	.339	.360	.388	.432	.514
10	.322	.342	.368	.410	.490
11	.307	.326	.352	.391	.468
12	.295	.313	.338	.375	.450
13	.284	.302	.325	.361	.433
14	.274	.292	.314	.349	.418
15	.266	.283	.304	.338	.404
16	.258	.274	.295	.328	.392
17	.250	.266	.286	.318	.381
18	.244	.259	.278	.309	.371
19	.237	.252	.272	.301	.363
20	.231	.246	.264	.294	.356
25	.21	.22	.24	.27	.32
30	.19	.20	.22	.24	.29
35	.18	.19	.21	.23	.27
Over 35	$\dfrac{1.07}{\sqrt{N}}$	$\dfrac{1.14}{\sqrt{N}}$	$\dfrac{1.22}{\sqrt{N}}$	$\dfrac{1.36}{\sqrt{N}}$	$\dfrac{1.63}{\sqrt{N}}$

Table K. Critical Values for K_D (Kolmogorov-Smirnov Two-Sample Test)

N	LEVEL OF SIGNIFICANCE FOR ONE-TAILED TEST		LEVEL OF SIGNIFICANCE FOR TWO-TAILED TEST	
	$\alpha = .05$	$\alpha = .01$	$\alpha = .05$	$\alpha = .01$
3	3	—	—	—
4	4	—	4	—
5	4	5	5	5
6	5	6	5	6
7	5	6	6	6
8	5	6	6	7
9	6	7	6	7
10	6	7	7	8
11	6	8	7	8
12	6	8	7	8
13	7	8	7	9
14	7	8	8	9
15	7	9	8	9
16	7	9	8	10
17	8	9	8	10
18	8	10	9	10
19	8	10	9	10
20	8	10	9	11
21	8	10	9	11
22	9	11	9	11
23	9	11	10	11
24	9	11	10	12
25	9	11	10	12
26	9	11	10	12
27	9	12	10	12
28	10	12	11	13
29	10	12	11	13
30	10	12	11	13
35	11	13	12	
40	11	14	13	

Table L.   Critical Values for K_D (Kolmogorov-Smirnov Two-Sample Test for Large Samples)

Level of Significance for Two-tailed Test

.10	$1.22\sqrt{\dfrac{N_1 + N_2}{N_1 N_2}}$
.05	$1.36\sqrt{\dfrac{N_1 + N_2}{N_1 N_2}}$
.025	$1.48\sqrt{\dfrac{N_1 + N_2}{N_1 N_2}}$
.01	$1.63\sqrt{\dfrac{N_1 + N_2}{N_1 N_2}}$
.005	$1.73\sqrt{\dfrac{N_1 + N_2}{N_1 N_2}}$
.001	$1.95\sqrt{\dfrac{N_1 + N_2}{N_1 N_2}}$

Table M. Critical Values of U and U' (Mann-Whitney Test)

ONE-TAILED TEST AT $\alpha = 0.005$ OR A TWO-TAILED TEST AT $\alpha = 0.01$

Each cell lists U (upper value) and U' (lower value), written here as U / U'.

N_2 \ N_1	1	2	3	4	5	6	7	8	9	10	11	12	13	14	15	16	17	18	19	20
1	—	—	—	—	—	—	—	—	—	—	—	—	—	—	—	—	—	—	—	—
2	—	—	—	—	—	—	—	—	—	—	—	—	—	—	—	—	—	—	0/38	0/40
3	—	—	—	—	—	—	—	—	0/27	0/30	0/33	1/35	1/38	1/41	2/43	2/46	2/49	2/52	3/54	3/57
4	—	—	—	—	—	0/24	0/28	1/31	1/35	2/38	2/42	3/45	3/49	4/52	5/55	5/59	6/62	6/66	7/69	8/72
5	—	—	—	—	0/25	1/29	1/34	2/38	3/42	4/46	5/50	6/54	7/58	7/63	8/67	9/71	10/75	11/79	12/83	13/87
6	—	—	—	0/24	1/29	2/34	3/39	4/44	5/49	6/54	7/59	9/63	10/68	11/73	12/78	13/83	15/87	16/92	17/97	18/102
7	—	—	—	0/28	1/34	3/39	4/45	6/50	7/56	9/61	10/67	12/72	13/78	15/83	16/89	18/94	19/100	21/105	22/111	24/116
8	—	—	—	1/31	2/38	4/44	6/50	7/57	9/63	11/69	13/75	15/81	17/87	18/94	20/100	22/106	24/112	26/118	28/124	30/130
9	—	—	0/27	1/35	3/42	5/49	7/56	9/63	11/70	13/77	16/83	18/90	20/97	22/104	24/111	27/117	29/124	31/131	33/138	36/144
10	—	—	0/30	2/38	4/46	6/54	9/61	11/69	13/77	16/84	18/92	21/99	24/106	26/114	29/121	31/129	34/136	37/143	39/151	42/158
11	—	—	0/33	2/42	5/50	7/59	10/67	13/75	16/83	18/92	21/100	24/108	27/116	30/124	33/132	36/140	39/148	42/156	45/164	48/172
12	—	—	1/35	3/45	6/54	9/63	12/72	15/81	18/90	21/99	24/108	27/117	31/125	34/134	37/143	41/151	44/160	47/169	51/177	54/186
13	—	—	1/38	3/49	7/58	10/68	13/78	17/87	20/97	24/106	27/116	31/125	34/135	38/144	42/153	45/163	49/172	53/181	56/191	60/200
14	—	—	1/41	4/52	7/63	11/73	15/83	18/94	22/104	26/114	30/124	34/134	38/144	42/154	46/164	50/174	54/184	58/194	63/203	67/213
15	—	—	2/43	5/55	8/67	12/78	16/89	20/100	24/111	29/121	33/132	37/143	42/153	46/164	51/174	55/185	60/195	64/206	69/216	73/227
16	—	—	2/46	5/59	9/71	13/83	18/94	22/106	27/117	31/129	36/140	41/151	45/163	50/174	55/185	60/196	65/207	70/218	74/230	79/241
17	—	—	2/49	6/62	10/75	15/87	19/100	24/112	29/124	34/136	39/148	44/160	49/172	54/184	60/195	65/207	70/219	75/231	81/242	86/254
18	—	—	2/52	6/66	11/79	16/92	21/105	26/118	31/131	37/143	42/156	47/169	53/181	58/194	64/206	70/218	75/231	81/243	87/255	92/268
19	—	0/38	3/54	7/69	12/83	17/97	22/111	28/124	33/138	39/151	45/164	51/177	56/191	63/203	69/216	74/230	81/242	87/255	93/268	99/281
20	—	0/40	3/57	8/72	13/87	18/102	24/116	30/130	36/144	42/158	48/172	54/186	60/200	67/213	73/227	79/241	86/254	92/268	99/281	105/295

(Dashes in the body of the table indicate that no decision is possible at the stated level of significance.)

Critical Values of U and U' (Continued)

One-tailed test at $\alpha = 0.01$ or a two-tailed test at $\alpha = 0.02$. Each cell gives U (upper) and U' (lower).

$N_2 \backslash N_1$	1	2	3	4	5	6	7	8	9	10	11	12	13	14	15	16	17	18	19	20
1	—	—	—	—	—	—	—	—	—	—	—	—	—	—	—	—	—	—	—	—
2	—	—	—	—	—	—	—	—	—	—	—	—	0/26	0/28	0/30	0/32	0/34	0/36	1/37	1/39
3	—	—	—	—	—	—	0/21	0/24	1/26	1/29	1/32	2/34	2/37	2/40	3/42	3/45	4/47	4/50	4/52	5/55
4	—	—	—	—	0/20	1/23	1/27	2/30	3/33	3/37	4/40	5/43	5/47	6/50	7/53	7/57	8/60	9/63	9/67	10/70
5	—	—	—	0/20	1/24	2/28	3/32	4/36	5/40	6/44	7/48	8/52	9/56	10/60	11/64	12/68	13/72	14/76	15/80	16/84
6	—	—	—	1/23	2/28	3/33	4/38	6/42	7/47	8/52	9/57	11/61	12/66	13/71	15/75	16/80	18/84	19/89	20/94	22/98
7	—	—	0/21	1/27	3/32	4/38	6/43	7/49	9/54	11/59	12/65	14/70	16/75	17/81	19/86	21/91	23/96	24/102	26/107	28/112
8	—	—	0/24	2/30	4/36	6/42	7/49	9/55	11/61	13/67	15/73	17/79	20/84	22/90	24/96	26/102	28/108	30/114	32/120	34/126
9	—	—	1/26	3/33	5/40	7/47	9/54	11/61	14/67	16/74	18/81	21/87	23/94	26/100	28/107	31/113	33/120	36/126	38/133	40/140
10	—	—	1/29	3/37	6/44	8/52	11/59	13/67	16/74	19/81	22/88	24/96	27/103	30/110	33/117	36/124	38/132	41/139	44/146	47/153
11	—	—	1/32	4/40	7/48	9/57	12/65	15/73	18/81	22/88	25/96	28/104	31/112	34/120	37/128	41/135	44/143	47/151	50/159	53/167
12	—	—	2/34	5/43	8/52	11/61	14/70	17/79	21/87	24/96	28/104	31/113	35/121	38/130	42/138	46/146	49/155	53/163	56/172	60/180
13	—	0/26	2/37	5/47	9/56	12/66	16/75	20/84	23/94	27/103	31/112	35/121	39/130	43/139	47/148	51/157	55/166	59/175	63/184	67/193
14	—	0/28	2/40	6/50	10/60	13/71	17/81	22/90	26/100	30/110	34/120	38/130	43/139	47/149	51/159	56/168	60/178	65/187	69/197	73/207
15	—	0/30	3/42	7/53	11/64	15/75	19/86	24/96	28/107	33/117	37/128	42/138	47/148	51/159	56/169	61/179	66/189	70/200	75/210	80/220
16	—	0/32	3/45	7/57	12/68	16/80	21/91	26/102	31/113	36/124	41/135	46/146	51/157	56/168	61/179	66/190	71/201	76/212	82/222	87/233
17	—	0/34	4/47	8/60	13/72	18/84	23/96	28/108	33/120	38/132	44/143	49/155	55/166	60/178	66/189	71/201	77/212	82/224	88/234	93/247
18	—	0/36	4/50	9/63	14/76	19/89	24/102	30/114	36/126	41/139	47/151	53/163	59/175	65/187	70/200	76/212	82/224	88/236	94/248	100/260
19	—	1/37	4/53	9/67	15/80	20/94	26/107	32/120	38/133	44/146	50/159	56/172	63/184	69/197	75/210	82/222	88/235	94/248	101/260	107/273
20	—	1/39	5/55	10/70	16/84	22/98	28/112	34/126	40/140	47/153	53/167	60/180	67/193	73/207	80/220	87/233	93/247	100/260	107/273	114/286

(Dashes in the body of the table indicate that no decision is possible at the stated level of significance.)

Critical Values of *U* and *U'* (Continued)

One-Tailed Test at $\alpha = 0.025$ or a Two-Tailed Test at $\alpha = 0.05$

(Cells show U / U'.)

N_2 \ N_1	1	2	3	4	5	6	7	8	9	10	11	12	13	14	15	16	17	18	19	20
1	—	—	—	—	—	—	—	—	—	—	—	—	—	—	—	—	—	—	—	—
2	—	—	—	—	—	—	—	0/16	0/18	0/20	0/22	1/23	1/25	1/27	1/29	1/31	2/32	2/34	2/36	2/38
3	—	—	—	—	0/15	1/17	1/20	2/22	2/25	3/27	3/30	4/32	4/35	5/37	5/40	6/42	6/45	7/47	7/50	8/52
4	—	—	—	0/16	1/19	2/22	3/25	4/28	4/32	5/35	6/38	7/41	8/44	9/47	10/50	11/53	11/57	12/60	13/63	13/67
5	—	—	0/15	1/19	2/23	3/27	5/30	6/34	7/38	8/42	9/46	11/49	12/53	13/57	14/61	15/65	17/68	18/72	19/76	20/80
6	—	—	1/17	2/22	3/27	5/31	6/36	8/40	10/44	11/49	13/53	14/58	16/62	17/67	19/71	21/75	22/80	24/84	25/89	27/93
7	—	—	1/20	3/25	5/30	6/36	8/41	10/46	12/51	14/56	16/61	18/66	20/71	22/76	24/81	26/86	28/91	30/96	32/101	34/106
8	—	0/16	2/22	4/28	6/34	8/40	10/46	13/51	15/57	17/63	19/69	22/74	24/80	26/86	29/91	31/97	34/102	36/108	38/111	41/119
9	—	0/18	2/25	4/32	7/38	10/44	12/51	15/57	17/64	20/70	23/76	26/82	28/89	31/95	34/101	37/107	39/114	42/120	45/126	48/132
10	—	0/20	3/27	5/35	8/42	11/49	14/56	17/63	20/70	23/77	26/84	29/91	33/97	36/104	39/111	42/118	45/125	48/132	52/138	55/145
11	—	0/22	3/30	6/38	9/46	13/53	16/61	19/69	23/76	26/84	30/91	33/99	37/106	40/114	44/121	47/129	51/136	55/143	58/151	62/158
12	—	1/23	4/32	7/41	11/49	14/58	18/66	22/74	26/82	29/91	33/99	37/107	41/115	45/123	49/131	53/139	57/147	61/155	65/163	69/171
13	—	1/25	4/35	8/44	12/53	16/62	20/71	24/80	28/89	33/97	37/106	41/115	45/124	50/132	54/141	59/149	63/158	67/167	72/175	76/184
14	—	1/27	5/37	9/47	13/57	17/67	22/76	26/86	31/95	36/104	40/114	45/123	50/132	55/141	59/151	64/160	67/171	74/178	78/188	83/197
15	—	1/29	5/40	10/50	14/61	19/71	24/81	29/91	34/101	39/111	44/121	49/131	54/141	59/151	64/161	70/170	75/180	80/190	85/200	90/210
16	—	1/31	6/42	11/53	15/65	21/75	26/86	31/97	37/107	42/118	47/129	53/139	59/149	64/160	70/170	75/181	81/191	86/202	92/212	98/222
17	—	2/32	6/45	11/57	17/68	22/80	28/91	34/102	39/114	45/125	51/136	57/147	63/158	67/171	75/180	81/191	87/202	93/213	99/224	105/235
18	—	2/34	7/47	12/60	18/72	24/84	30/96	36/108	42/120	48/132	55/143	61/155	67/167	74/178	80/190	86/202	93/213	99/225	106/236	112/248
19	—	2/36	7/50	13/63	19/76	25/89	32/101	38/111	45/126	52/138	58/151	65/163	72/175	78/188	85/200	92/212	99/224	106/236	113/248	119/261
20	—	2/38	8/52	13/67	20/80	27/93	34/106	41/119	48/132	55/145	62/158	69/171	76/184	83/197	90/210	98/222	105/235	112/248	119/261	127/273

(Dashes in the body of the table indicate that no decision is possible at the stated level of significance.)

Critical Values of U and U' (Continued)

ONE-TAILED TEST AT $\alpha = 0.05$ OR A TWO-TAILED TEST AT $\alpha = 0.10$

Each cell lists U (upper) and U' (lower).

N_2 \ N_1	1	2	3	4	5	6	7	8	9	10	11	12	13	14	15	16	17	18	19	20
1	—	—	—	—	—	—	—	—	—	—	—	—	—	—	—	—	—	—	0 / 19	0 / 20
2	—	—	—	—	0 / 10	0 / 12	0 / 14	1 / 15	1 / 17	1 / 19	1 / 21	2 / 22	2 / 24	2 / 26	3 / 27	3 / 29	3 / 31	4 / 32	4 / 34	4 / 36
3	—	—	0 / 9	0 / 12	1 / 14	2 / 16	2 / 19	3 / 21	3 / 24	4 / 26	5 / 28	5 / 31	6 / 33	7 / 35	7 / 38	8 / 40	9 / 42	9 / 45	10 / 47	11 / 49
4	—	—	0 / 12	1 / 15	2 / 18	3 / 21	4 / 24	5 / 27	6 / 30	7 / 33	8 / 36	9 / 39	10 / 42	11 / 45	12 / 48	14 / 50	15 / 53	16 / 56	17 / 59	18 / 62
5	—	0 / 10	1 / 14	2 / 18	4 / 21	5 / 25	6 / 29	8 / 32	9 / 36	11 / 39	12 / 43	13 / 47	15 / 50	16 / 54	18 / 57	19 / 61	20 / 65	22 / 68	23 / 72	25 / 75
6	—	0 / 12	2 / 16	3 / 21	5 / 25	7 / 29	8 / 34	10 / 38	12 / 42	14 / 46	16 / 50	17 / 55	19 / 59	21 / 63	23 / 67	25 / 71	26 / 76	28 / 80	30 / 84	32 / 88
7	—	0 / 14	2 / 19	4 / 24	6 / 29	8 / 34	11 / 38	13 / 43	15 / 48	17 / 53	19 / 58	21 / 63	24 / 67	26 / 72	28 / 77	30 / 82	33 / 86	35 / 91	37 / 96	39 / 101
8	—	1 / 15	3 / 21	5 / 27	8 / 32	10 / 38	13 / 43	15 / 49	18 / 54	20 / 60	23 / 65	26 / 70	28 / 76	31 / 81	33 / 87	36 / 92	39 / 97	41 / 103	44 / 108	47 / 113
9	—	1 / 17	3 / 24	6 / 30	9 / 36	12 / 42	15 / 48	18 / 54	21 / 60	24 / 66	27 / 72	30 / 78	33 / 84	36 / 90	39 / 96	42 / 102	45 / 108	48 / 114	51 / 120	54 / 126
10	—	1 / 19	4 / 26	7 / 33	11 / 39	14 / 46	17 / 53	20 / 60	24 / 66	27 / 73	31 / 79	34 / 86	37 / 93	41 / 99	44 / 106	48 / 112	51 / 119	55 / 125	58 / 132	62 / 138
11	—	1 / 21	5 / 28	8 / 36	12 / 43	16 / 50	19 / 58	23 / 65	27 / 72	31 / 79	34 / 87	38 / 94	42 / 101	46 / 108	50 / 115	54 / 122	57 / 130	61 / 137	65 / 144	69 / 151
12	—	2 / 22	5 / 31	9 / 39	13 / 47	17 / 55	21 / 63	26 / 70	30 / 78	34 / 86	38 / 94	42 / 102	47 / 109	51 / 117	55 / 125	60 / 132	64 / 140	68 / 148	72 / 156	77 / 163
13	—	2 / 24	6 / 33	10 / 42	15 / 50	19 / 59	24 / 67	28 / 76	33 / 84	37 / 93	42 / 101	47 / 109	51 / 118	56 / 126	61 / 134	65 / 143	70 / 151	75 / 159	80 / 167	84 / 176
14	—	2 / 26	7 / 35	11 / 45	16 / 54	21 / 63	26 / 72	31 / 81	36 / 90	41 / 99	46 / 108	51 / 117	56 / 126	61 / 135	66 / 144	71 / 153	77 / 161	82 / 170	87 / 179	92 / 188
15	—	3 / 27	7 / 38	12 / 48	18 / 57	23 / 67	28 / 77	33 / 87	39 / 96	44 / 106	50 / 115	55 / 125	61 / 134	66 / 144	72 / 153	77 / 163	83 / 172	88 / 182	94 / 191	100 / 200
16	—	3 / 29	8 / 40	14 / 50	19 / 61	25 / 71	30 / 82	36 / 92	42 / 102	48 / 112	54 / 122	60 / 132	65 / 143	71 / 153	77 / 163	83 / 173	89 / 183	95 / 193	101 / 203	107 / 213
17	—	3 / 31	9 / 42	15 / 53	20 / 65	26 / 76	33 / 86	39 / 97	45 / 108	51 / 119	57 / 130	64 / 140	70 / 151	77 / 161	83 / 172	89 / 183	96 / 193	102 / 204	109 / 214	115 / 225
18	—	4 / 32	9 / 45	16 / 56	22 / 68	28 / 80	35 / 91	41 / 103	48 / 114	55 / 123	61 / 137	68 / 148	75 / 159	82 / 170	88 / 182	95 / 193	102 / 204	109 / 215	116 / 226	123 / 237
19	0 / 19	4 / 34	10 / 47	17 / 59	23 / 72	30 / 84	37 / 96	44 / 108	51 / 120	58 / 132	65 / 144	72 / 156	80 / 167	87 / 179	94 / 191	101 / 203	109 / 214	116 / 226	123 / 238	130 / 250
20	0 / 20	4 / 36	11 / 49	18 / 62	25 / 75	32 / 88	39 / 101	47 / 113	54 / 126	62 / 138	69 / 151	77 / 163	84 / 176	92 / 188	100 / 200	107 / 213	115 / 225	123 / 237	130 / 250	138 / 262

(Dashes in the body of the table indicate that no decision is possible at the stated level of significance.)

Table N. Critical Values of *H* (Kruskal-Wallis Test)

N_1	N_2	N_3	H	p	N_1	N_2	N_3	H	p
2	1	1	2.7000	.500	4	3	2	6.4444	.008
								6.3000	.011
2	2	1	3.6000	.200				5.4444	.046
								5.4000	.051
2	2	2	4.5714	.067				4.5111	.098
			3.7143	.200				4.4444	.102
3	1	1	3.2000	.300	4	3	3	6.7455	.010
								6.7091	.013
3	2	1	4.2857	.100				5.7909	.046
			3.8571	.133				5.7273	.050
3	2	2	5.3572	.029				4.7091	.092
			4.7143	.048				4.7000	.101
			4.5000	.067					
			4.4643	.105	4	4	1	6.6667	.010
								6.1667	.022
3	3	1	5.1429	.043				4.9667	.048
			4.5714	.100				4.8667	.054
			4.0000	.129				4.1667	.082
								4.0667	.102
3	3	2	6.2500	.011					
			5.3611	.032	4	4	2	7.0364	.006
			5.1389	.061				6.8727	.011
			4.5556	.100				5.4545	.046
			4.2500	.121				5.2364	.052
								4.5545	.098
3	3	3	7.2000	.004				4.4455	.103
			6.4889	.011					
			5.6889	.029	4	4	3	7.1439	.010
			5.6000	.050				7.1364	.011
			5.0667	.086				5.5985	.049
			4.6222	.100				5.5758	.051
								4.5455	.099
4	1	1	3.5714	.200				4.4773	.102
4	2	1	4.8214	.057	4	4	4	7.6538	.008
			4.5000	.076				7.5385	.011
			4.0179	.114				5.6923	.049
								5.6538	.054
4	2	2	6.0000	.014				4.6539	.097
			5.3333	.033				4.5001	.104
			5.1250	.052					
			4.4583	.100	5	1	1	3.8571	.143
			4.1667	.105					
4	3	1	5.8333	.021	5	2	1	5.2500	.036
			5.2083	.050				5.0000	.048
			5.0000	.057				4.4500	.071
			4.0556	.093				4.2000	.095
			3.8889	.129				4.0500	.119

N_1	N_2	N_3	H	p	N_1	N_2	N_3	H	p
5	2	2	6.5333	.008				4.5487	.099
			6.1333	.013				4.5231	.103
			5.1600	.034					
			5.0400	.056	5	4	4	7.7604	.009
			4.3733	.090				7.7440	.011
			4.2933	.122				5.6571	.049
								5.6176	.050
5	3	1	6.4000	.012				4.6187	.100
			4.9600	.048				4.5527	.102
			4.8711	.052					
			4.0178	.095	5	5	1	7.3091	.009
			3.8400	.123				6.8364	.011
								5.1273	.046
5	3	2	6.9091	.009				4.9091	.053
			6.8218	.010				4.1091	.086
			5.2509	.049				4.0364	.105
			5.1055	.052					
			4.6509	.091	5	5	2	7.3385	.010
			4.4945	.101				7.2692	.010
								5.3385	.047
5	3	3	7.0788	.009				5.2462	.051
			6.9818	.011				4.6231	.097
			5.6485	.049				4.5077	.100
			5.5152	.051					
			4.5333	.097	5	5	3	7.5780	.010
			4.4121	.109				7.5429	.010
								5.7055	.046
5	4	1	6.9545	.008				5.6264	.051
			6.8400	.011				4.5451	.100
			4.9855	.044				4.5363	.102
			4.8600	.056					
			3.9873	.098	5	5	4	7.8229	.010
			3.9600	.102				7.7914	.010
								5.6657	.049
5	4	2	7.2045	.009				5.6429	.050
			7.1182	.010				4.5229	.099
			5.2727	.049				4.5200	.101
			5.2682	.050					
			4.5409	.098	5	5	5	8.0000	.009
			4.5182	.101				7.9800	.010
								5.7800	.049
5	4	3	7.4449	.010				5.6600	.051
			7.3949	.011				4.5600	.100
			5.6564	.049				4.5000	.102
			5.6308	.050					

Answers to Selected Exercises

Chapter 2

1. (a) Interval
   (b) Nominal
   (c) Ordinal
   (d) Interval
   (e) Nominal

   (f) Ordinal
   (g) Interval
   (h) Nominal
   (i) Nominal
   (j) Nominal

2. (a) Nominal/Discrete
   (b) Nominal/Discrete
   (c) Interval/Continuous
   (d) Nominal/Discrete
   (e) Interval/Continuous
   (f) Nominal/Discrete
   (g) Nominal/Discrete
   (h) Interval/Continuous
   (i) Nominal/Discrete
   (j) Interval/Continuous

   (k) Nominal/Discrete
   (l) Nominal/Discrete
   (m) Nominal/Discrete
   (n) Nominal/Discrete
   (o) Interval/Continuous
   (p) Nominal/Discrete
   (q) Ordinal/Discrete
   (r) Nominal/Discrete
   (s) Interval/Continuous
   (t) Interval/Continuous

8. (a) 10.005 to 10.015
   (b) −9.5 to −10.5
   (c) −0.5 to 0.5
   (d) 1.55 to 1.65
   (e) 3.05 to 3.15

   (f) 0.45 to 0.55
   (g) 0.50 to 0.60
   (h) 0.45 to 0.55
   (i) 15.5 to 16.5
   (j) −34.5 to −35.5

9. (a) $\dfrac{\displaystyle\sum_{i=1}^{N} X_i}{2}$ (b) $\left[\displaystyle\sum_{i=3}^{8} (X_i - Y_i)\right] 7$ (c) $\displaystyle\sum_{i=1}^{N} X_i k = k\sum_{i=1}^{N} X_i$

(d) $\displaystyle\sum_{i=1}^{3} (X_i + Y_i + Z_i)$ (e) $\bar{X} = \Sigma X/N$

10. (a) $\displaystyle\sum_{1}^{7} X$ (b) ΣX (c) $\displaystyle\sum_{1}^{4}(X + Y)$ (d) $\displaystyle\sum_{2}^{N}(X + a)$

(e) $\displaystyle\sum_{6}^{9}(X + Y - k)$ (f) $\displaystyle\sum_{1}^{6} X^2 Y^2$ (g) $\displaystyle\sum_{1}^{8}(X - Y)$

(h) $\Sigma(X + Y)$ (i) $\displaystyle\sum_{2}^{4}(X - Z)$ (j) $\Sigma[X - (a + b)]$

11. (a) $(X_2 + Y_2) + (X_3 + Y_3) + (X_4 + Y_4) + (X_5 + Y_5) + (X_6 + Y_6)$
$+ (X_7 + Y_7) + (X_8 + Y_8)$

(b) $(X_1 + k) + (X_2 + k) + (X_3 + k) + \ldots + (X_N + k)$

(c) $(X_2 + Y_2) + (X_3 + Y_3) + (X_4 + Y_4)$

(d) $(X_2 + 2) + (X_3 + 2) + (X_4 + 2) + (X_5 + 2) + (X_6 + 2)$
$+ (X_7 + 2) + (X_8 + 2) + (X_9 + 2) + (X_{10} + 2)$

(e) $(X_4 + Y_4 + k) + (X_5 + Y_5 + k) + (X_6 + Y_6 + k) + (X_7 + Y_7 + k)$

(f) $(X_1 + Y_1 - k) + (X_2 + Y_2 - k) + (X_3 + Y_3 - k)$

(g) $X_5^2 + X_6^2 + X_7^2 + X_8^2 + X_9^2$

(h) $X_1^2 Y_1^2 Z_1^2 + X_2^2 Y_2^2 Z_2^2 + X_3^2 Y_3^2 Z_3^2$

(i) $(X_2 + a^2) + (X_3 + a^2)$

(j) $(X_7 - Z_7 + a) + (X_8 - Z_8 + a) + (X_9 - Z_9 + a)$
$+ (X_{10} - Z_{10} + a) + (X_{11} - Z_{11} + a) + (X_{12} - Z_{12} + a)$
$+ (X_{13} - Z_{13} + a) + (X_{14} - Z_{14} + a) + (X_{15} - Z_{15} + a)$

12. (a) 20 (b) 6 (c) 38.2 (d) 37 (e) -12.3 (f) -112.3 (g) -32.3
(h) 32.7 (i) 14 (j) 16

Chapter 3

2. (b) *Class Interval* *f*

Class Interval	f
0–19	32
20–39	17
40–59	17
60–79	17
80–89	17

3.

(a) Class Interval	(b) Class Interval	(c) Class Interval
20–28	21–64	1,201–2,060
29–37	65–108	2,061–2,920
38–46	109–152	2,921–3,780
47–55	153–196	3,781–4,640
56–64	197–240	4,641–5,500
65–73	241–284	5,501–6,360
74–82	285–328	6,361–7,220
83–91	329–372	7,221–8,080
	373–416	8,081–8,940
	417–460	8,941–9,800

4. (a) 19.5, 91.5   (d) 20.5, 460.5
      (e) 1,200.5, 9800.5   (f) .5, 100.5

5.

(a) Class Interval	Cum f	(b) Class Interval	Cum %
0–9	15	0–9	15
10–19	32	10–19	32
20–29	40	20–29	40
30–39	49	30–39	49
40–49	58	40–49	58
50–59	66	50–59	66
60–69	75	60–69	75
70–79	83	70–79	83
80–89	93	80–89	93
90–99	100	90–99	100

6.

(a) Class Interval	f
40–44	4
45–49	1
50–54	3
55–59	10
60–64	24
65–69	33
70–74	20
75–79	4
80–84	1

9.

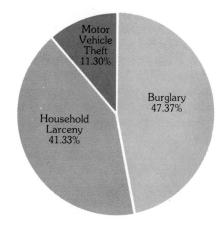

12.

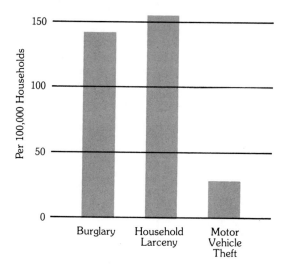

17.

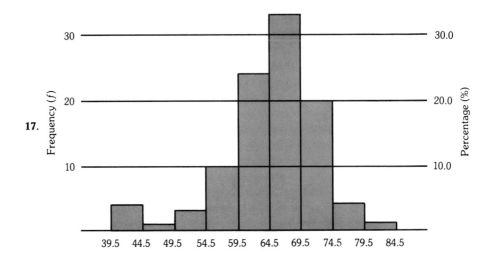

18.

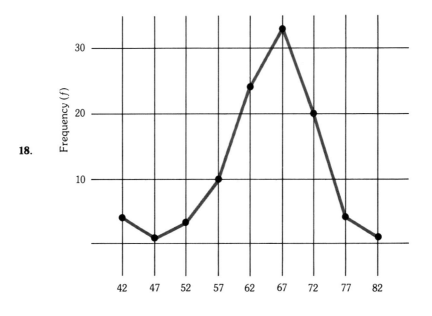

Chapter 4

3. (a) 27.67 (b) 22.27 (c) 2.33 (d) 97.87 (e) 90.46

4. (a) 135.79 (b) 124.00 (c) 108.74 (d) 112.57 (e) 108.93

7. (a) answer for Chapter 3, Exercise 2(c)

Class Interval	Cum %
0–9	15
10–19	32
20–29	40
30–39	49
40–49	58
50–59	66
60–69	75
70–79	83
80–89	93
90–99	100

(b)

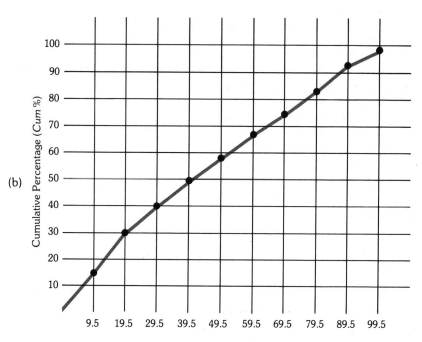

(c) 14.79, 24.50, 43.94, 63.94, 86.50

10. (a) 6 (b) 8.33 (c) 9.58 (d) 10.42 (e) 13.30

Chapter 5

	$\bar{X}$	Median	Mode(s)
1. (a)	5.2	4.5	3
(b)	9.9	10.5	18
(c)	6.42	6.3	6.3
(d)	113.09	105	104 and 105
(e)	2.53	2	2

(f)	1.627	1.624	all
(g)	0.0051	0.004	0.001 and 0.002
(h)	0.48	1.15	1.3
(i)	37.67	46.00	51.00
(j)	98.6	98.6	98.6

2. (a) Positively skewed
   (b) Symmetrical
   (c) Negatively skewed
   (d) Positively skewed
   (e) Positively skewed
   (f) No skew, multimodal
   (g) Positively skewed
   (h) No skew, all scores the same
   (i) Symmetrical
   (j) No skew, bimodal

6. Figure 3.1: Oppose
   Figure 3.2: Psychology
   Figure 3.3: Highest Ability
   Figure 3.4: For Not Cleared, Auto Thefts; for Cleared, Robbery
   Figure 3.5: 109
   Figure 3.6: 109
   Figure 3.7: For blacks, approximately $4,000; for whites, approximately $12,500

10. (a) $\bar{X}$ = 60.03; Median = 62.50
    (b) $\bar{X}$ = 75.26; Median = 79.50
    (c) $\bar{X}$ = 66.74; Median = 77.23
    (d) Mode = 64.5
    (e) Mode = 74.5

11. $\bar{X}$ = 52.96

12. (a) $\bar{X}$ = 24.44   (b) $\bar{X}$ = 40   (c) $\bar{X}$ = 35

Chapter 6

3. s = 3.09

4. (a) 1.63   (b) 1.29   (c) 2.35   (d) 4.47   (e) 10.65

5. e

6. 6; 5; 8; 16; 35

9. 2.6; 1.33; 1.00; 2.00; 3.00; 7.00

10. (a) 7.79   (b) 60.68   (c) 35   (d) 12.5   (e) 6.25

13. All scores are the same.

14. 10; 3.16

16. 0.72

17. 0.81

18. No

Chapter 7

2. Yes

3. (a) 49.62   (c) 44.84
   (b) 48.12   (d) 11.03

(e) 11.03		(h) 36.21	
(f) 46.93		(i) 49.88	
(g) 49.53		(j) 19.85	

4. (a) 93.57   (f) 9.29
    (b) 95.05   (g) 77.96
    (c) 47.83   (h) 23.84
    (d) 12.71   (i) 27.43
    (e) 97.29   (j) 92.57

5. (a) −0.30   (f) 0.19
    (b) −3.26   (g) 1.17
    (c) −3.57   (h) 0.07
    (d) −3.50   (i) 0.47
    (e) −4.45   (j) −2.06

6. (a) 60.80   (f) 84.95
    (b) 72.20   (g) 87.80
    (c) 69.95   (h) 93.80
    (d) 80.00   (i) 100.10
    (e) 90.05   (j) 101.15

7. (a) 68.02   (b) 115.54   (c) 100   (d) 100   (e) 100

8. 80.78

9. 89.2; 66.25

10. (a) 0.50   (b) 69.15   (c) 30.85

12. 30.5

13. (a) 2 in every 100 applicants
    (b) 102.6
    (c) 906.4; 93.6
    (d) 103.55 (same number of applicants); 105.9 (twice as many applicants)

Chapter 8

2. (a) 1/13   (b) 1/52   (c) 21/52   (d) 15/52   (e) 7/13

3. (a) 5/36   (b) 2/9   (c) 1/6   (d) 1/2   (e) 2/3

4. d, e

5. c, d, e

6. (a) 1/156   (b) 3/156   (c) 2/39   (d) 2/39

8. (a) 0.80   (b) 0.15   (c) 0.30   (d) 0.50

10. (a) 11/850   (b) 11/1,105   (c) 1/5,525   (d) 4,324/5,525   (e) 27,417/66,300

12. (a) 1/4   (b) 3/4   (c) 1/2   (d) 3/4   (e) 1/4

13. (a) .1587   (f) .1932
    (b) .1409   (g) .3183
    (c) .3312   (h) .2207
    (d) .8772   (i) .2879
    (e) .3446   (j) .2767

Chapter 9

3. (a) Accept H_0 at $\alpha = 0.01$ and at $\alpha = 0.05$
   (b) Reject H_0 at $\alpha = 0.01$ and at $\alpha = 0.05$
   (c) Reject H_0 at $\alpha = 0.01$ and at $\alpha = 0.05$
   (d) Accept H_0 at $\alpha = 0.01$ and at $\alpha = 0.05$
   (e) Accept H_0 at $\alpha = 0.01$ and at $\alpha = 0.05$

4. (a) 0.265   (b) 0.290   (c) 0.623   (d) 0.001   (e) 0.377

6. (a) H_1 (e) H_1
   (b) H_0 (f) H_0
   (c) H_1 (g) H_0
   (d) H_0 (h) H_1

7. (a) One-tailed
   (b) One-tailed
   (c) One-tailed
   (e) One-tailed
   (h) One-tailed

8. (b) Two-tailed
   (d) Two-tailed
   (f) Two-tailed
   (g) Two-tailed

9. a and e; d and b; c and f

Chapter 10

3. $t = 2.18$; reject H_0: $\mu = 100$(one-tailed test) at $\alpha = 0.05$

4. $z = 0.71$; accept H_0: $\mu = 13,800$(two-tailed test) at $\alpha = 0.05$
   $z = 0.71$; accept H_0: $\mu = 13,800$(one-tailed test) at $\alpha = 0.01$

5. $t = 3.21$; reject H_0: $\mu = 30$ (one-tailed test) at $\alpha = 0.05$ and at $\alpha = 0.01$

6. $z = -11.31$; reject H_0: $\mu = 48$ (one-tailed test) at $\alpha = 0.01$

7. $t = -2.55$; reject H_0: $\mu = 48$ (one-tailed test) at $\alpha = 0.05$

12. (a) $\bar{X} = 50 \pm 1.94$
    (b) $\bar{X} = 50 \pm 2.61$

13. (a) $\bar{X} = 50 \pm 0.62$
    (b) $\bar{X} = 50 \pm 0.82$

14. $\bar{X} = 14,000 \pm 554.38$

15. $\bar{X} = 33 \pm 2.58$

16. $\bar{X} = 110 \pm 1.50$

17. $N = 1,600$

18. 95 percent

Chapter 11

2. (a) Yes (b) No (c) No (d) Yes

3. $t = 8.78$, $df = 14$; reject H_0: $\mu_1 = \mu_2$ at $\alpha = 0.05$ and at $\alpha = 0.01$

4. Yes ($z = 19.84$)

5. Yes ($z = 1.04$)

7. (a) H_0: $\mu_1 = \mu_2$; H_1: $\mu_1 \neq \mu_2$
    (b) $t = 1.50$, $df = 36$; accept H_0

8. (a) $t = -1.94$, $df = 16$, accept H_0: $\mu_1 = \mu_2$ at $\alpha = 0.05$ and at $\alpha = 0.01$
    (b) $t = -1.07$, $df = 15.23$; accept H_0: $\mu_1 = \mu_2$ at $\alpha = 0.05$ and at $\alpha = 0.01$

9. (a) $\bar{X}_1 - \bar{X}_2 = 3 \pm 0.34$ (assuming homoscedasticity)
    (b) $\bar{X}_1 - \bar{X}_2 = 3 \pm 0.09$ (assuming homoscedasticity)
    (c) When $N \leqslant 30$, then:

$$\mu_{\bar{x}_1 - \bar{x}_2} = \bar{X}_1 - \bar{X}_2 \pm t s_{\bar{x}_1 - \bar{x}_2}$$

where:

$$s_{\bar{x}_1 - \bar{x}_2} = \sqrt{\frac{s_1^2}{N_1 - 1} + \frac{s_2^2}{N_2 - 1}}$$

and, if pooled:

$$s_{\bar{x}_1 - \bar{x}_2} = \sqrt{\frac{N_1 s_1^2 + N_2 s_2^2}{N_1 + N_2 - 2}\left(\frac{1}{N_1} + \frac{1}{N_2}\right)}$$

When $N > 30$, then:

$$\mu_{\bar{x}_1 - \bar{x}_2} = \bar{X}_1 - \bar{X}_2 \pm z s_{\bar{x}_1 - \bar{x}_2}$$

where:

$$s_{\bar{x}_1 - \bar{x}_2} = \sqrt{\frac{s_1^2}{N_1} + \frac{s_2^2}{N_2}}$$

and, if pooled:

$$s_{\bar{x}_1 - \bar{x}_2} = \sqrt{\frac{N_1 s_1^2 + N_2 s_2^2}{N_1 + N_2 - 2}\left(\frac{1}{N_1} + \frac{1}{N_2}\right)}$$

11. $t = -1.7$, $df = 9$; accept H_0: $\mu_1 = \mu_2$ at $\alpha = 0.01$

15. Exercise 11: A = 0.413, $df = 9$

Chapter 12

1. (a) $SS_w = 48.00$
   (b) $SS_b = 8.40$
   (c) $SS_t = 56.40$

2. No

4. Not significant at either the 0.01 or the 0.05 level

6. $F = 9.31$, $df = 3,12$; reject H_0 at $\alpha = 0.05$

7. (a) H_0: $\mu_1 = \mu_2 = \mu_3 = \mu_4 = \mu_5$; H_1: $\mu_1 \neq \mu_2 \neq \mu_3 \neq \mu_4 \neq \mu_5$
   (b) $F = 8.07$, $df = 4,20$; reject H_0 at $\alpha = 0.05$. At $\alpha = 0.05$, $q_{0.05} = 3.96$;
   therefore, HSD of 17.76 is significant: $\bar{X}_1$ vs. $\bar{X}_3$, $\bar{X}_1$ vs. $\bar{X}_5$, $\bar{X}_3$ vs. $\bar{X}_4$, and $\bar{X}_4$
   vs. $\bar{X}_5$ all exceed the critical value of 17.76

8. $F = 11$, $df = 3, 7$; reject H_0 at $\alpha = 0.05$

9. $F = 5$, $df = 3, 8$; accept H_0 at $\alpha = 0.05$

10. $F = 1.6$

Chapter 13

2. $1.00, -0.95, 0.75$ and $-0.75, -0.74, 0.68, 0.36, -0.30, -0.01, 0$

4. (a) $r = 0.75$ (b) $r = 0.70$ (c) $r = 0.52$

5. $r = 0.79$

6. (a)

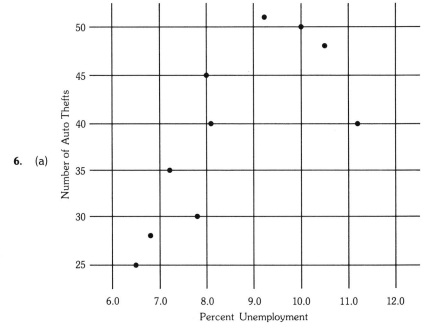

(c) $r = 0.73$
(d) $r_s = 0.78$

7. No

8. $r = 0.49$

12. (a) Not significant
    (b) Significant
    (c) Not significant
    (d) Not significant
    (e) Significant

13. $r = \pm 0.4438$

14. $r = \pm 0.7646$

15. (a) 1, 2, 3, 4, 5
    (b) 1, 3, 4, 5, 2
    (c) 2, 1, 5, 4, 3
    (d) 5, 3, 2, 1, 4
    (e) 5, 4, 3, 2, 1

16. $r_s = 0.85$

17. (a) $r_s = 0.49$ (b) accept H_0

18. (a) $r_s = 0.48$ (b) accept H_0

Chapter 14

1. (a) $Y' = 3.00$

(b)

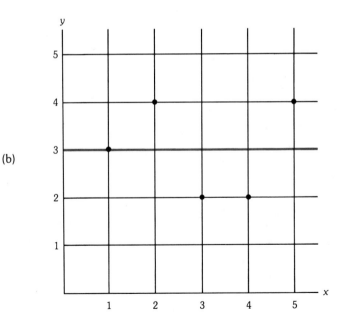

(c) 0%
(d) 100%
(e) $F = 0$

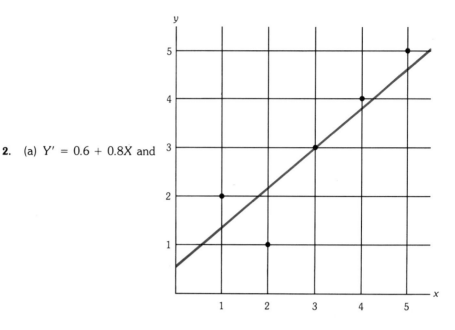

2.   (a) $Y' = 0.6 + 0.8X$ and

   (b) 2.2
   (c) 3.0
   (d) 0.64 and 0.36
   (e) $F = 5.33$, not significant

4.   (a) $Y' = 10.32 + 0.02X$
   (b) $Y' = -12.41 + 6.36X$
   (c) Income

6.   Illiteracy; $r^2 = 0.68$

7.   (a) $Y' = 19.43 - 0.09X$
   (b) $r^2 = 0.57$ and $k^2 = 0.43$

8.   Yes

9.   (a) $5,400
   (b) $7,900
   (c) Yes

10.   (a) 54.8
   (b) 3.2
   (c) 34.75
   (d) 8
   (e) Yes

12. (a) $Y' = 0.14 + 0.002X$
   (b) 2.09
   (c) 0.31
   (d) 2.44; 2.44 ± 0.61
   (e) $F = 13.28$, $df = 1, 8$; significant
13. (a) $8,000
   (b) 16%

Chapter 15

2. (a)

	Low	Medium	High
Low	2	4	2
Medium	1	3	1
High	1	3	3

(c)

above $500

0	0	0
0	1	0
1	1	3

$250–$500

1	3	0
1	2	1
0	2	0

below $250

1	1	2
0	0	0
0	0	0

   (e) $r_{12.3} = 0.08$
4. $r_{12.3} = 0.54$, $r_{13.2} = 0.42$; alternative causes
6. (a) $r_{12.3} = -0.14$
   (b) The correlation between education and number of books read overwhelms the original bivariate relationship.
7. $r_{12.3}$ (Note: The control variable must be correlated with the dependent variable and the other independent variable to have any impact.)
8. (a) $r_{12.3} = 0.59$
   (b) $r_{23.1} = -0.50$
   (c) $F = 51.80$; significant at $\alpha = 0.05$ and at $\alpha = 0.01$

9. c

10. a and d

11. $r_{23} = 0$

12. $Y'_1 = 19.17 + 2.04X_2 + 0.018X_3$

13. (a) 3.10 (b) 2.69 (c) 3.07 (d) 3.00 (e) 2.74

14. (a) 0.63
     (b) −0.01
     (c) $Y'_1 = 4.02 + 4.32X_2 + .0004X_3$
     (d) 56%

15. β_3

17. $R_{1.23} = 0.65$

18. (a) $R = 0.69$
     (b) $R^2_{1.23} = 0.47$
     (c) $F = 20.84$, $df = 2$, 47; significant at $\alpha = 0.05$ and at $\alpha = 0.01$

19. For exercise 4, $R^2_{1.23} = 0.47$; $K^2_{1.23} = 0.53$

Chapter 16

1. $\chi^2 = 0.40$, not significant

2. (a) $\chi^2 = 82.44$, significant (b) $\chi^2 = 52.00$, significant

3. (a) $\chi^2 = 12.91$, significant
     (b) $\chi^2 = 19.36$, significant for New York City Democrats;
         $\chi^2 = 16.87$, significant for Upstate New York Democrats

4. (a) $\chi^2 = 110.28$, significant
     (b) $\chi^2 = 2.57$, not significant
     (c) $\chi^2 = 108.00$, significant

5. $\chi^2 = 7.70$, significant

6. $\phi = 0.09$

7. 2(a). V = 0.32   4(a). V = 0.30   4(b). V = 0.09   4(c). V = 0.42
     5. V = 0.14

9. $\lambda_a = 0.29$ $\tau_r = 0.19$

10. $\gamma = 0.51$

11. $Q = 0.80$

12. $\chi^2 = 19.17$, significant; $D = 0.05$, not significant

13. $K_D = 0.29$, significant

14. $U = 20$, $U = 61$; significant

15. $H = 4.61$, not significant

Index